LANDMARK COLLECTOR'S LIBRARY

Collieries in the North Staffordshire Coalfield

Paul Deakin

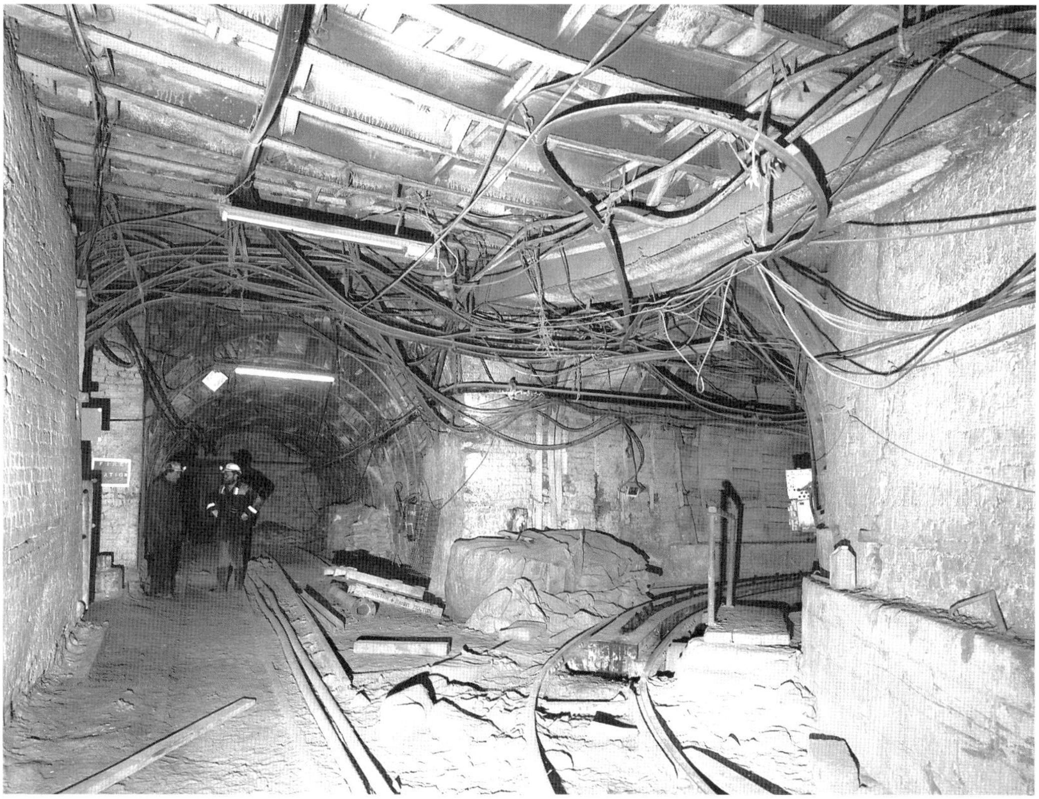

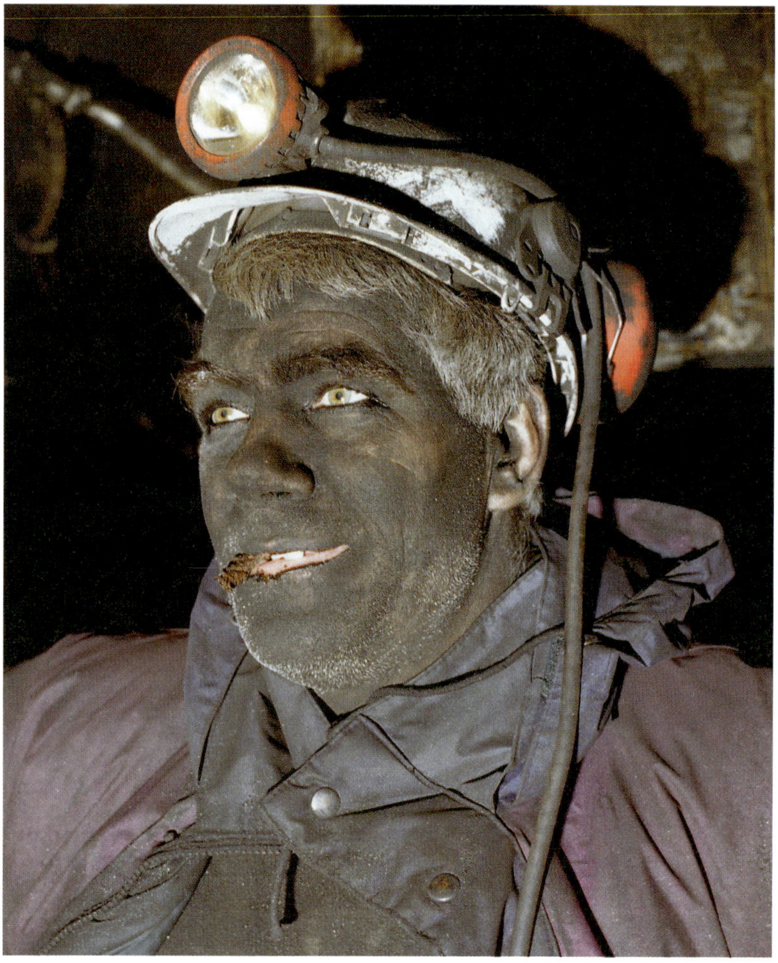

LANDMARK COLLECTOR'S LIBRARY

COLLIERIES IN THE NORTH STAFFORDSHIRE COALFIELD

Paul Deakin

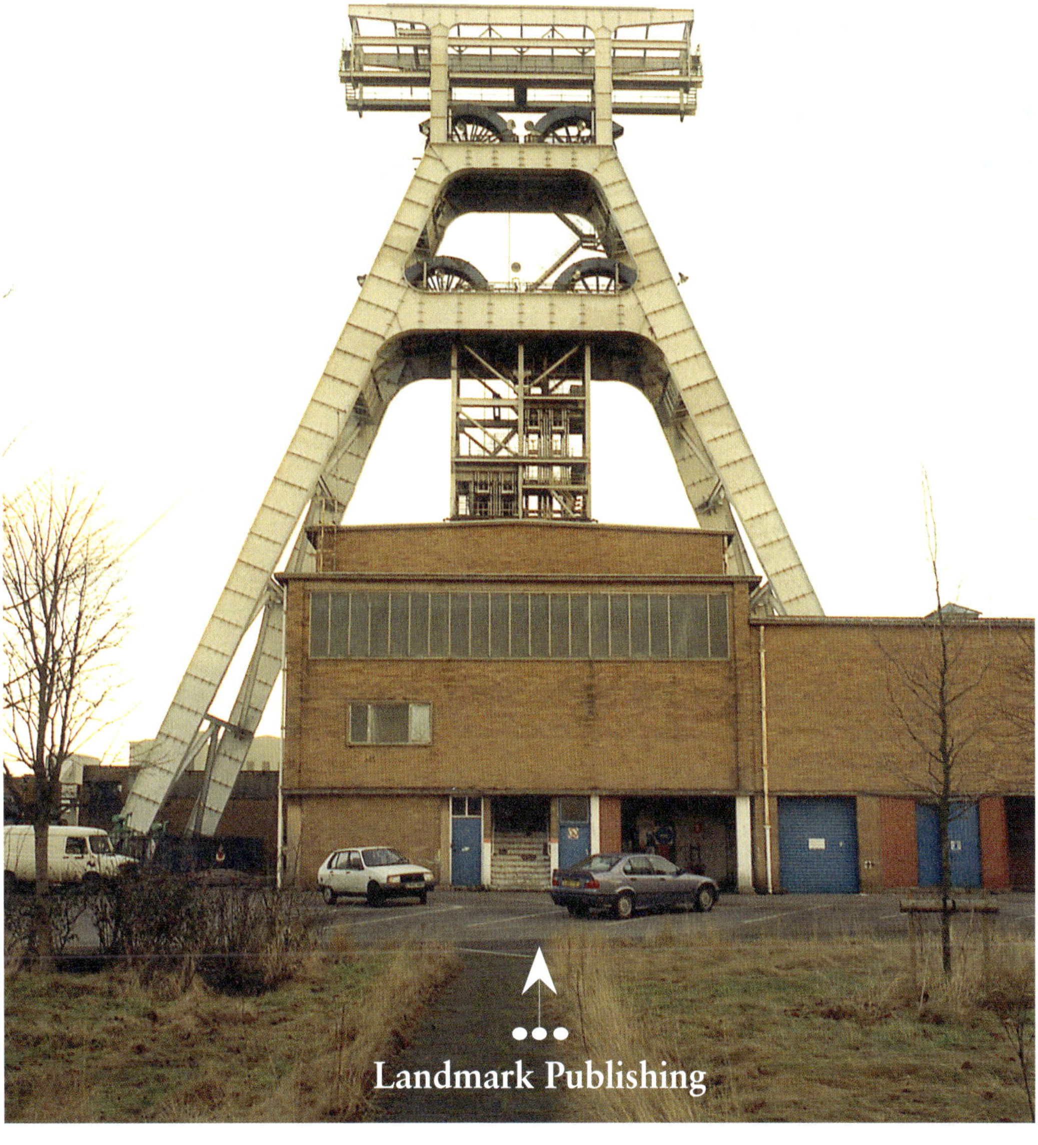

Landmark Publishing

Published by

LANDMARK
Publishing Ltd

Ashbourne Hall, Cokayne Ave
Ashbourne, Derbyshire DE6 1EJ England
Tel: (01335) 347349 Fax: (01335) 347303
e-mail: landmark@clara.net
web site: www.landmarkpublishing.co.uk

1st edition

ISBN 1 84306 138 4

British Library Cataloguing in Publication Data: a catalogue
record for this book is available from the British Library.

Printed by Cromwell Press Ltd, Trowbridge, Wiltshire

Design & reproduction by James Allsopp

Cover captions:

Front cover: The headstocks at Hem Heath No 2 Shaft (PD)

Back cover Top: 1062 Yard horizons, Hem Heath
Back cover bottom left: Monument to mining in Silverdale
Back cover bottom middle: Demolition of No 2 Headstocks, Hem Heath
Back cover bottom right: Level in Haying Wood No 2 colliery

Page 1: The 612 yard pit bottom, Hem Heath, looking north (PD)
Page 2: Rudi Leteller chewing tobacco, Silverdale (PD)
Page 3:
The headstocks at Hem Heath No 2 Shaft (PD)

CONTENTS

Preface

This book records the passing of one of the two most dominant industries of the Potteries: coal mining. Many of the photographs are from 1951 or from the 1980s-90s. Some 29 collieries are covered. This is not an historical account of coal mining in the North Staffordshire coalfield; it is a photographic record of most of the pits which were Nationalised in 1947. During the 1980s and 1990s, I was fortunate to be able to photograph some of the surviving pits underground, both in deep mines and in footrails, which were relatively shallow private mines.

This collection of photographs have come mainly from my own collection and the Metcalfe Collection (MC), with additional scenes from the collection of Albert Baines-Davies (ABD).

Although Chatterley Whitfield opened as a mining museum after its working life was over, it is currently closed, its future uncertain. Surely this record breaking mine site should be reopened; this former integral part of daily life for so many men and boys needs to be kept for future generations to understand just what it was all about. At so many mine sites, no vestige of mining now remains.

All is not quite lost, however, for Apedale Colliery now has a museum and underground visits are included as part of the tour. It deserves support; do go and visit it. More details of this venture are on page 159.

In the text accompanying the illustrations, shaft depths are taken from the NCB's statutory mine plans. The Company name given below the name of a colliery is that of the owner at the time the mine passed to the National Coal Board on 1st January 1947.

Acknowledgements

I wish to acknowledge, with thanks, the assistance I have received from: Mike Arthur; Albert Baines-Davies; Ray Hammersley; Allan Hand; Eddie Hand; Bob Lewis; Barry, George and Graham McCreedy; Bob Metcalfe and Andy Mifflin.

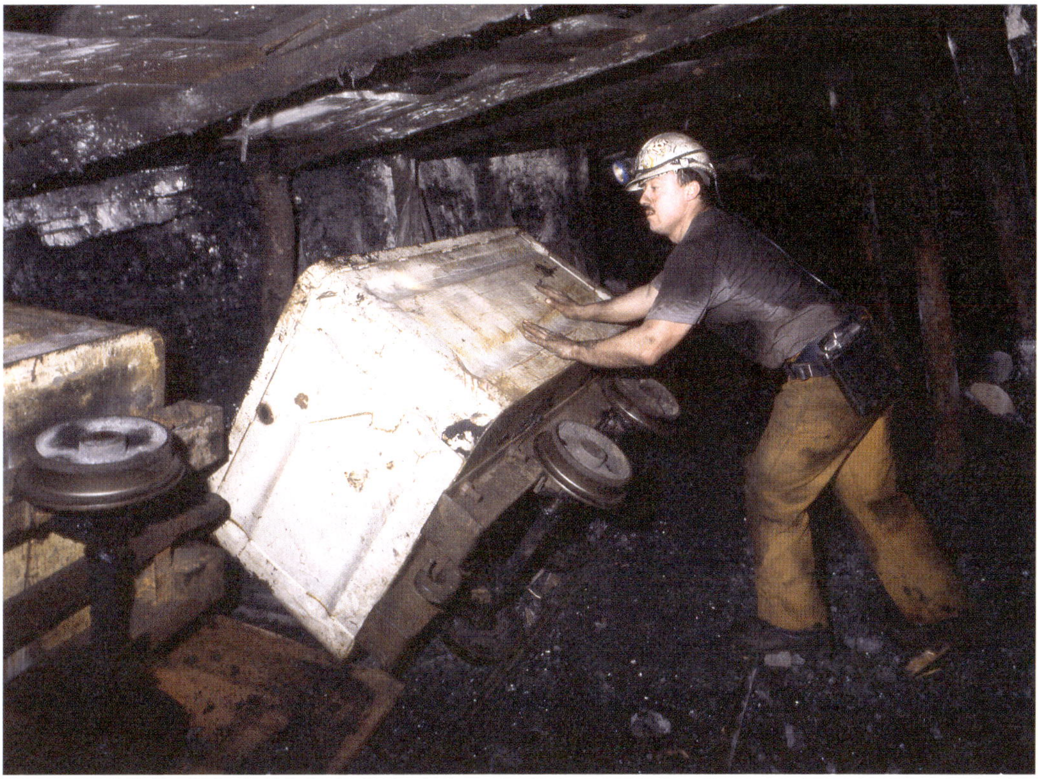

Above: Siding over a tub in Haying Wood No 2 Colliery. PD February 1993.
Below: Kevin Pritchard driving the shearer on Ragman 3s Face, Silverdale Colliery. PD Dec. 1998

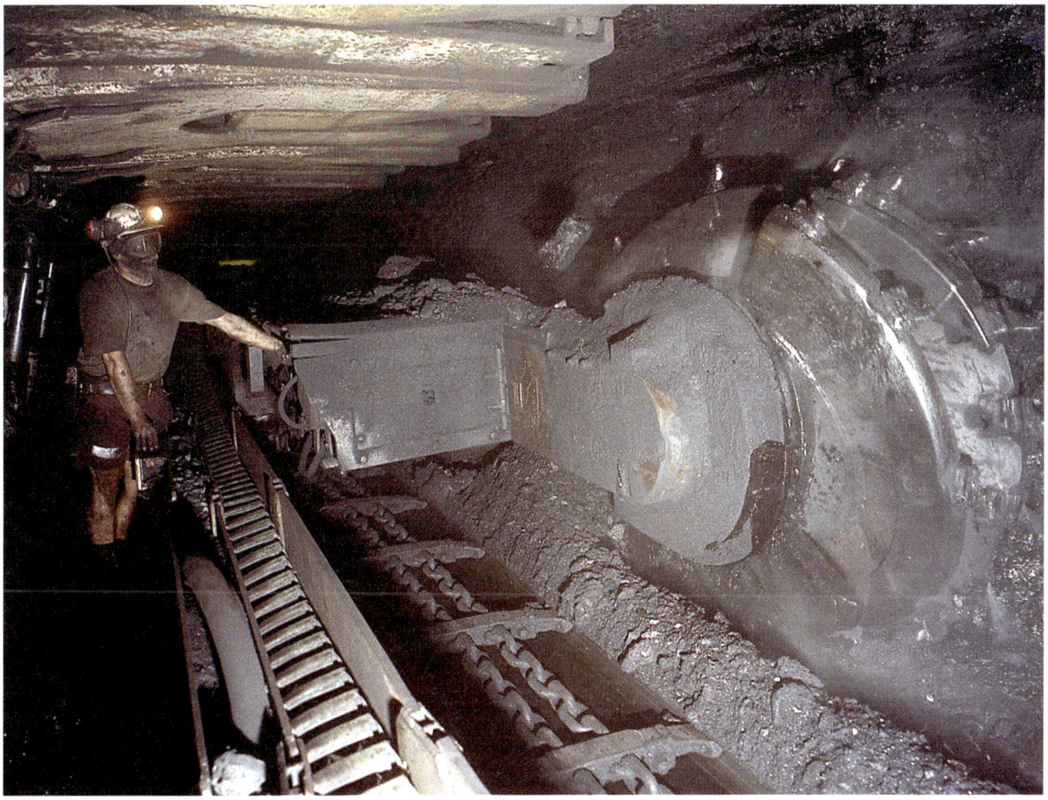

APEDALE

Operated by Holditch Mines Ltd. at Nationalisation (1st January 1947), later by Great Row Colliery Co. and then by Aurora Mining Co. Ltd. The present mine started in the 1920s to work the coal that had previously been sterilised by the Apedale Ironworks and ceased working in September 1998. There were nine adits and two shallow ventilation shafts.

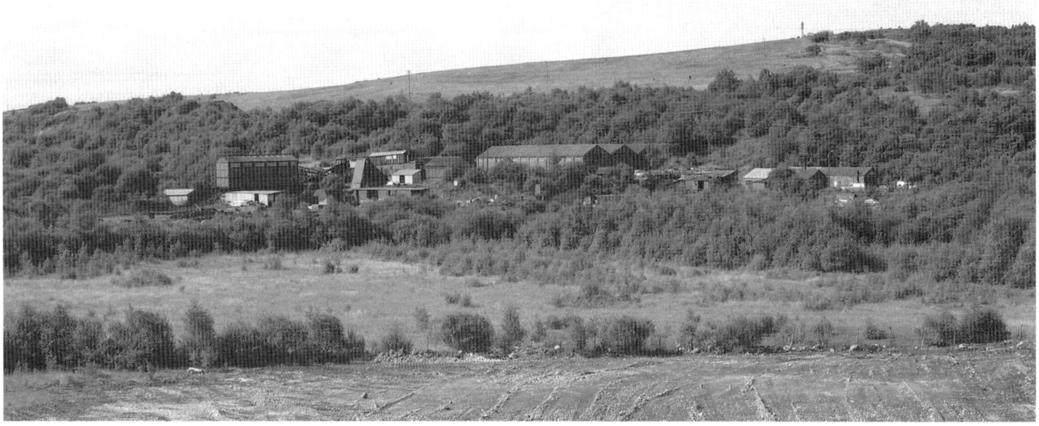

Above: A general view of the colliery, taken from the 'White Rock'. PD

Below: Numbers 1 & 2 Pits; No. 1 is on the left and is a steep adit (35° or 1 in 4) driven all the way in the Great Row seam. The No. 2 adit was initially a shallow-dipping cross measures drift until contacting the Spencroft Seam which it then followed. Both scenes taken in 1987. PD

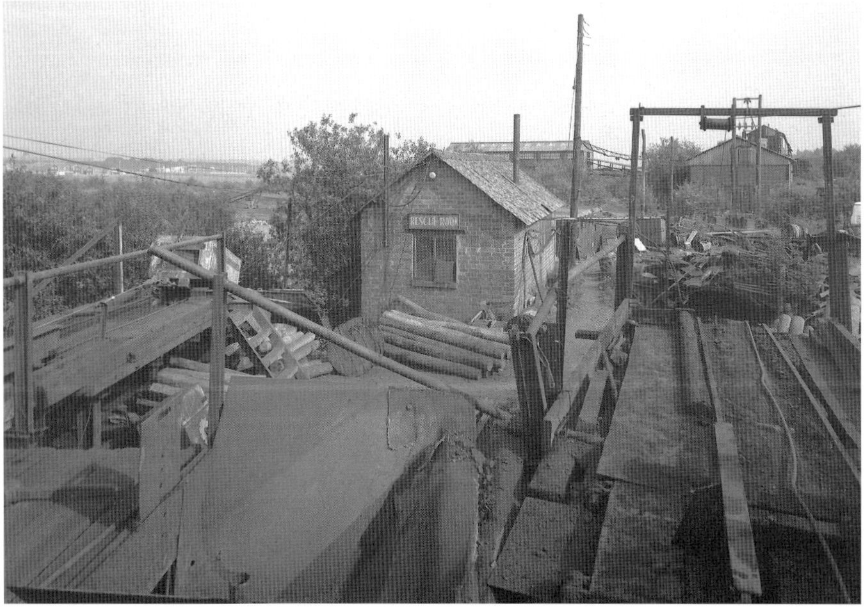

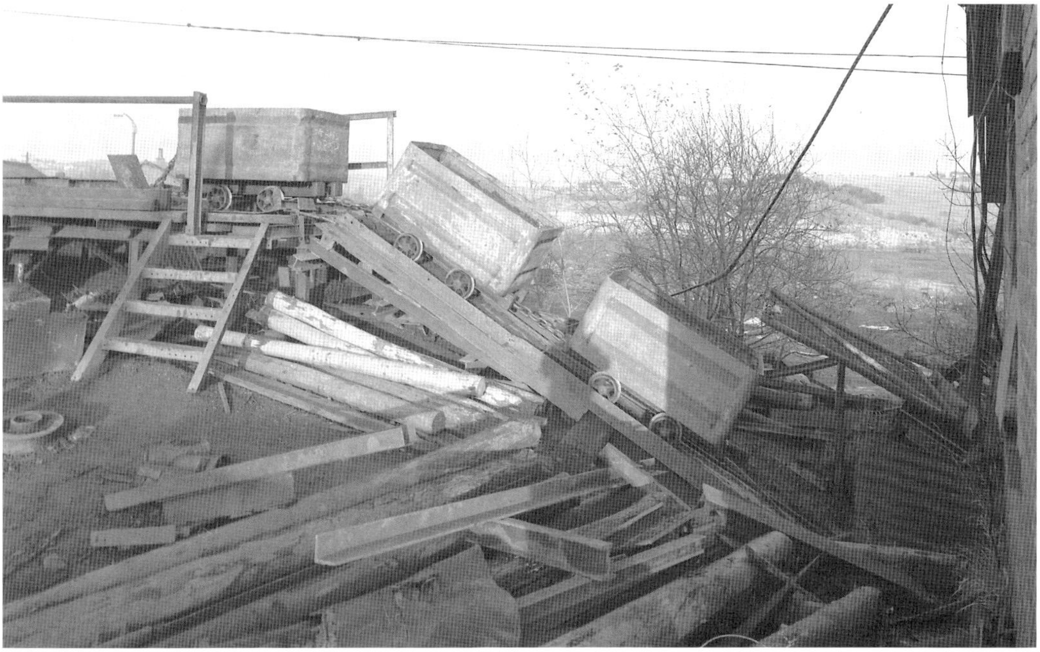

Above: A journey of three tubs entering the No. 1 adit. PD

Below: The tipping arrangements at Nos. 1 & 2 Pits with Len Goulding, the engine man, tipping a tub of Great Row coal. Taken in 1987. PD

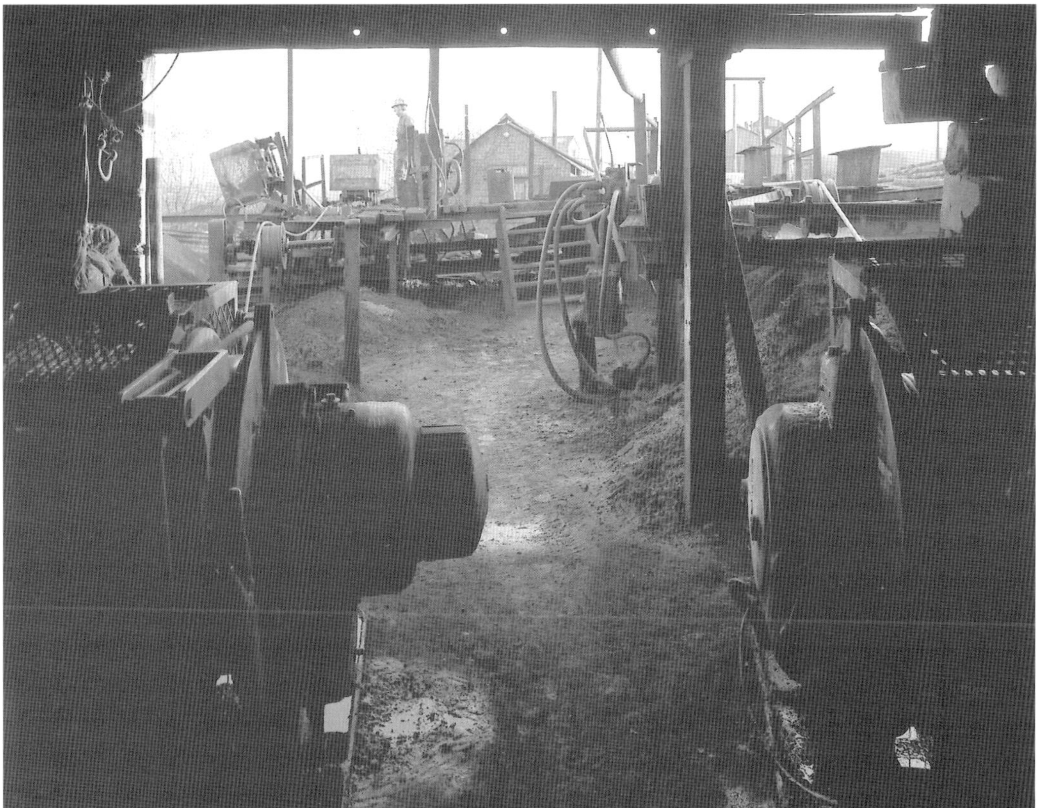

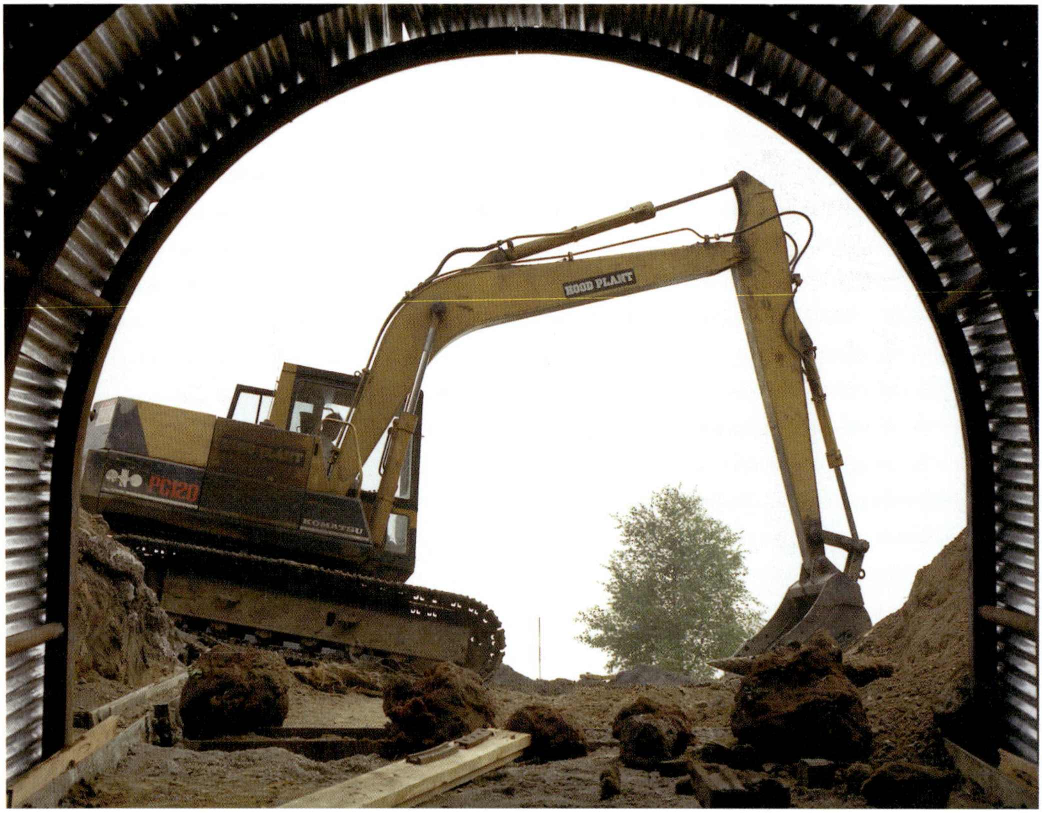
HOOD PLANT
PC120
KOMATSU

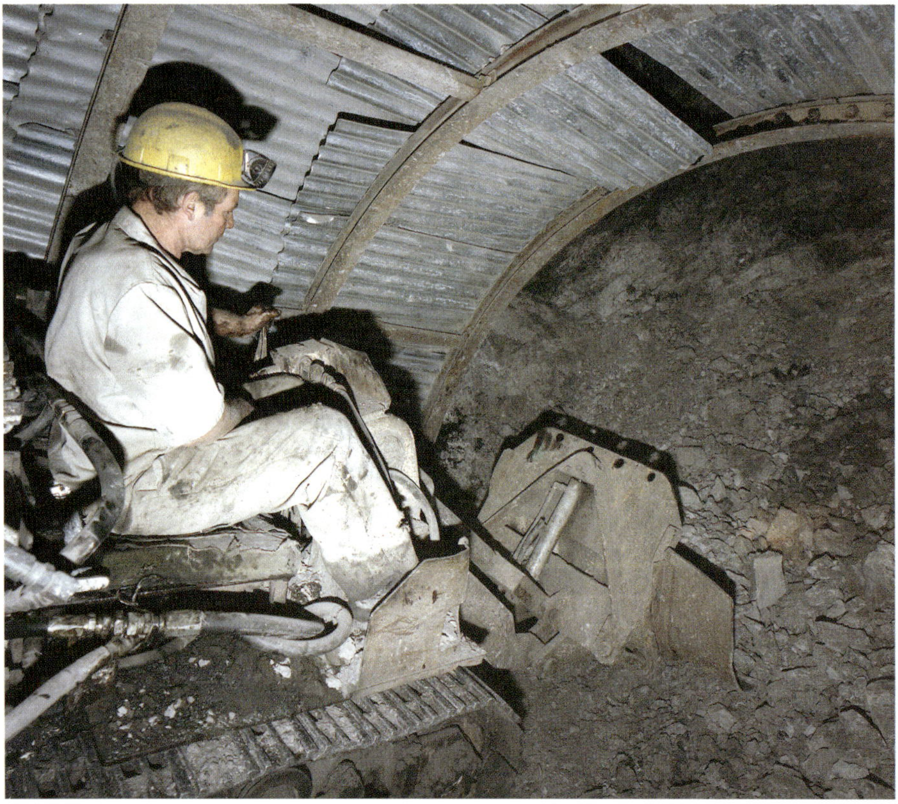

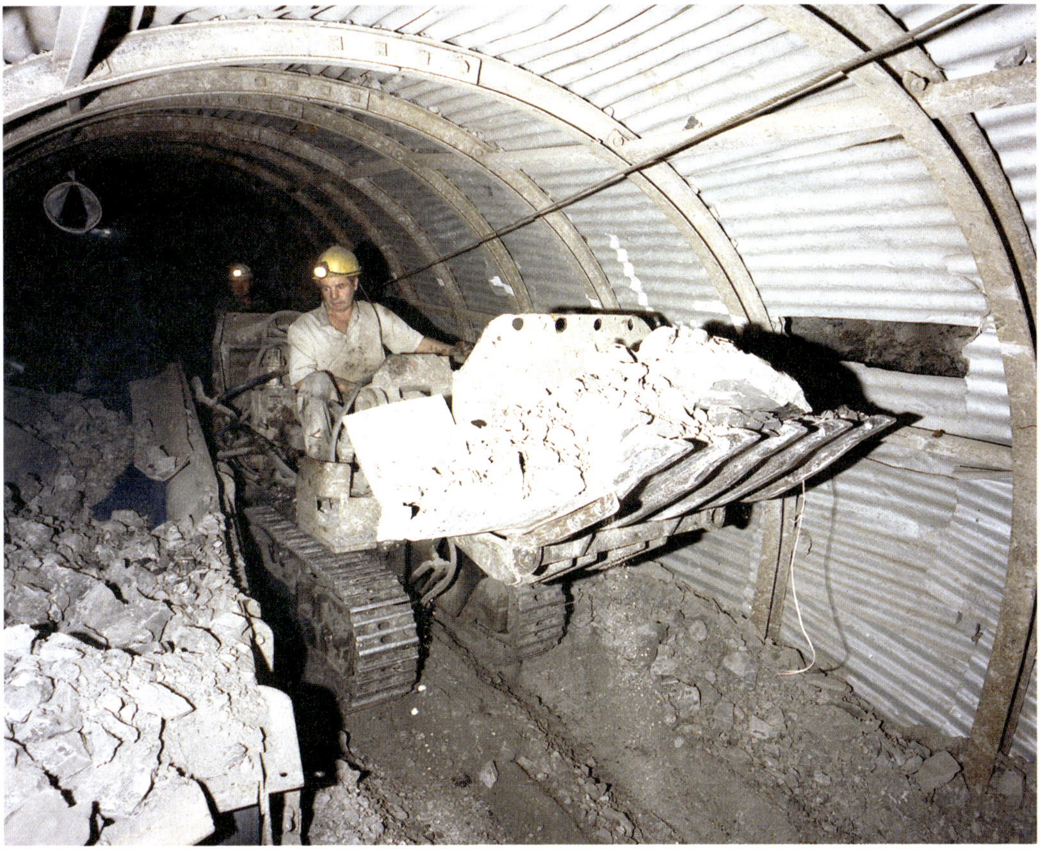

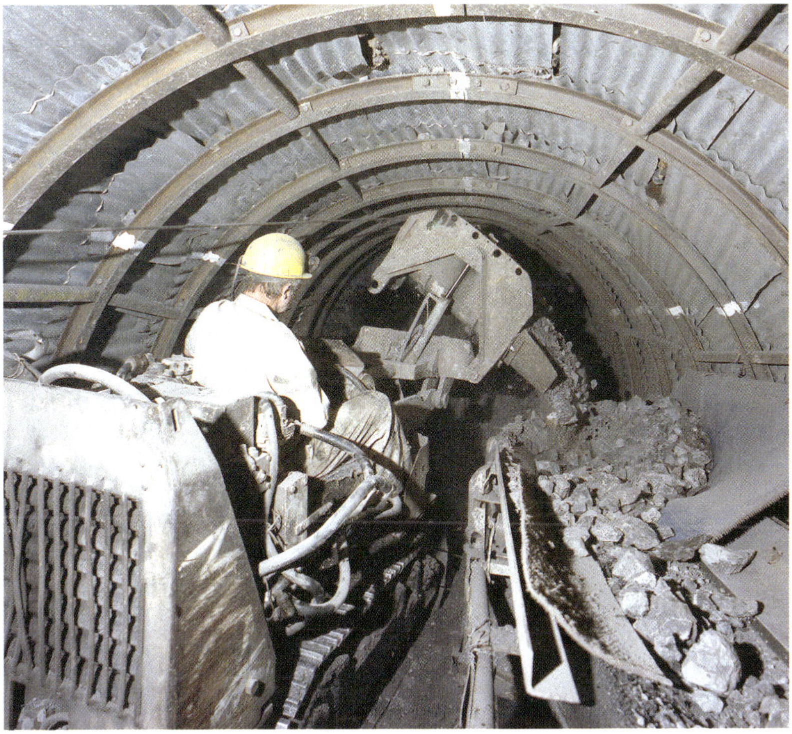

Opposite page top: No. 7 Pit during construction in 1988-89. PD

Opposite page bottom & this page: Lindsay Tooth driving a Mindev 90 side loader during the driving of No. 7 Pit. PD

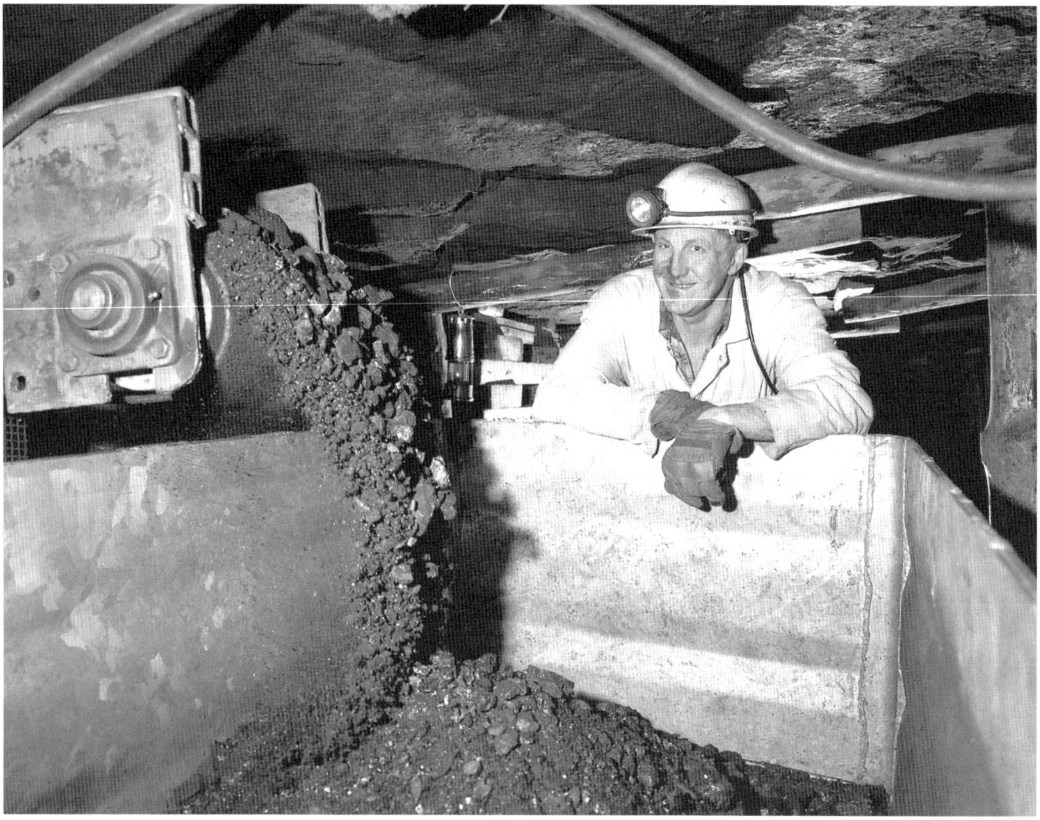

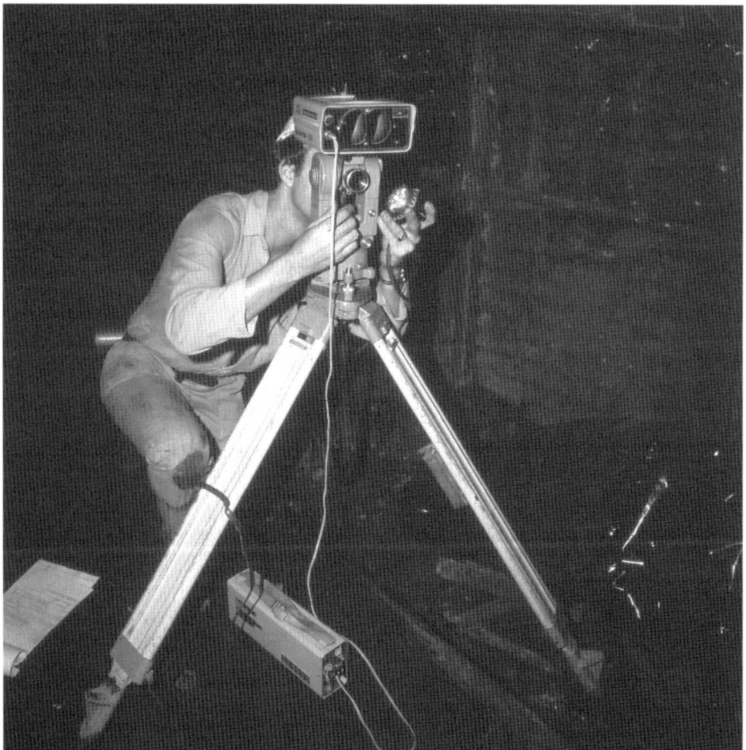

Above:
Arthur Cartwright, the Colliery manager, watching a tub being filled in the No. 4 Pit (Bassey Mine) workings, in 1991. PD

Left:
Allan Hand undertaking a theodolite survey in Apedale No. 5 Pit in 1983. PD

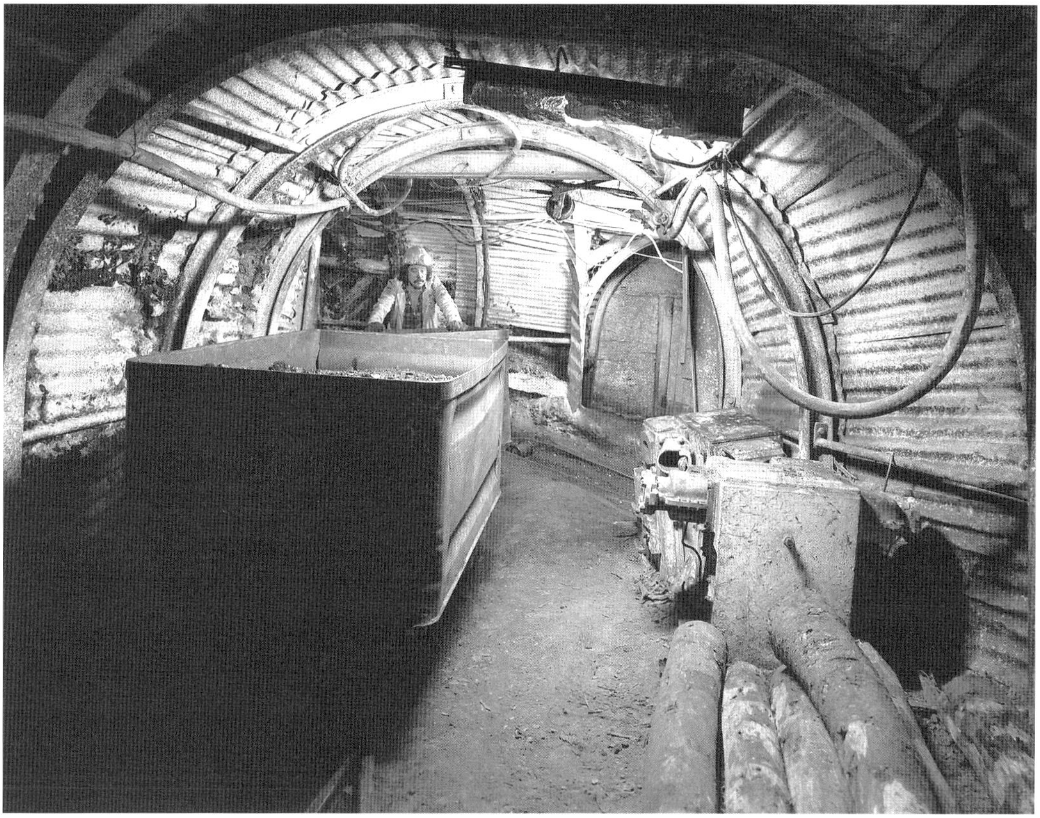

Above:
Alan Deaville pushing a tub of coal on the No. 5 Peacock Landing in 1983. PD

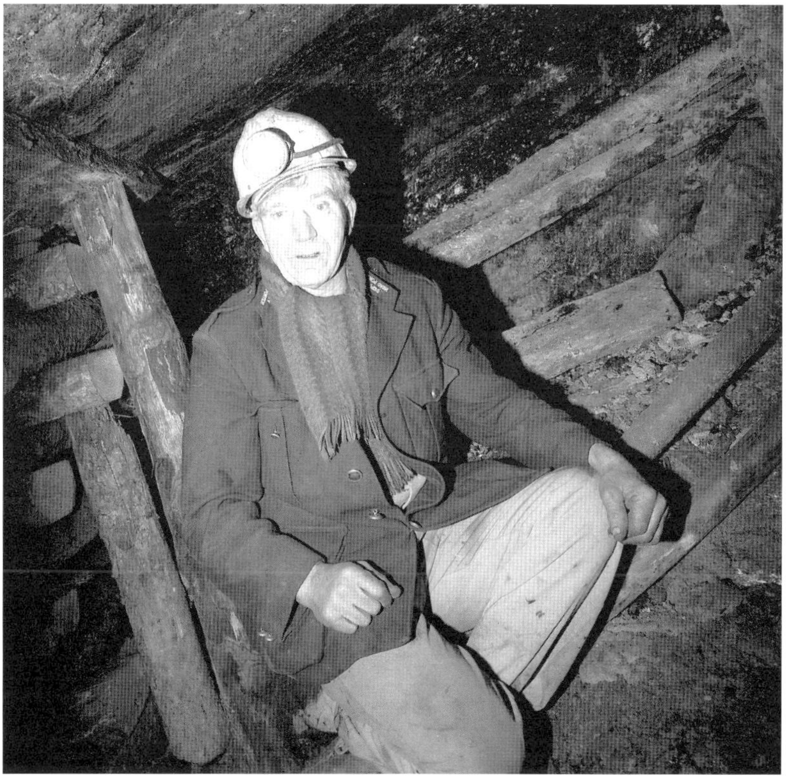

Right:
Joey Thomasson taking a break in No. 1 Pit, Great Row Dip, 1984. PD

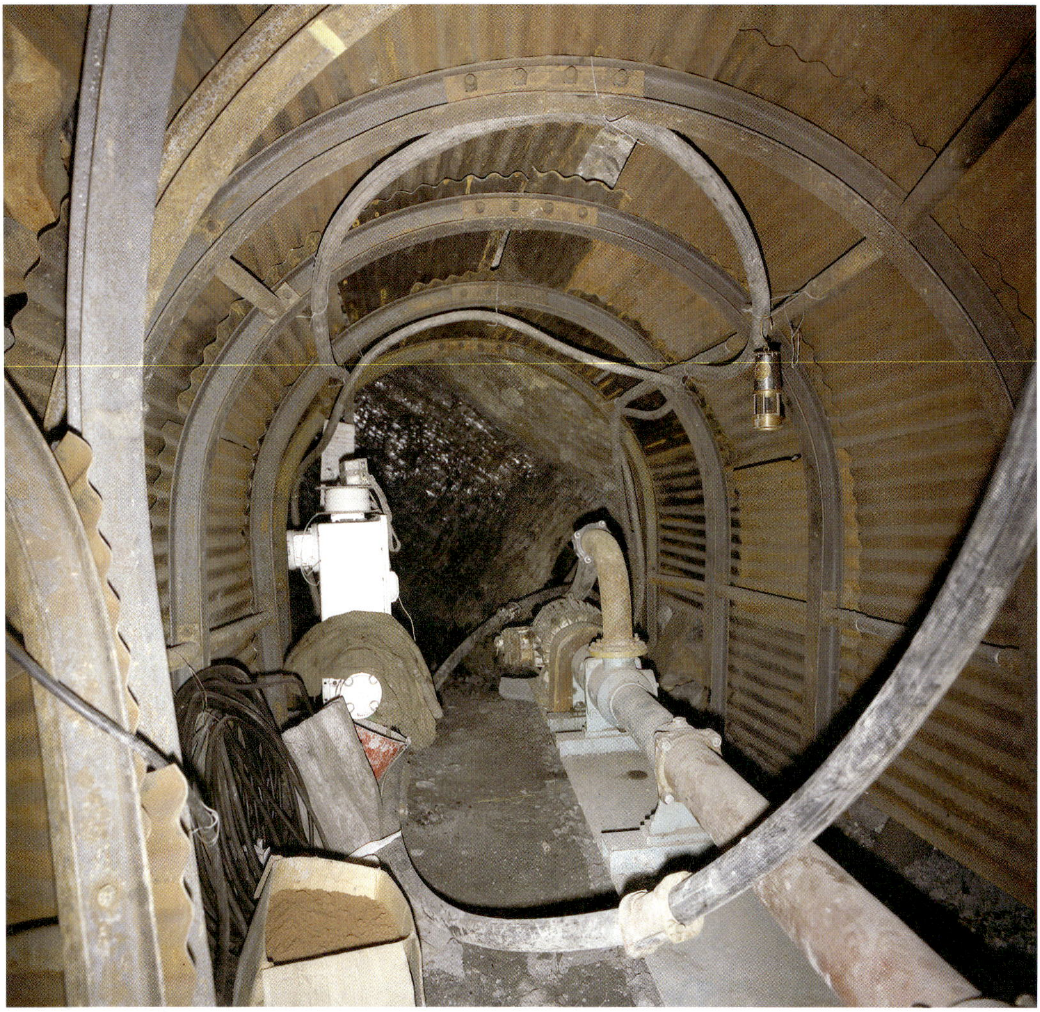

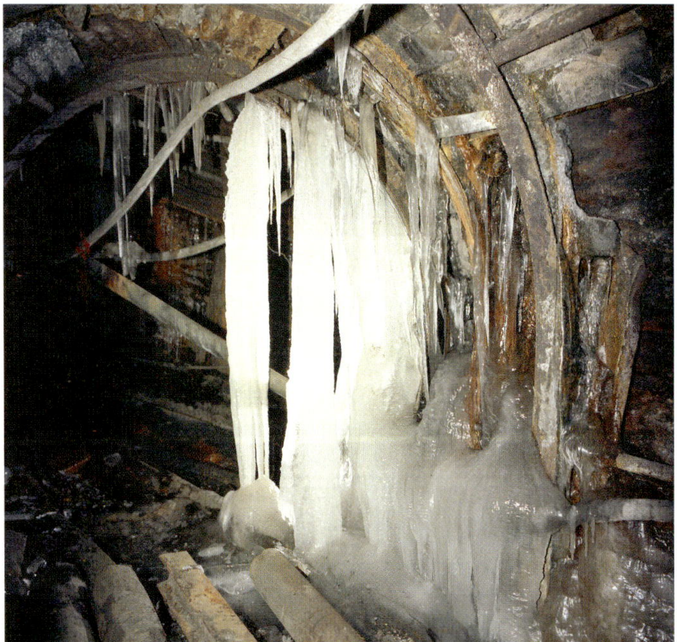

Above: No. 6 Pit, at the Peacock Pump lodge. PD

Left: Severe ice formations at Apedale No. 5 Pit, Great Row Landing, during the winter of 1984. PD

Opposite page;
Top: Arthur Cartwright in the No. 4 Pit (Bassey Mine) workings, 1991. PD

Below: A collier drilling shot holes in the same workings, 1991. PD

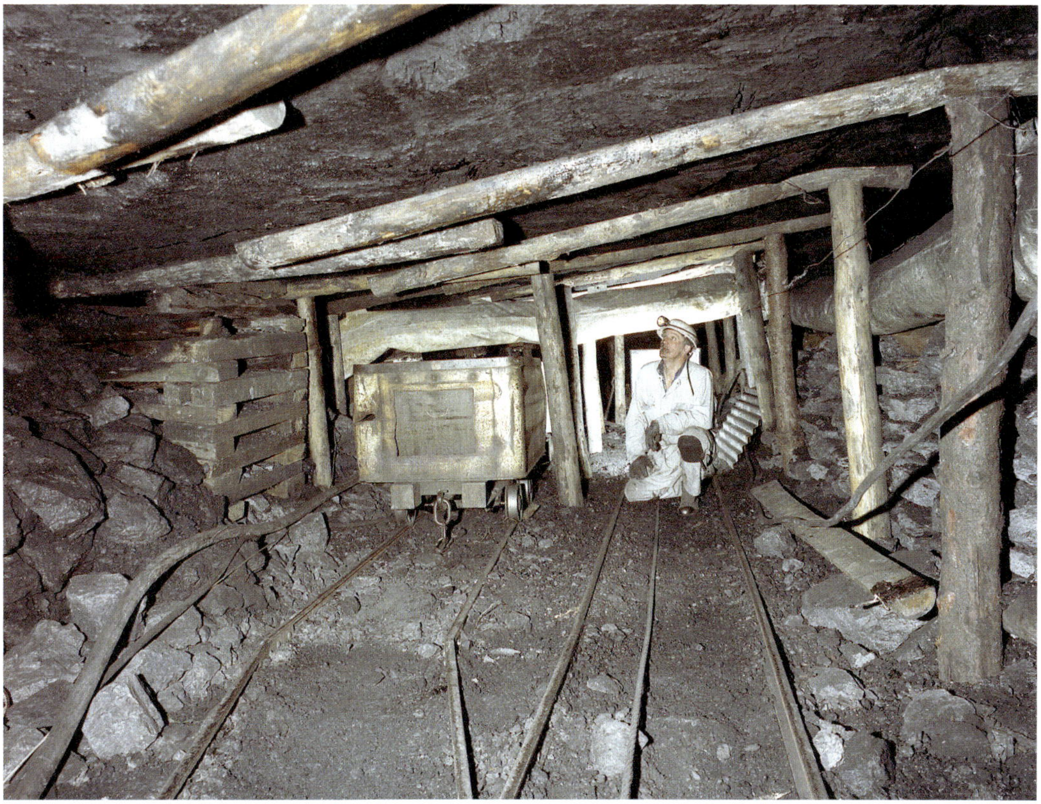

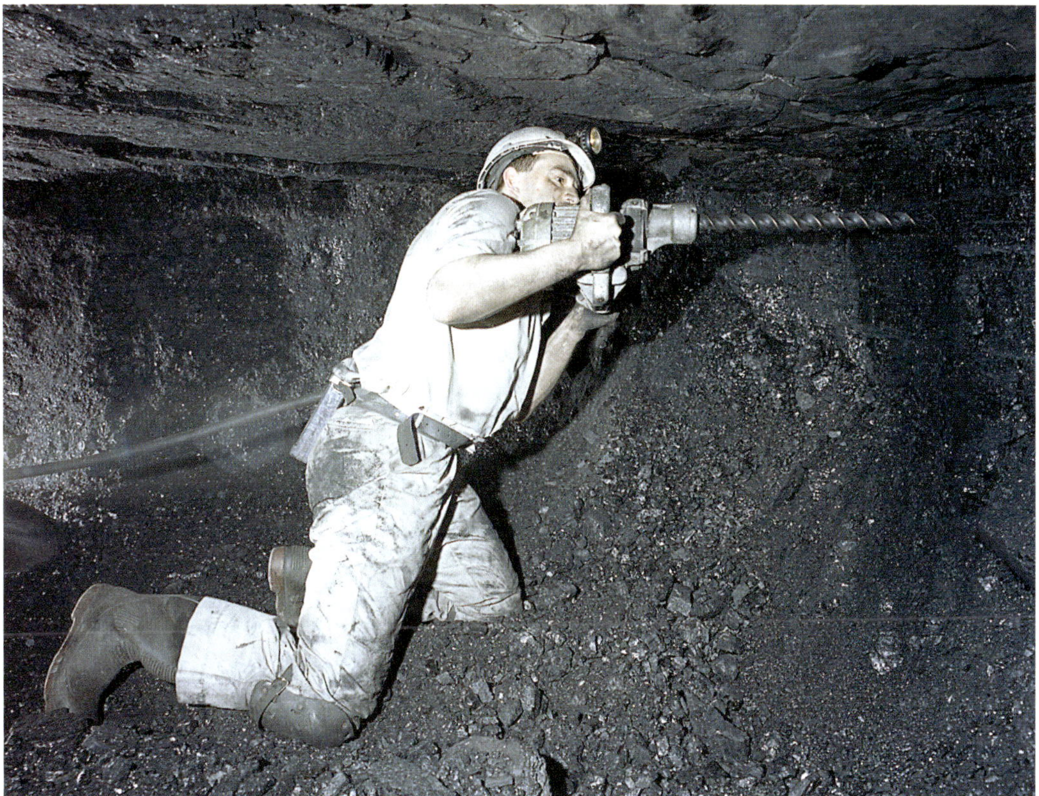

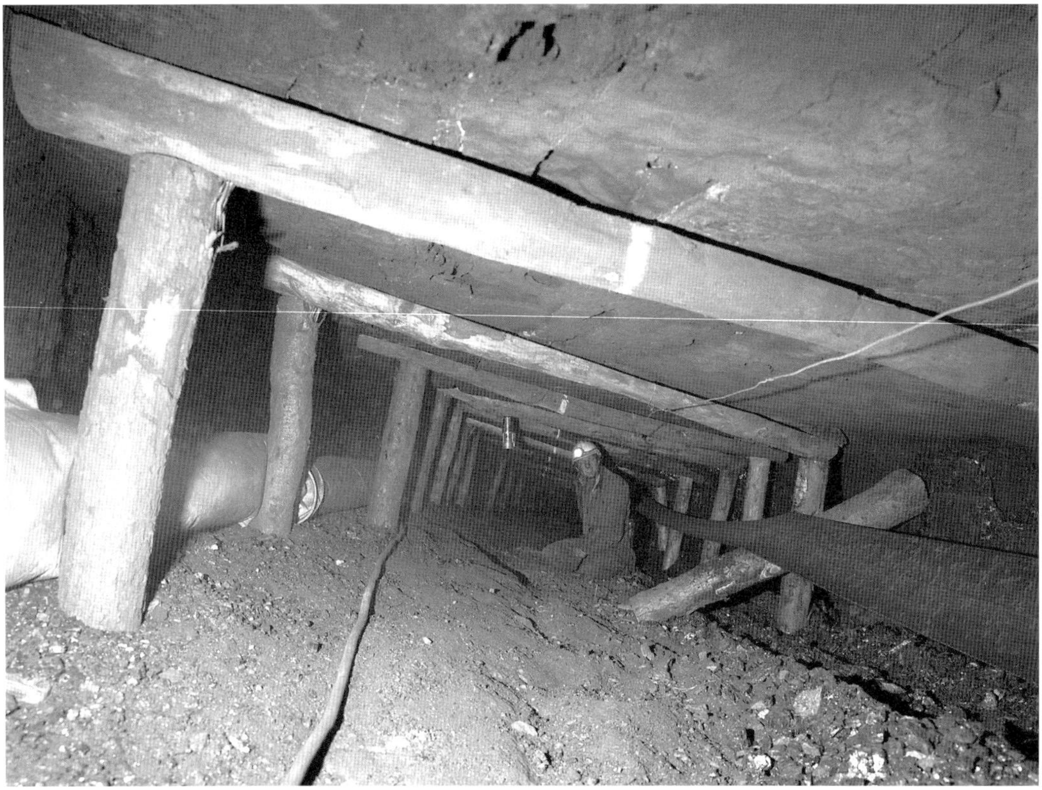

Above: No. 4 Pit (Bassey Mine) showing a view of a typical level. Note the conveyor belt on the right-hand side and the air bagging (ducting) on the left. 1991 PD

Below: The pit deputy scaling the roof in a No. 4 Pit (Bassey Mine) heading. 1991 PD

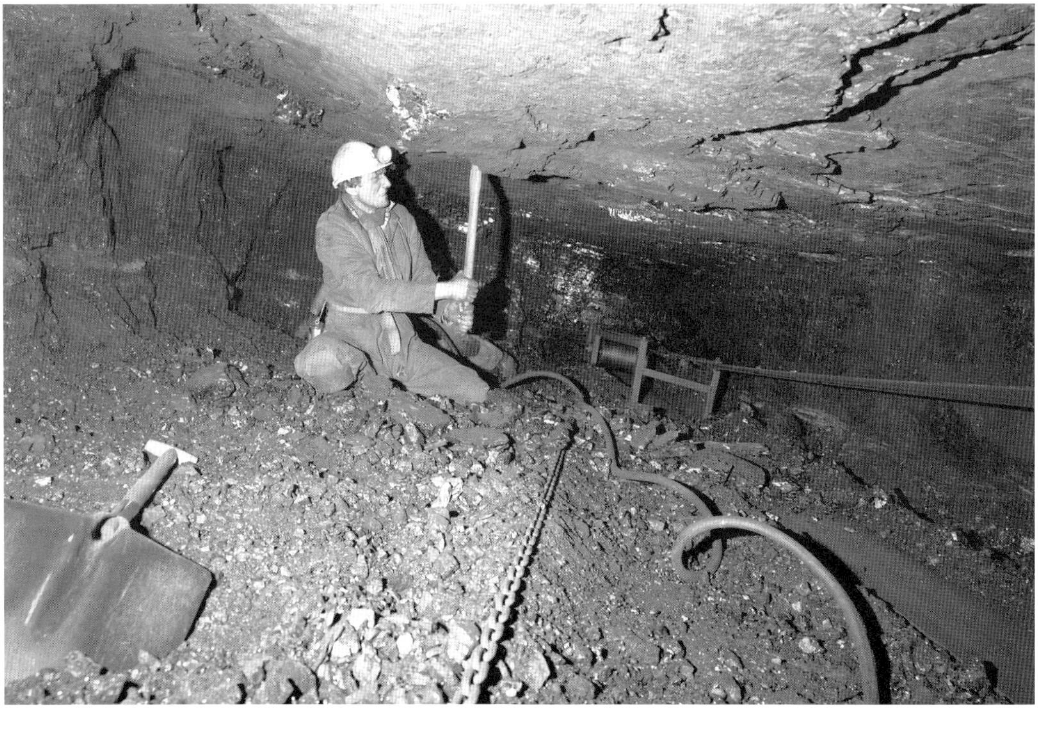

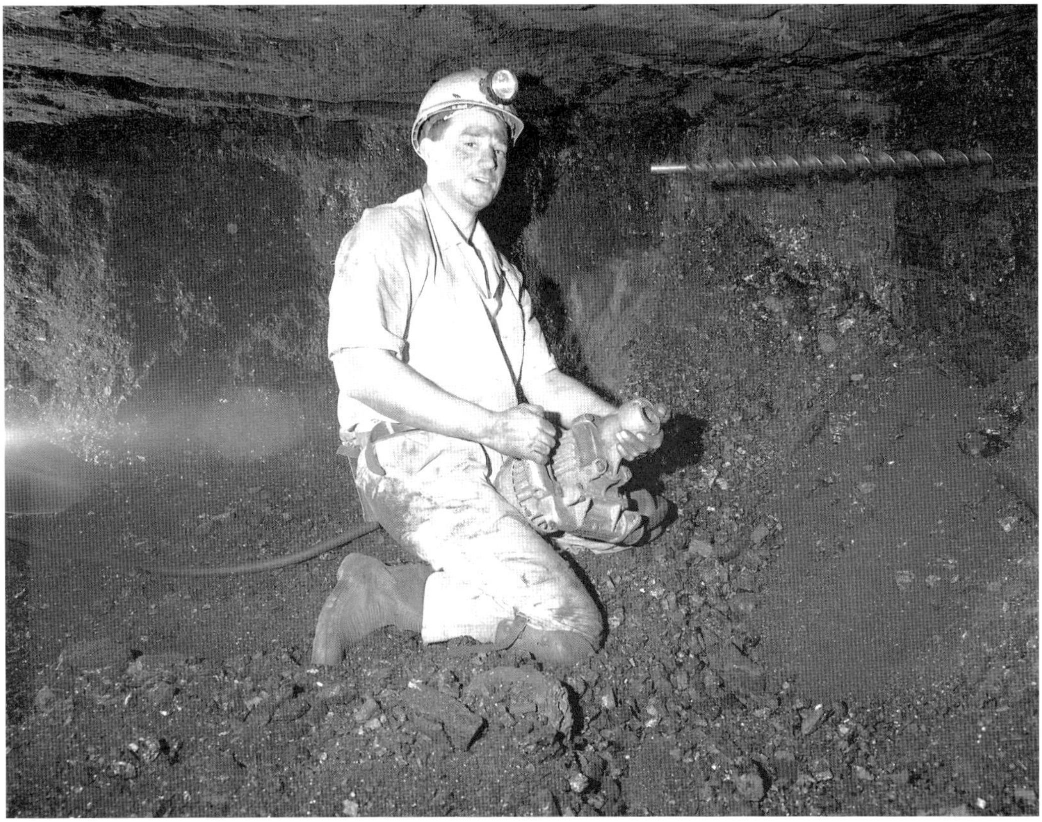

Above: A collier taking a break between drilling shot holes in No. 4 Pit (Bassey Mine).
Below: Hand loading onto a conveyor belt in the same mine. Both 1991. PD

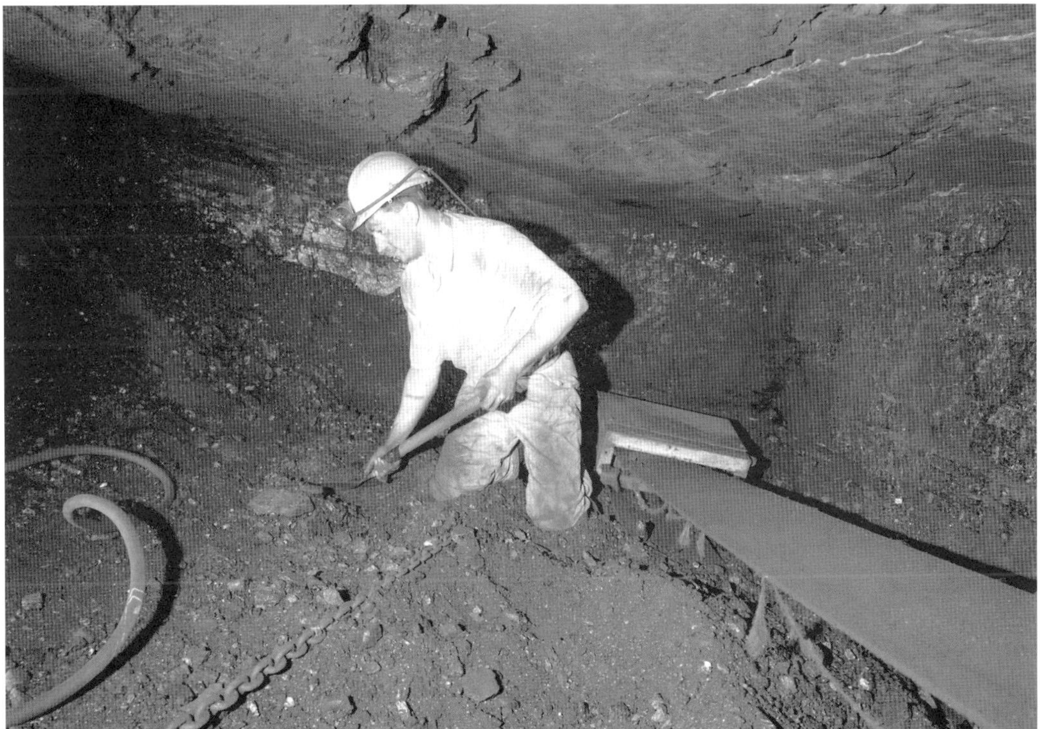

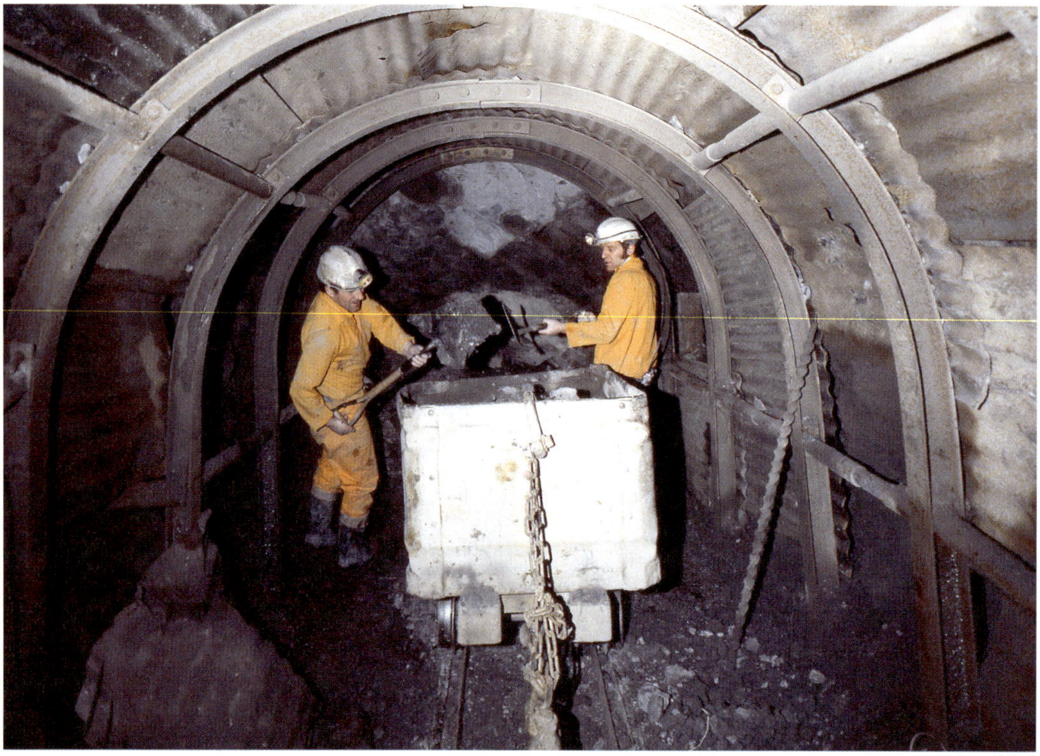

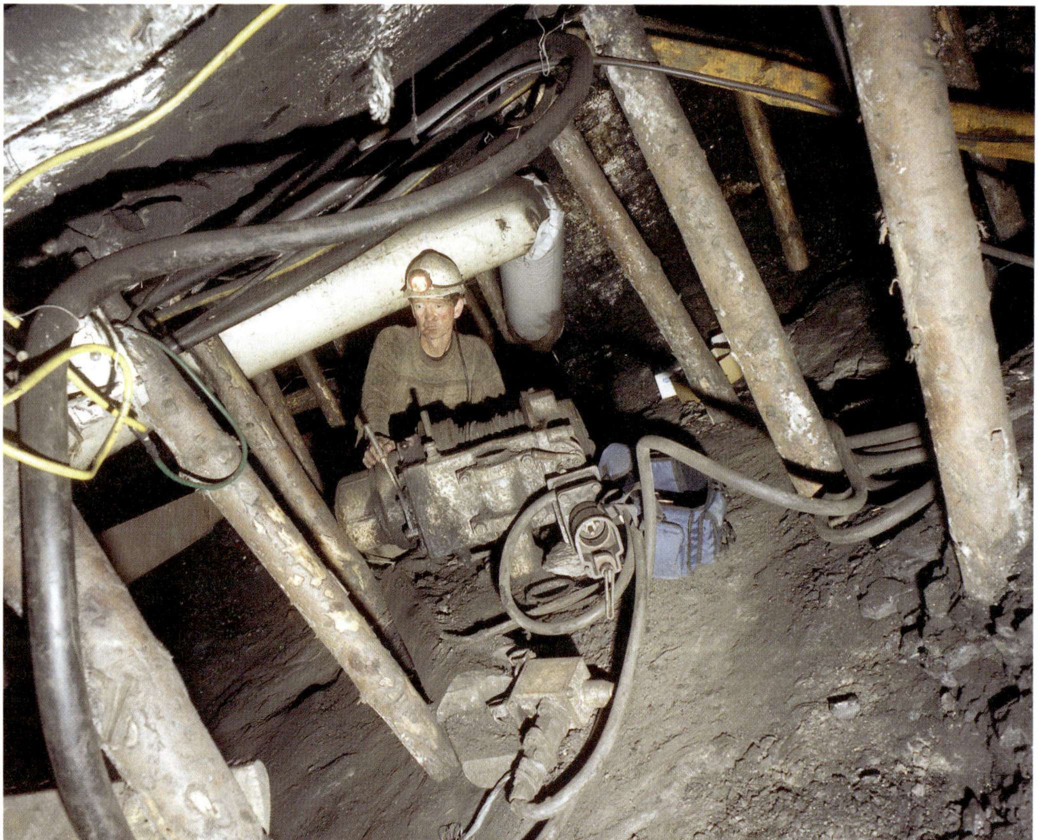

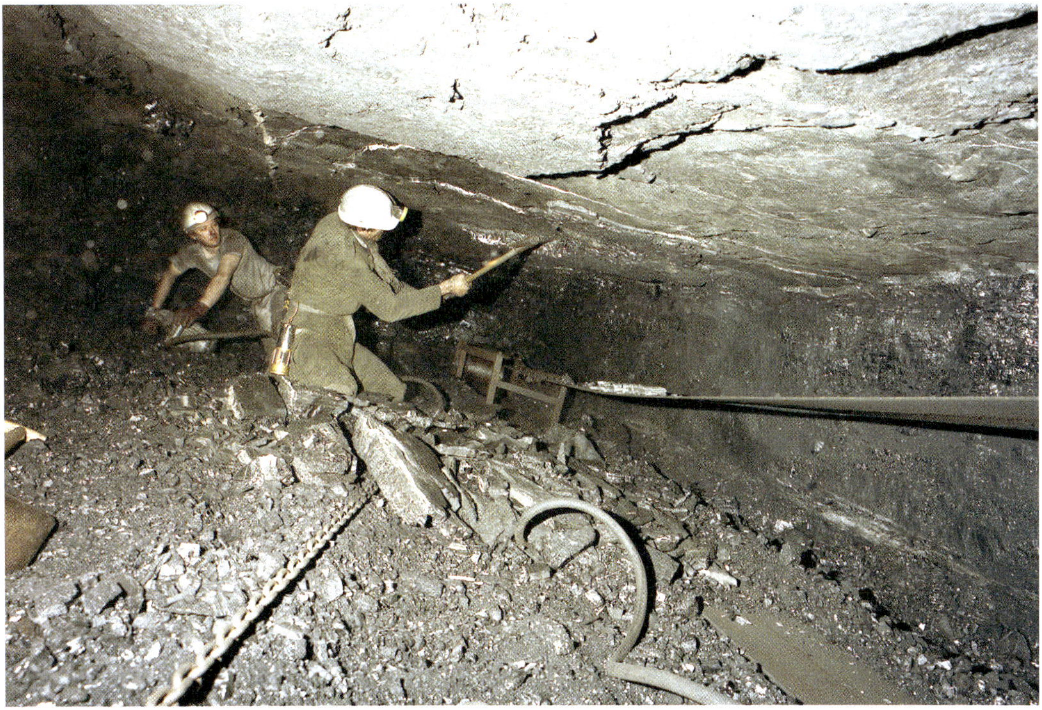

Above: No. 4 Pit with colliers scaling the roof in a Bassey Mine heading. 1991 PD

Below: No.6 Pit adit being driven through a near-surface sand deposit with Harry Hallam and Arthur Cartwright, 1983. PD

Opposite page; Above: Pat Keilly and Billy Smith hand loading a tub during the drivage of the No. 6 Pit crut. These men drove 220m of 1:2 ½ cross measures drift (or crut) in one year. They did all their own hand-filling too. 1983, PD

Below: A collier driving a small Pikrose winch in the No. 5 Pit, Chalky Seam workings. 1998, PD

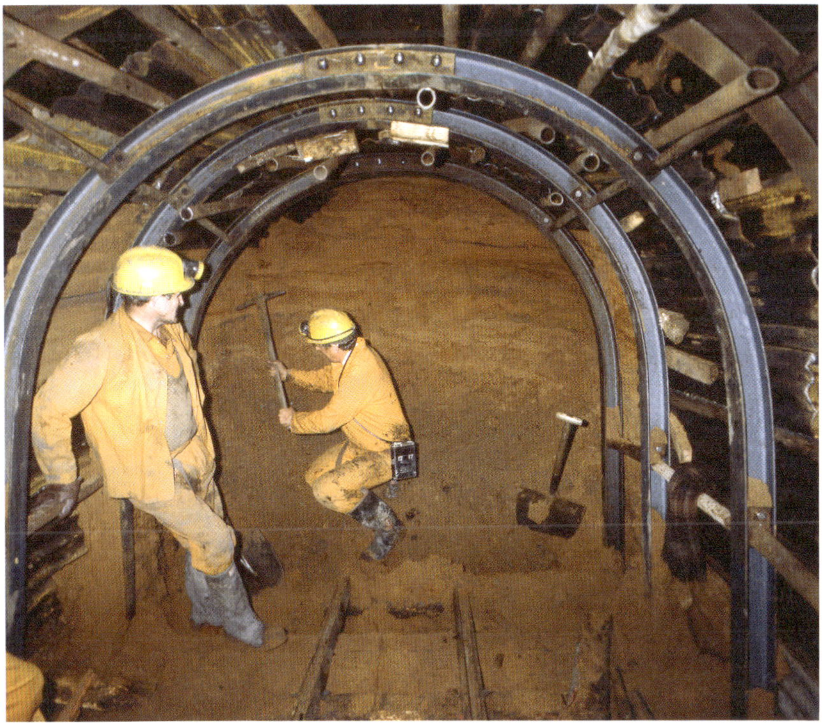

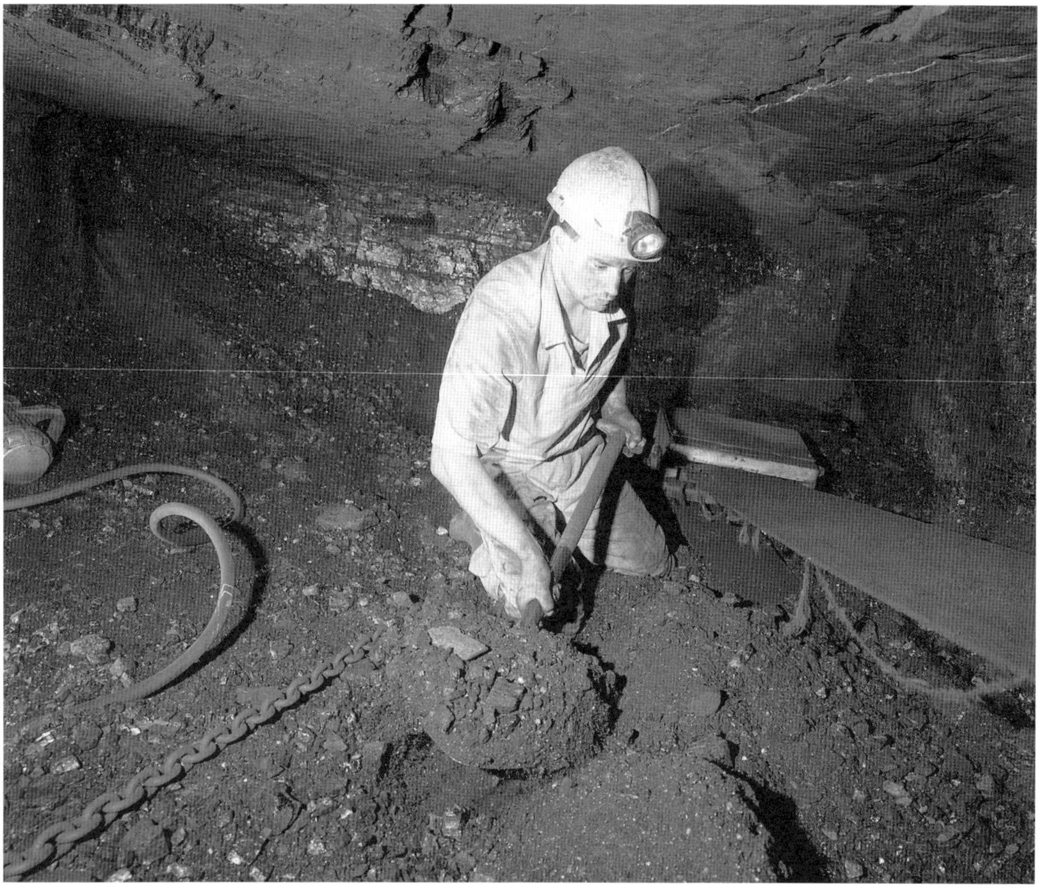

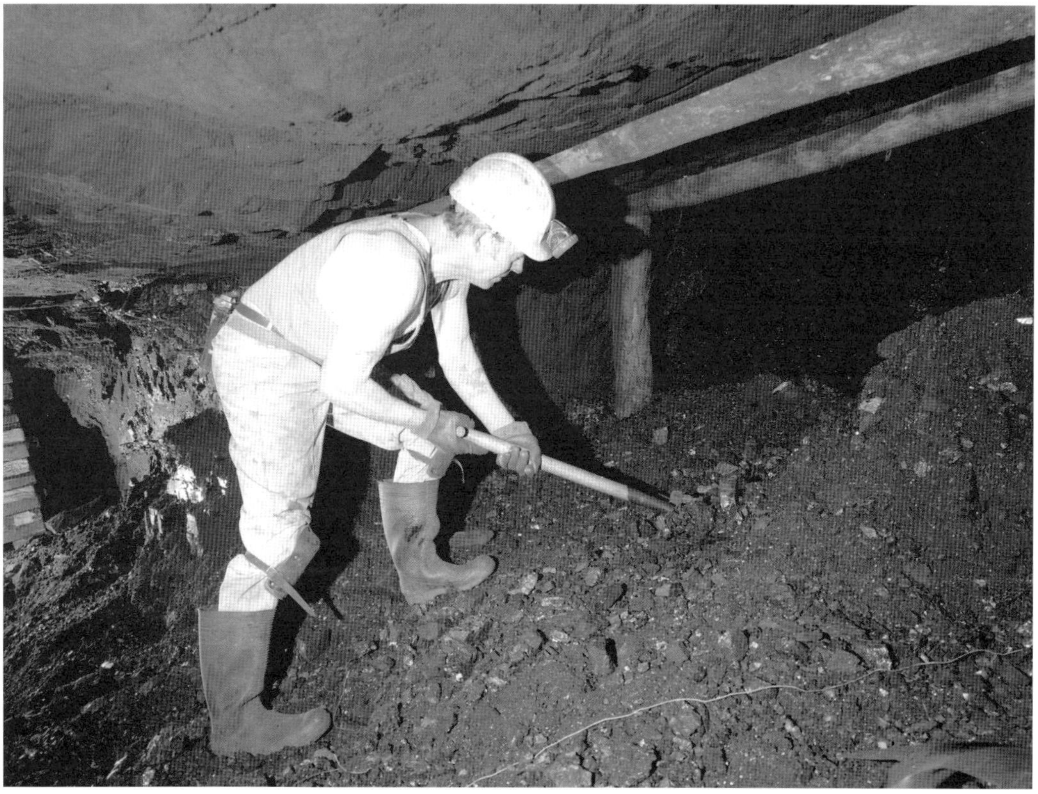

Opposite page;
Top: Another view of hand loading in No. 4 Pit (Bassey Mine), 1991. PD

Below: Roland Barnet hand loading coal at No. 4 Pit (Bassey Mine) 1991. PD

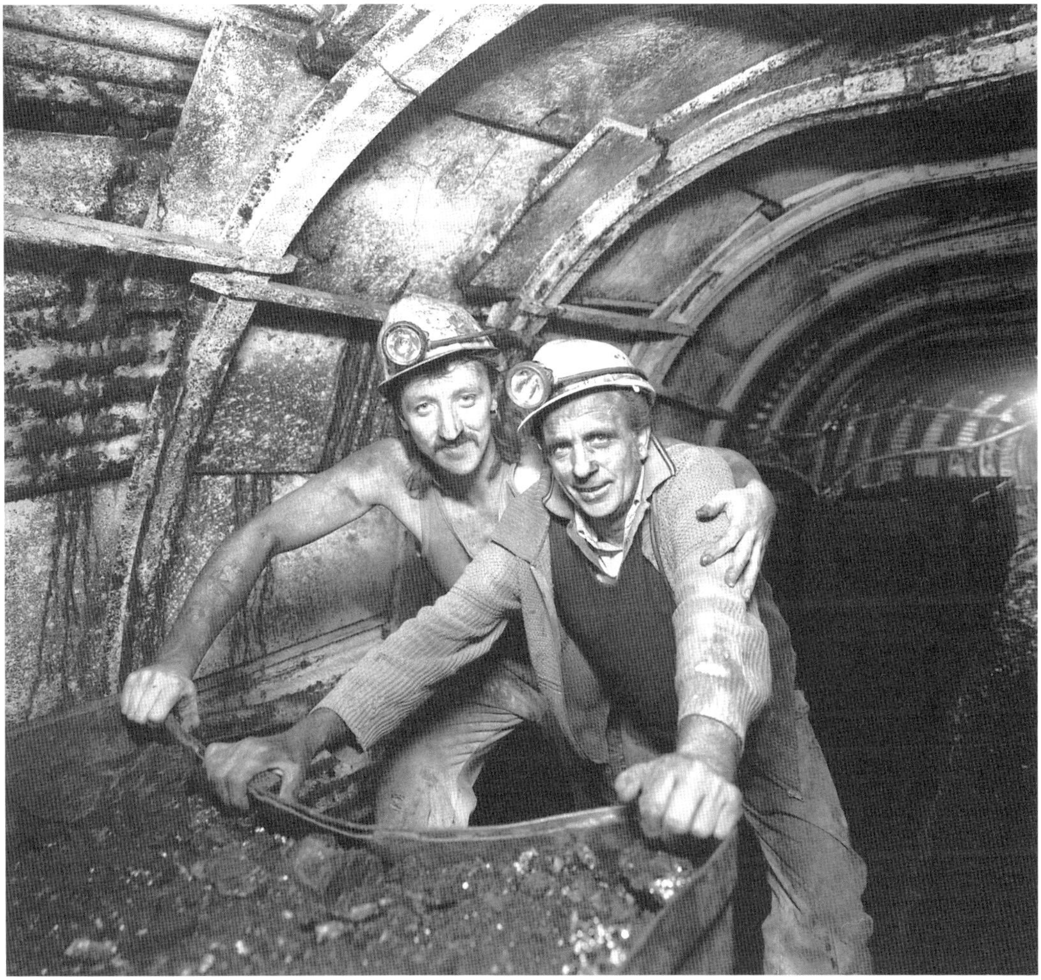

Jimmy Proudlock and Frankie Kirkham, Apedale No. 6 Pit, Cannel Row Landing, c. 1986. PD

The site is now preserved as the Apedale Mining Museum and Heritage Centre – see also page **159**.

Bank Top Colliery

Great Row Colliery Co.

Situated 1.6km west of Silverdale Church.

Two footrails operated on this site working the Winghay Seam from 1971-83 and the Rowhurst Seam from 1982-91. This view shows the surface engine house. Photo: PD. The site has been restored to agriculture.

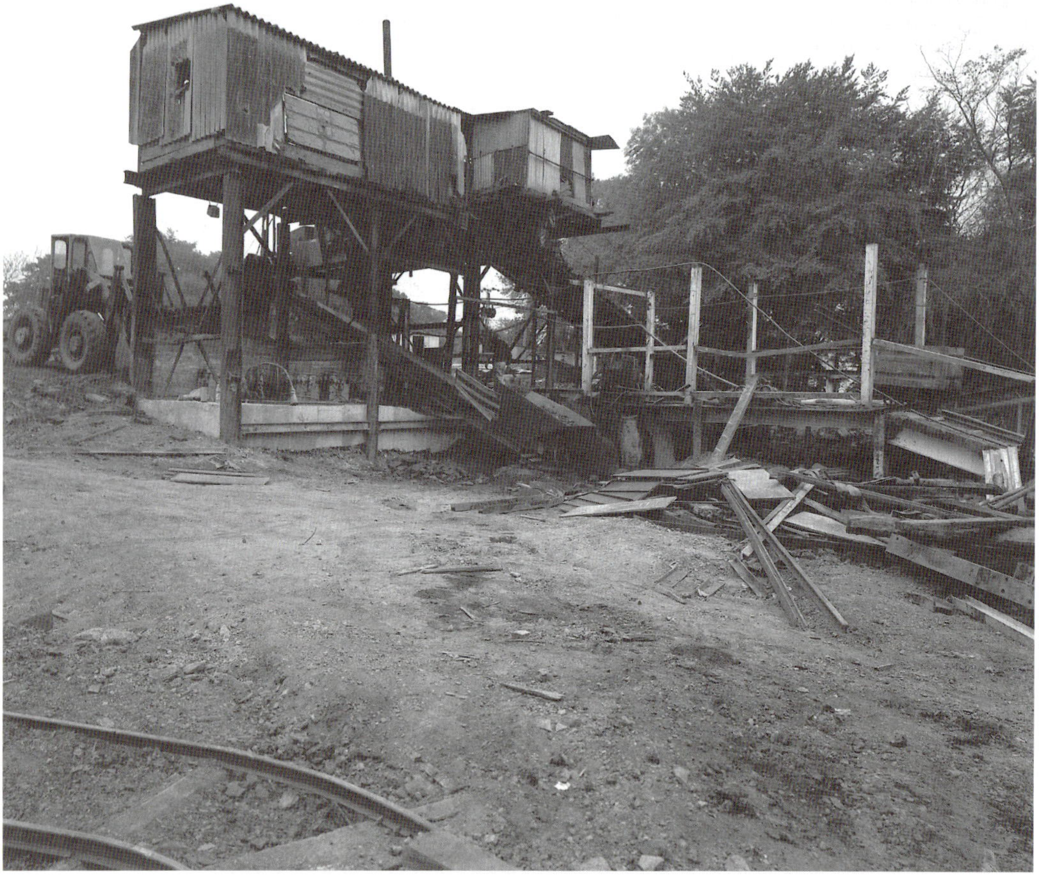

Berry Hill Colliery

Berry Hill Colliery Ltd.

Situated 1,600m north of Fenton and 400m east of Victoria Road. The colliery was on two sites and at the time of closure had four shafts: No. 1 Pit (New Pit), sunk to 684m in 1932; Knowles Pit, 390.4m; Bush Pit, 311.8m; Top Pit, 509.3m. Top, or Deep, Pit lay about 800m to the north-east. These three shafts were sunk in 1863.

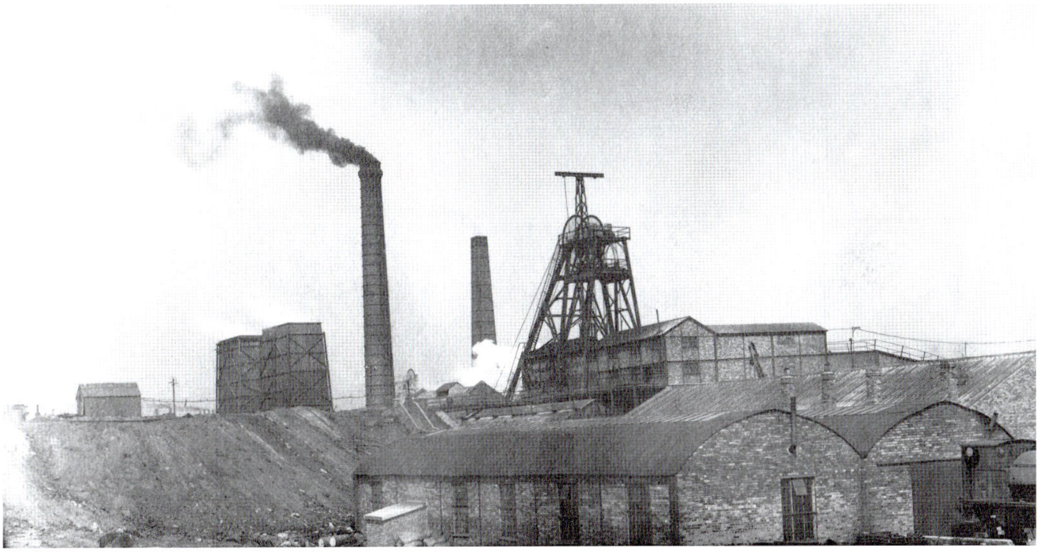

Above: No. 1 Pit in 1951. MC

Below: No. 1 Pit during demolition. The mine closed in April 1960 and the photograph was taken on 14th February 1961. MC

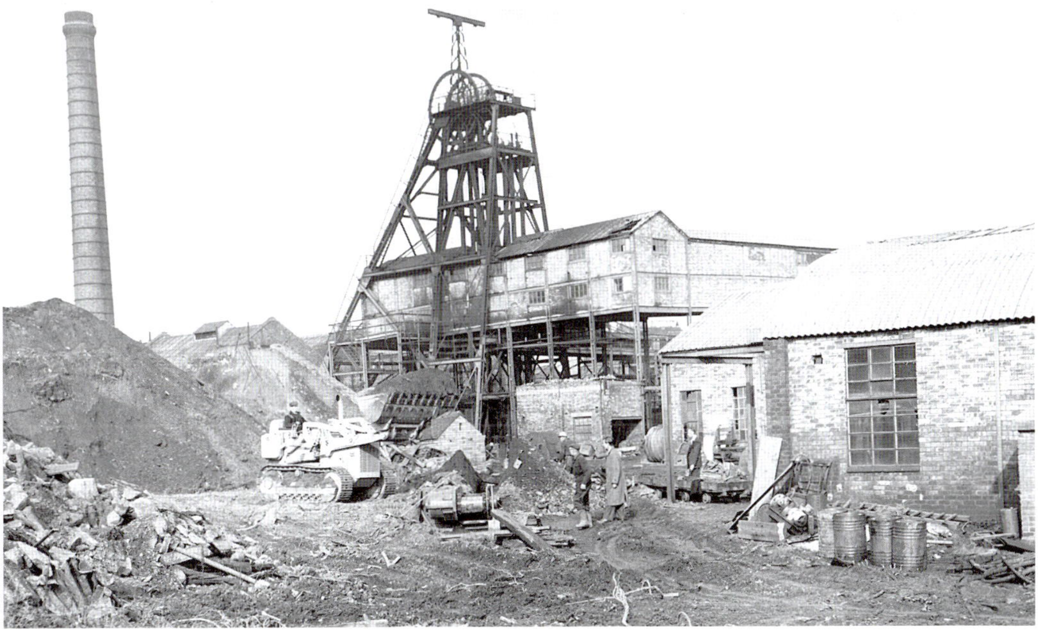

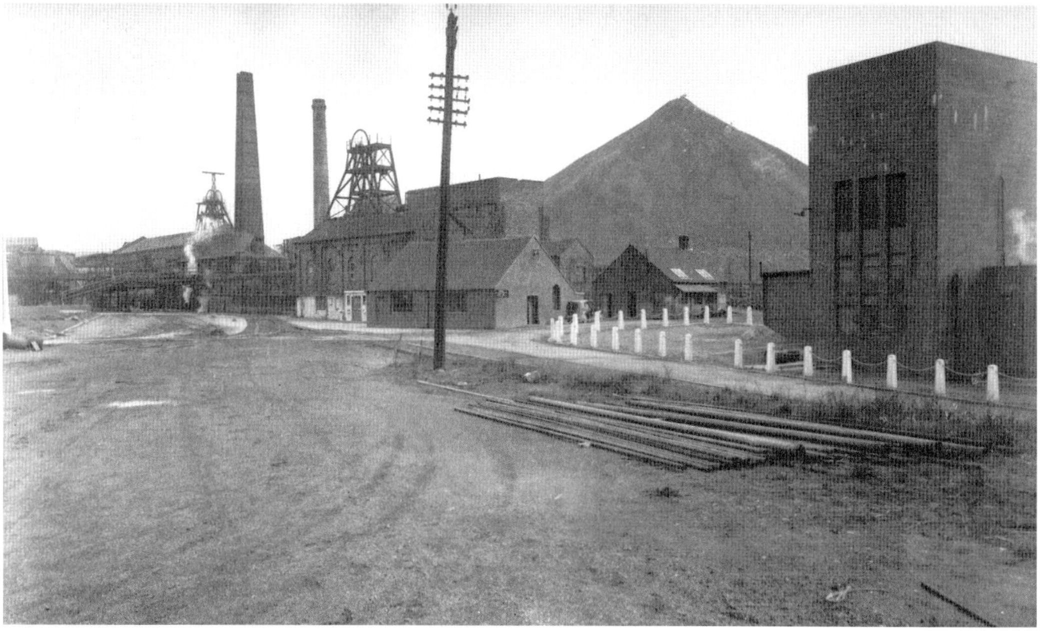

Above: View of the Colliery in 1951. No. 1 Pit is on the left and Knowles Shaft to its right. Note the conical tip, that is now largely removed and landscaped. MC
Below: No. 1 Pithead (shaft top) during demolition. The site subsequently became the NCB Area office and has now been cleared for industrial development. MC
Both scenes were taken on 14th February 1961.

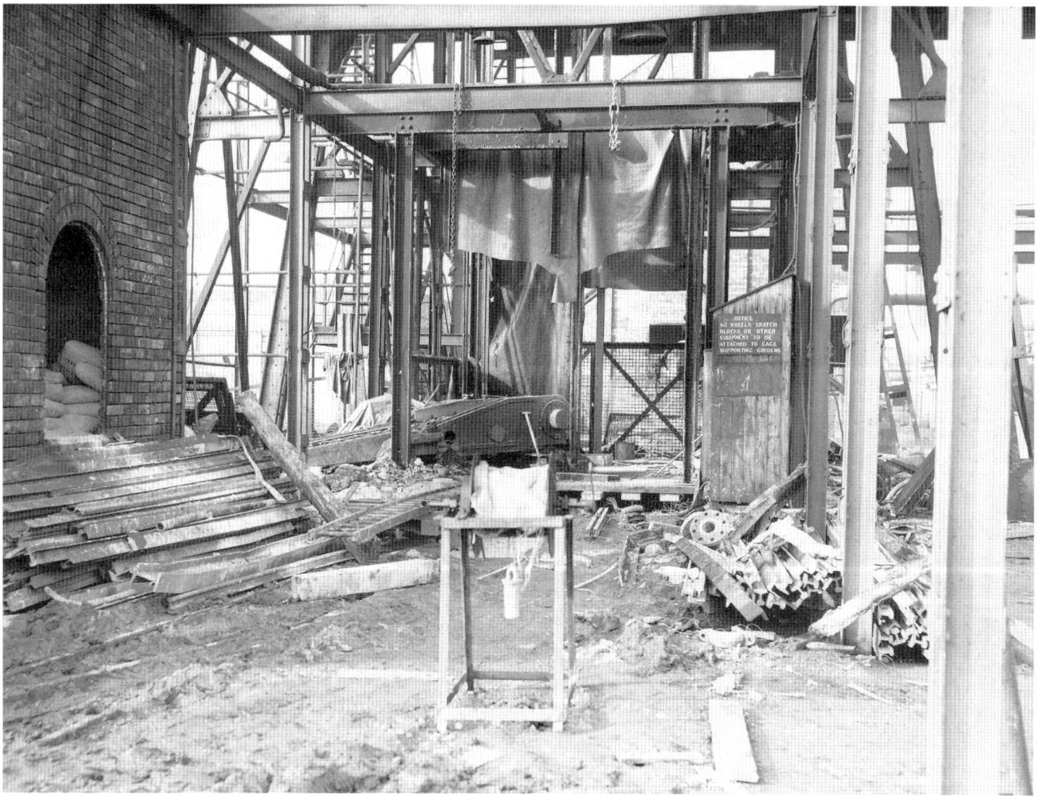

Bignall Hill Colliery

Bignall Hill Colliery Co. Ltd.

Like Berry Hill, this mine occupied two sites, The Jamage Site (that closed in the late 1930s) and The Rookery Pit, that is shown in the photograph below. Rookery is situated 400m south of the A500 (D-Road), 1km north-west of Bignall End. The Jamage site was a further 600m to the east. The site is now derelict land. There were two shafts on each site, Rookery being 240.2m deep. The Rookery downcast shaft is shown in this photograph. The Jamage shafts were slightly deeper at 245.2m deep. The mine was closed in November 1947, shortly after Nationalisation. MC

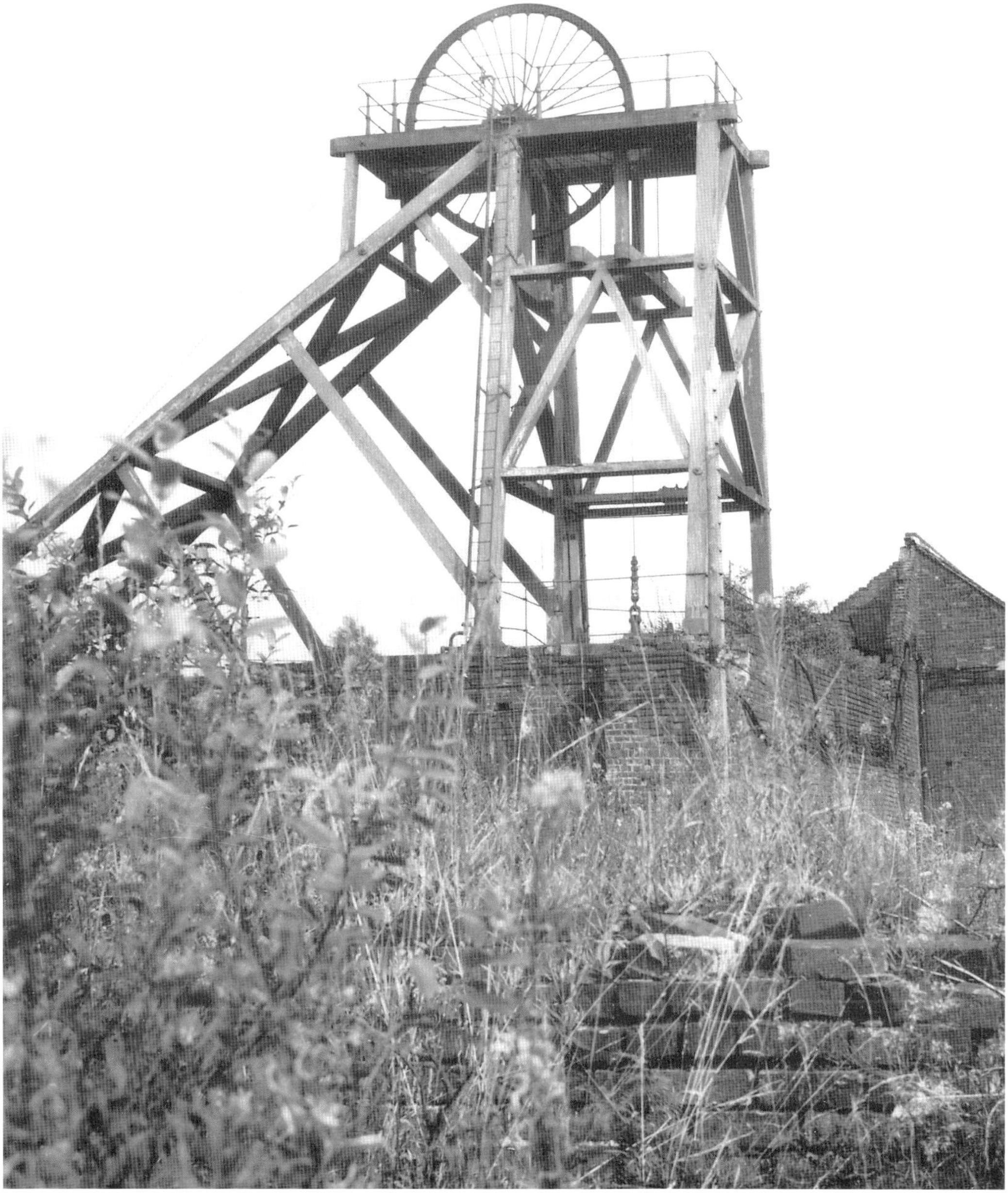

BIRCHENWOOD COLLIERY

Birchenwood Coal & Coke Co. Ltd.

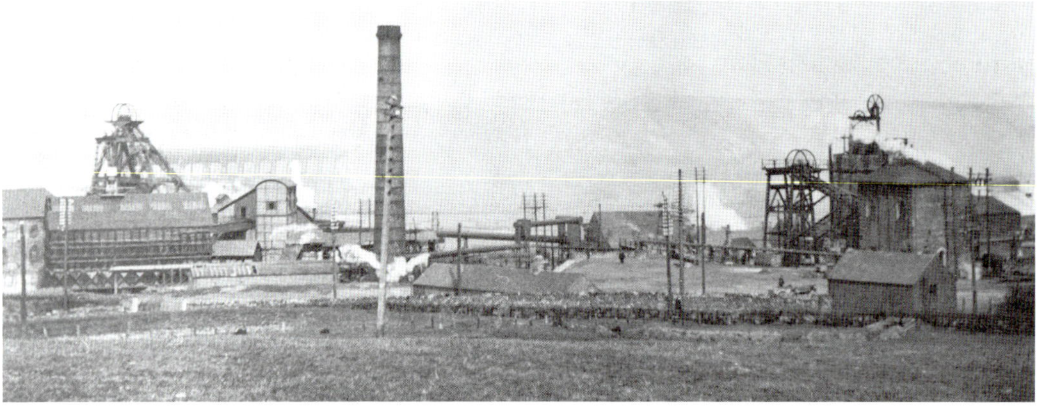

Above: General view of the colliery. It operated between 1891-1932, but the associated coke works continued until 1973 with coal supplied from other mines in the area. MC

Below: The underground pumphouse at No. 4 Pit, Birchenwood Colliery, showing the two Mather and Platt turbine pumps, 750 bhp delivering 1,800 gallons per minute at 1,000 feet head. MC

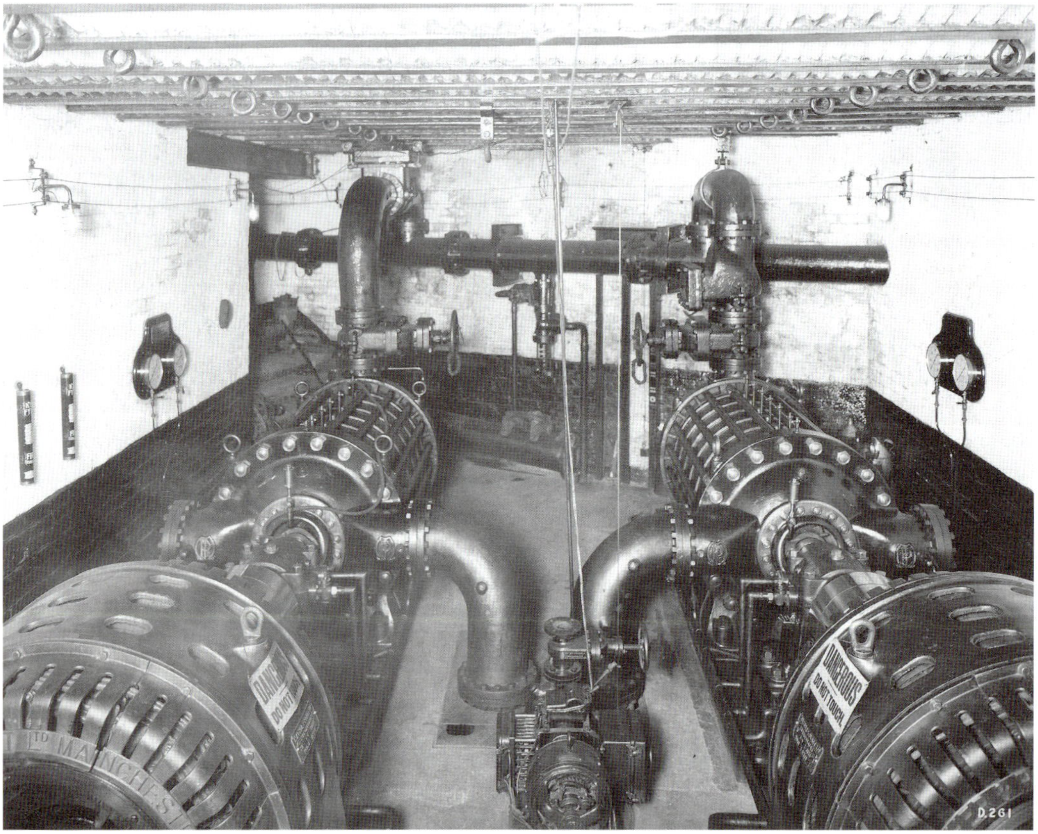

CHATTERLEY WHITFIELD

Chatterley Whitfield Colliery Ltd.

The mine was situated at Fegg Hayes, about 3km north off Tunstall town centre. It was sunk in 1839, with five shafts: Hesketh, 595.3m; Middle, 237.7m; Winstanley, 227.7m; Institute, 393.2m and Platt, 391.4m. It ceased coal production in March 1977, when it merged with Wolstanton. It subsequently became a mining museum with an underground experience. Most of the buildings and headgear survive in situ, whilst its future is debated at length.

It was a very large mine, employing over 4,000 people in the 1930s. It was the first colliery in Britain to produce over 1 million tons of saleable coal in one year.

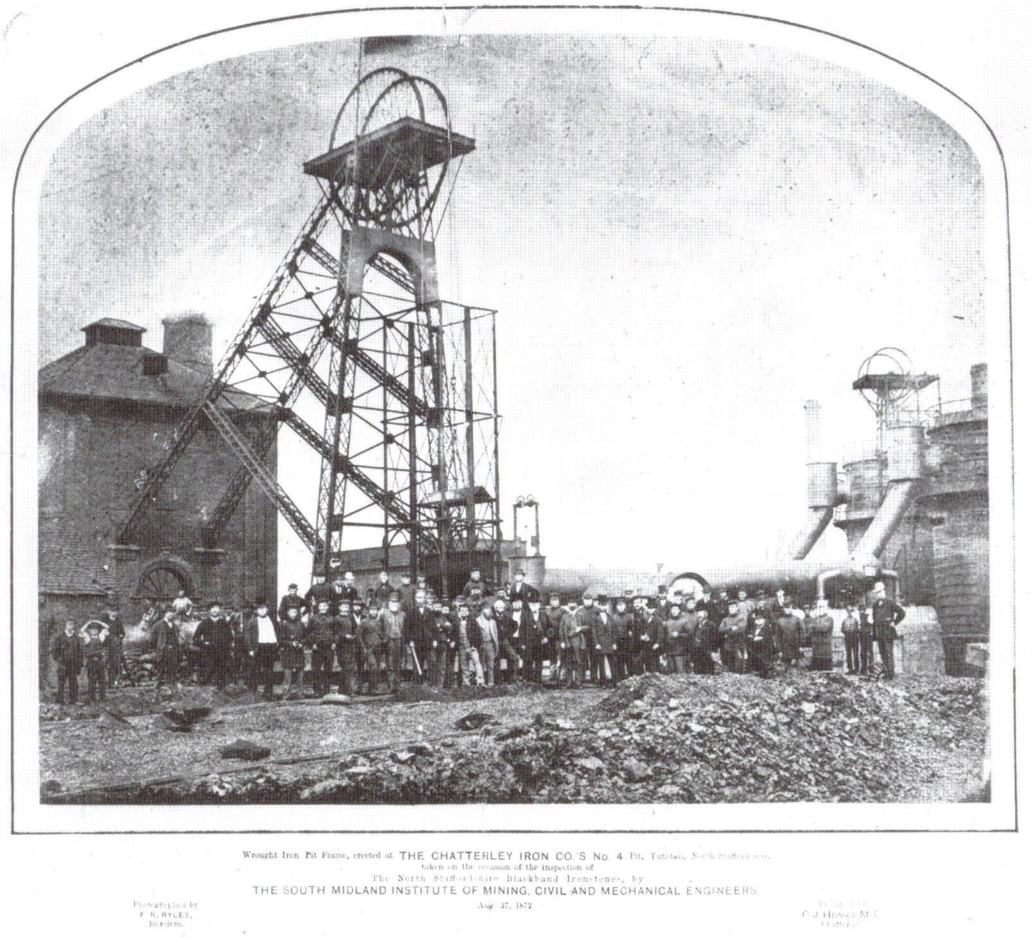

Above: No. 4 pit taken on the occasion of the visit of the South Midlands Institute of Mining, Civil and Mechanical Engineers on 27th August 1872. Note the two blast furnaces for iron production on the right hand side. MC

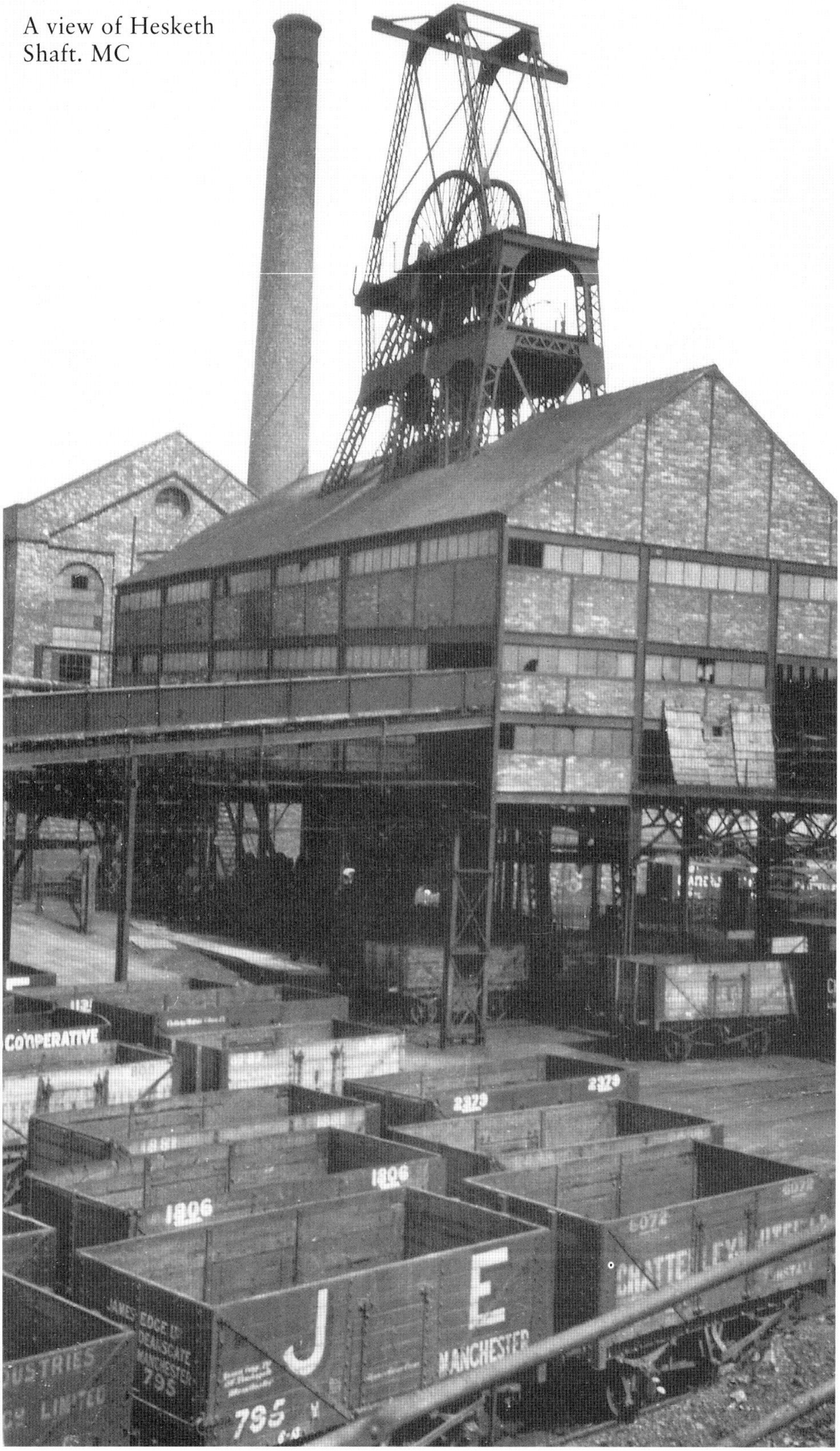

A view of Hesketh Shaft. MC

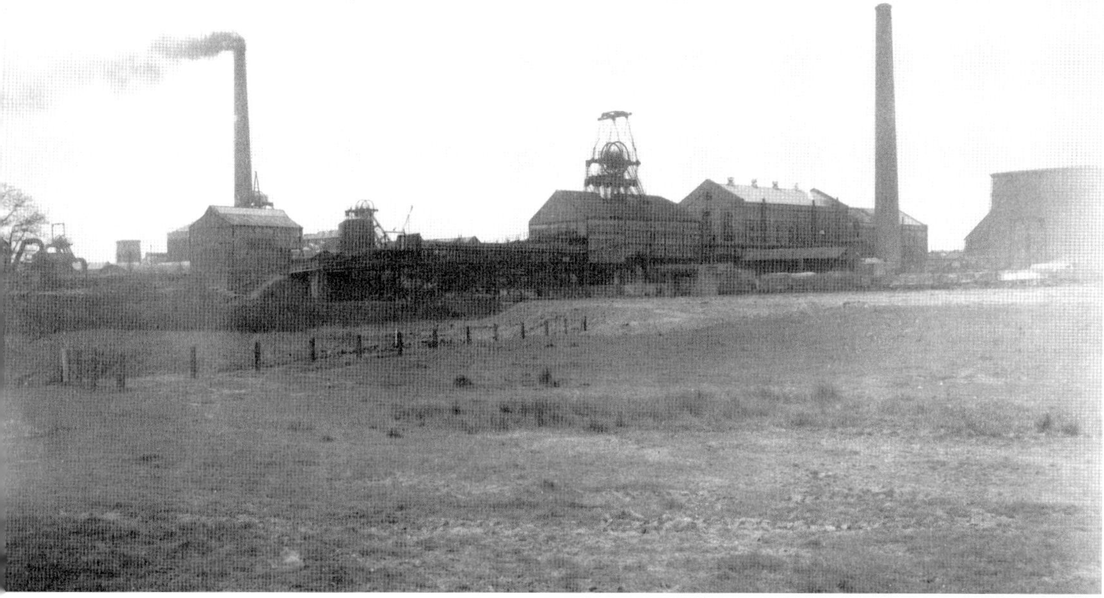

Above: View of the mine in 1951, with Hesketh Shaft prominent in the centre of the photograph. MC

Below: A similar view. Note that four of the headstocks are visible. MC

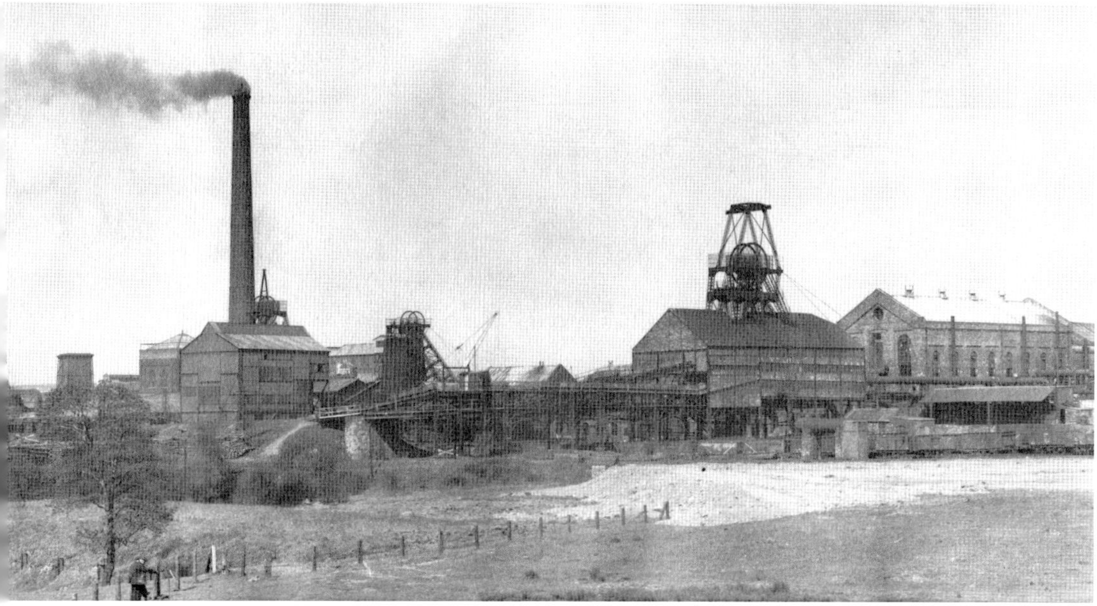

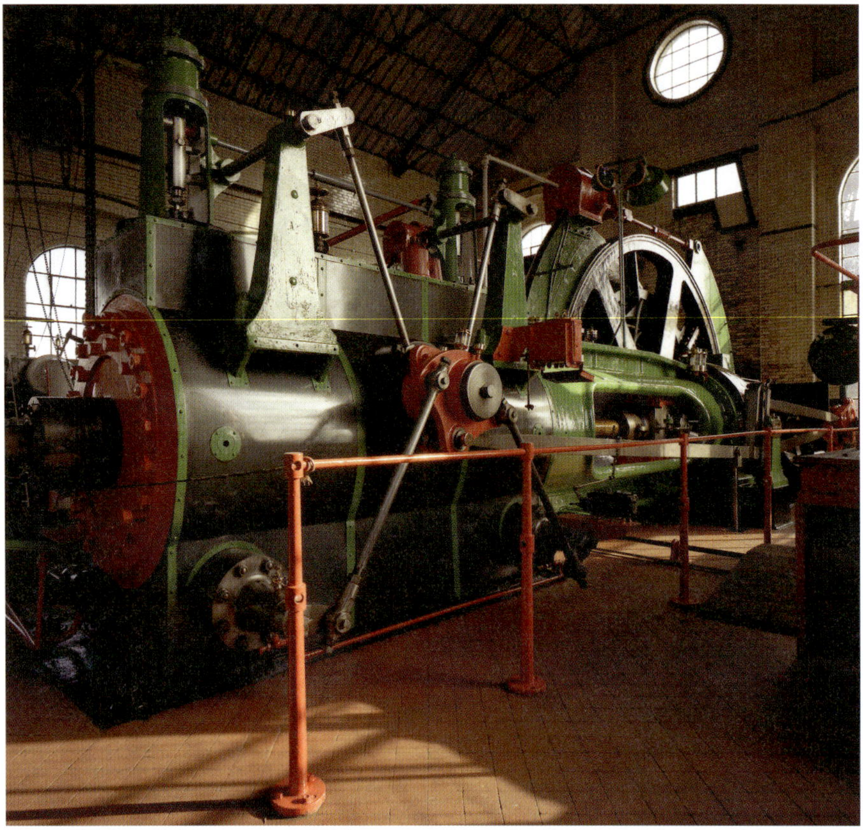

Two views of the steam winding engine on the Hesketh Shaft. It is now preserved. PD

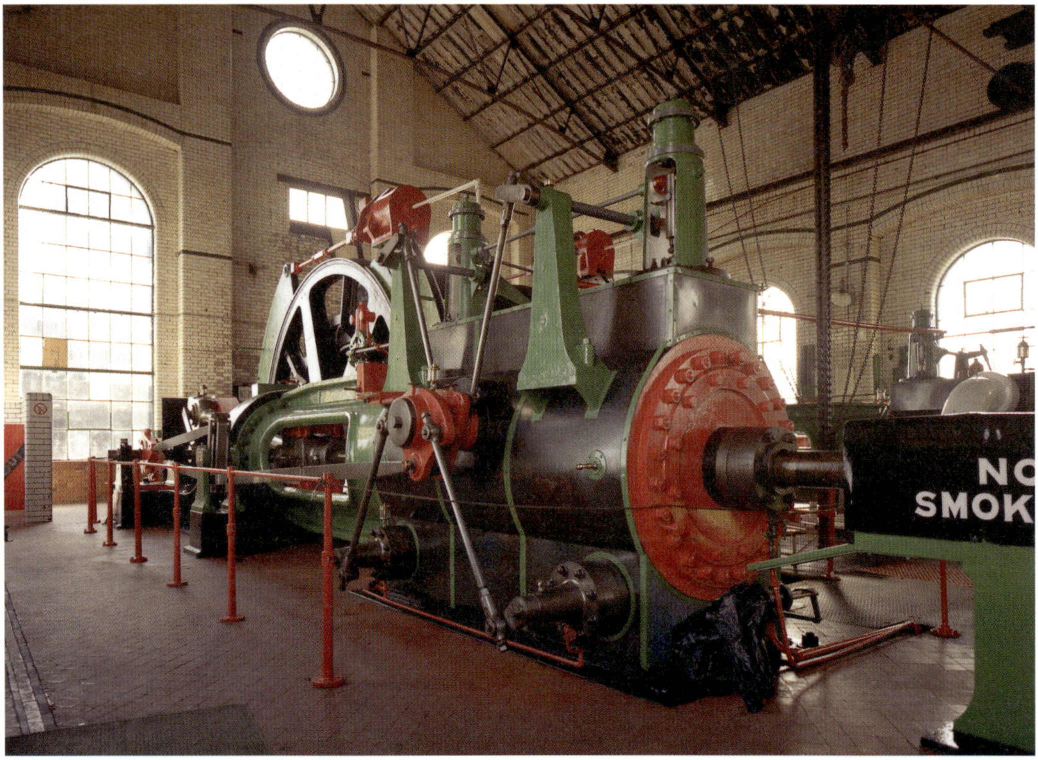

Fenton Colliery

Fenton Collieries Ltd.

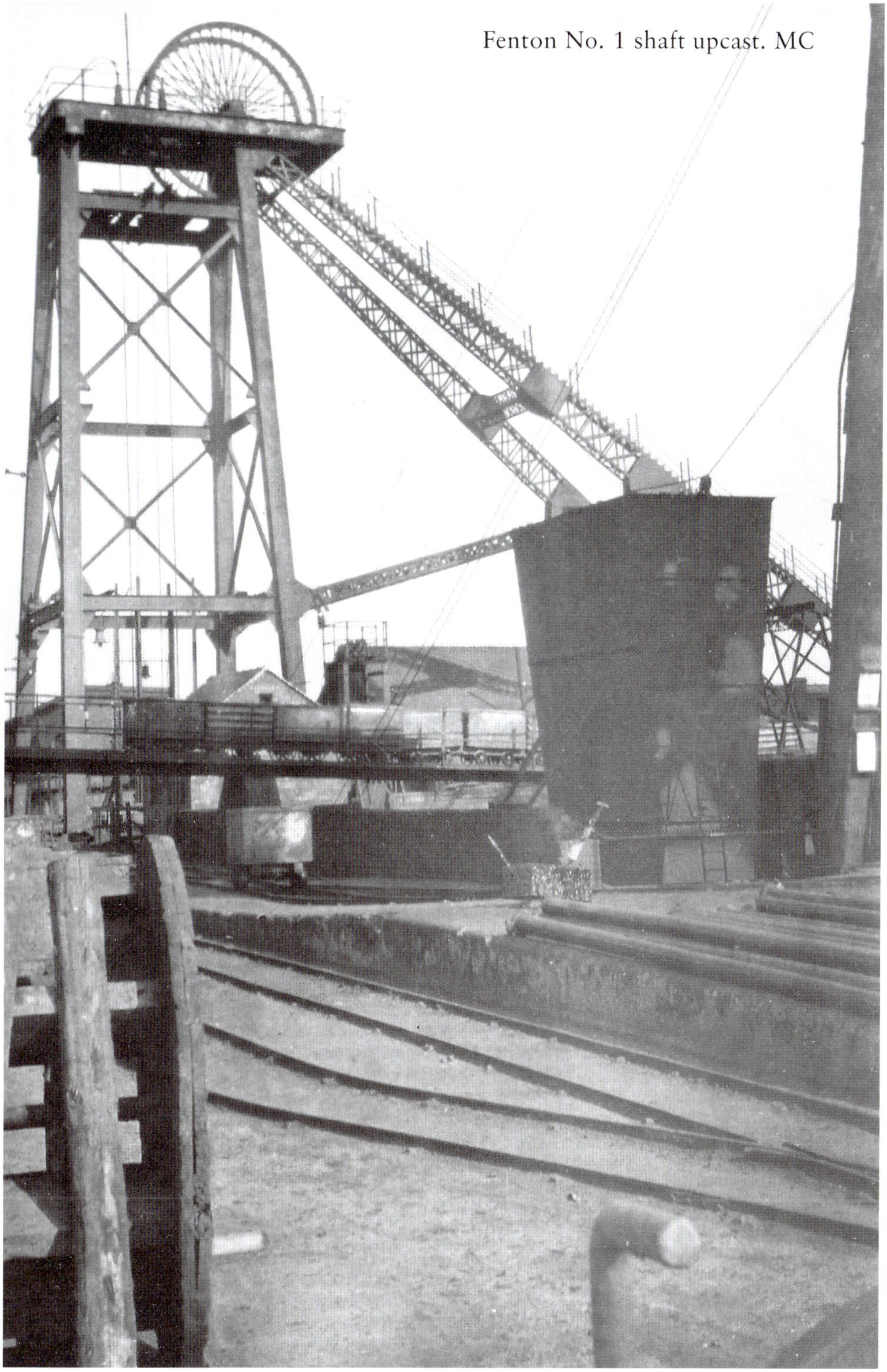

Fenton No. 1 shaft upcast. MC

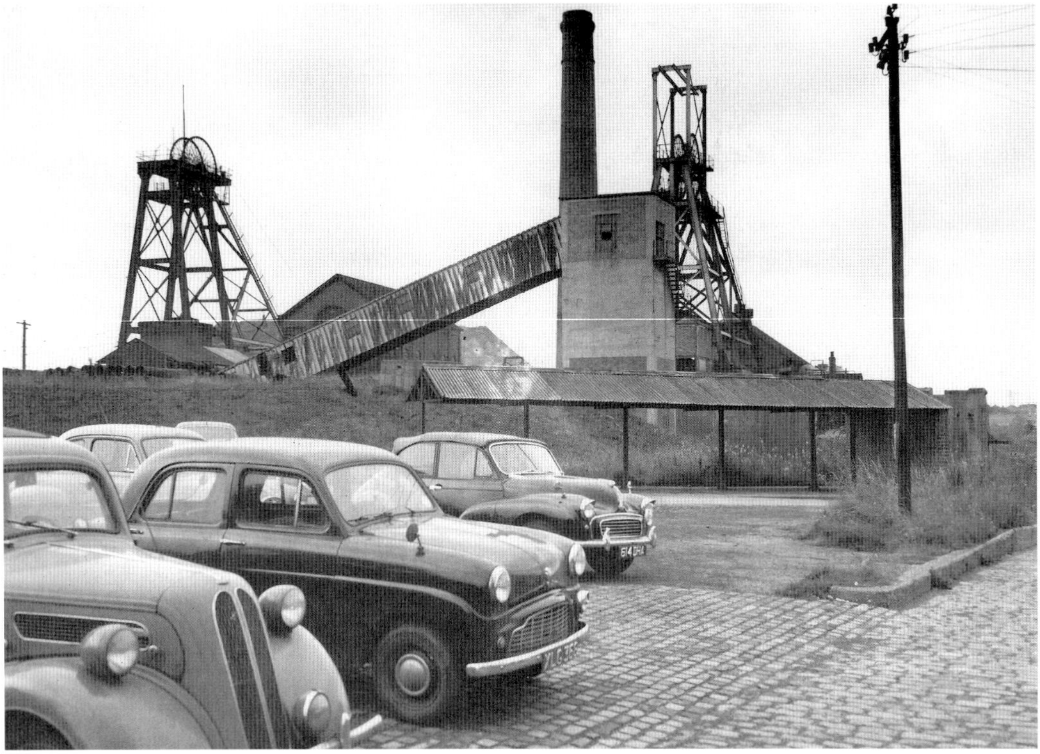

Above: This mine was sometimes known as Glebe or Fenton Glebe. It was sunk in 1865 and closed in October 1964. This view was taken on 10 August 1964. There were two shafts in use at the time of closure; No. 1 being 826.7m deep and No. 3, 715.1m deep. The site is now an industrial estate and the tips have been made into the Glebedale Park. MC

Below: A view of the mine in 1951. No. 1 shaft on the left of the picture and No. 2 on the right. MC

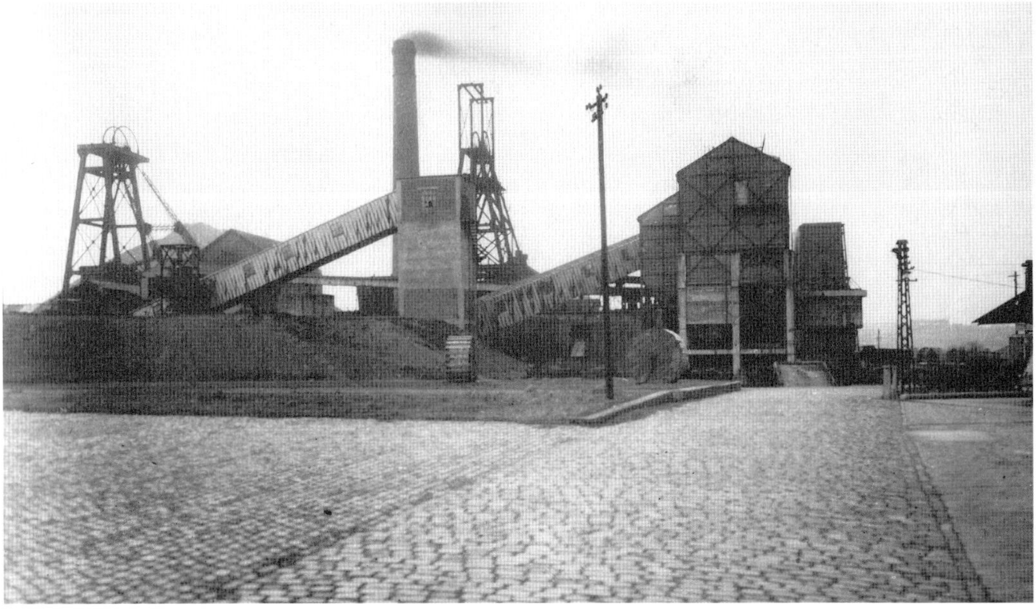

Florence Colliery

Florence Coal & Iron Co. Ltd.

Originally sunk in 1876, there are three shafts – No. 1, 792.9m deep; No. 2, 814.5m deep; and No. 3, 791m deep. The mine was later connected to Hem Heath and was abandoned in April 1994. Situated 1.5km south east of Longton.

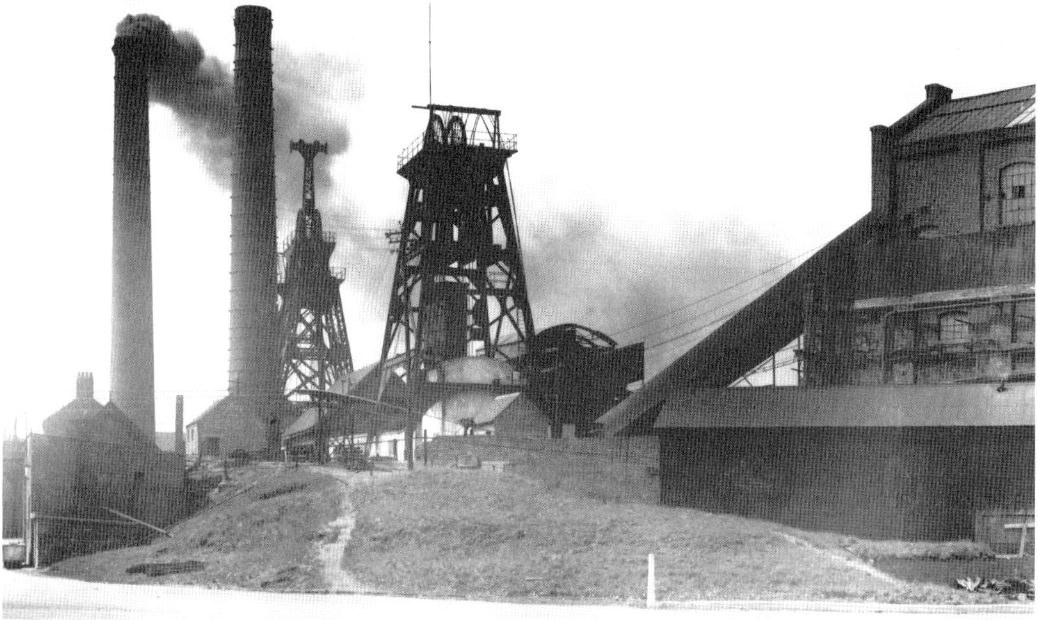

Above & below: The colliery in 1951. MC

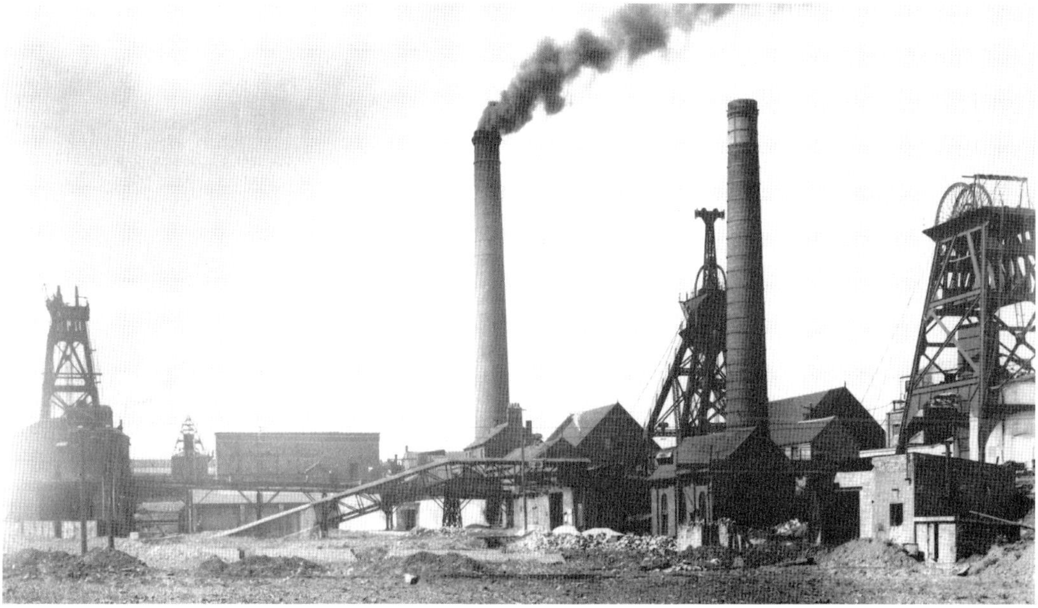

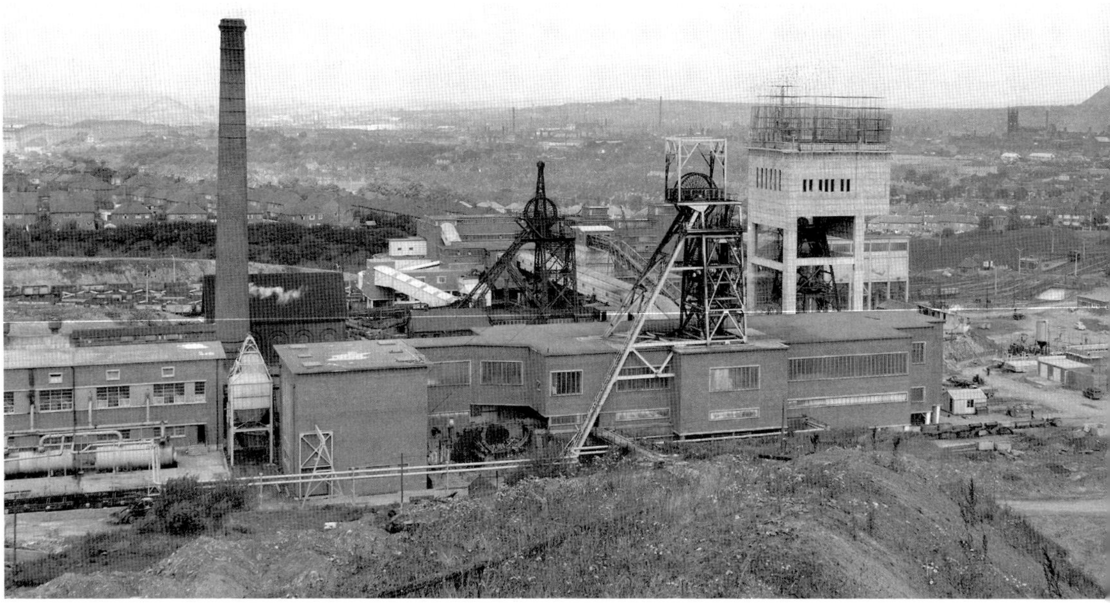

Above: A general view of the colliery taken during reconstruction work on 15 August 1960. Note the old headgear situated beneath the new winding tower under construction. MC

Below: Another view taken on the same day. MC

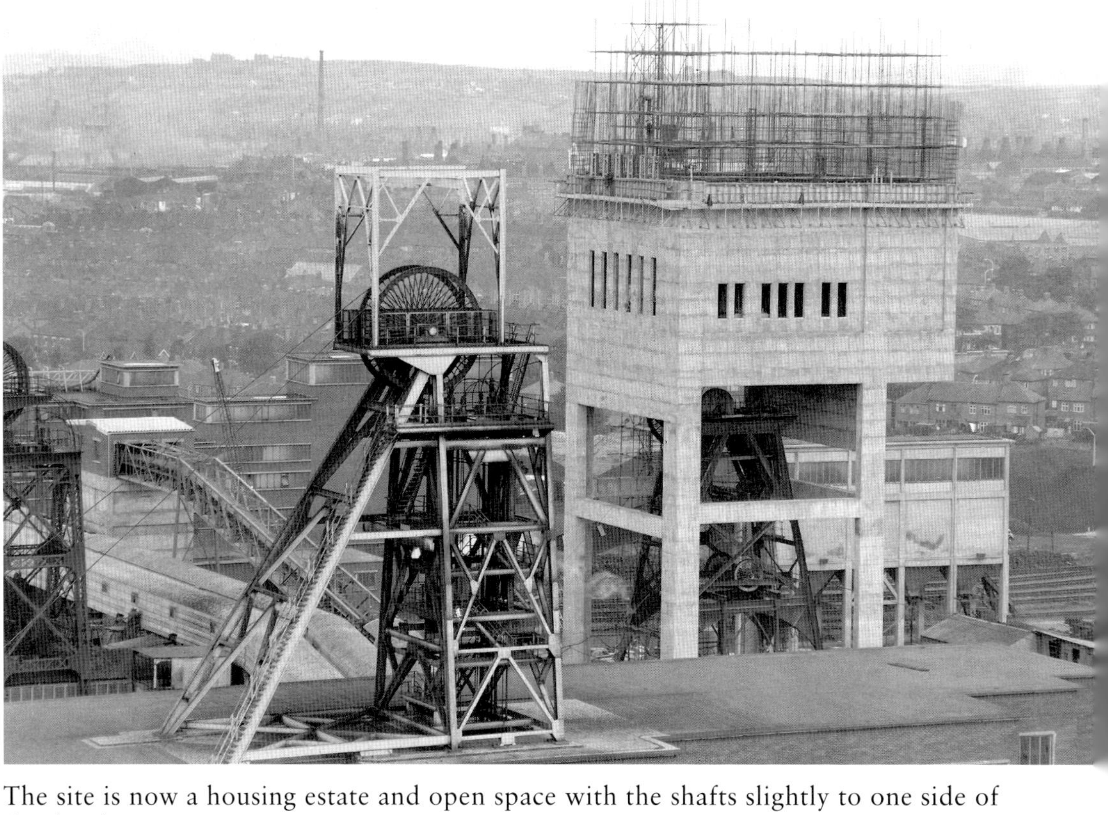

The site is now a housing estate and open space with the shafts slightly to one side of the development.

Foxfield Colliery

Foxfield Colliery Co. Ltd.

Situated on the Cheadle Coalfield 1km north of Dilhorne. There are two shafts, the Upcast being 215.4m deep and the Downcast 224.5m deep. The mine was sunk in 1880 and closed in October 1965. The colliery headstocks are made of reinforced concrete and are still standing, see page 148. There are proposals for their long-term preservation. The Foxfield Steam Railway terminates at the mine and the rest of the site is an industrial estate.

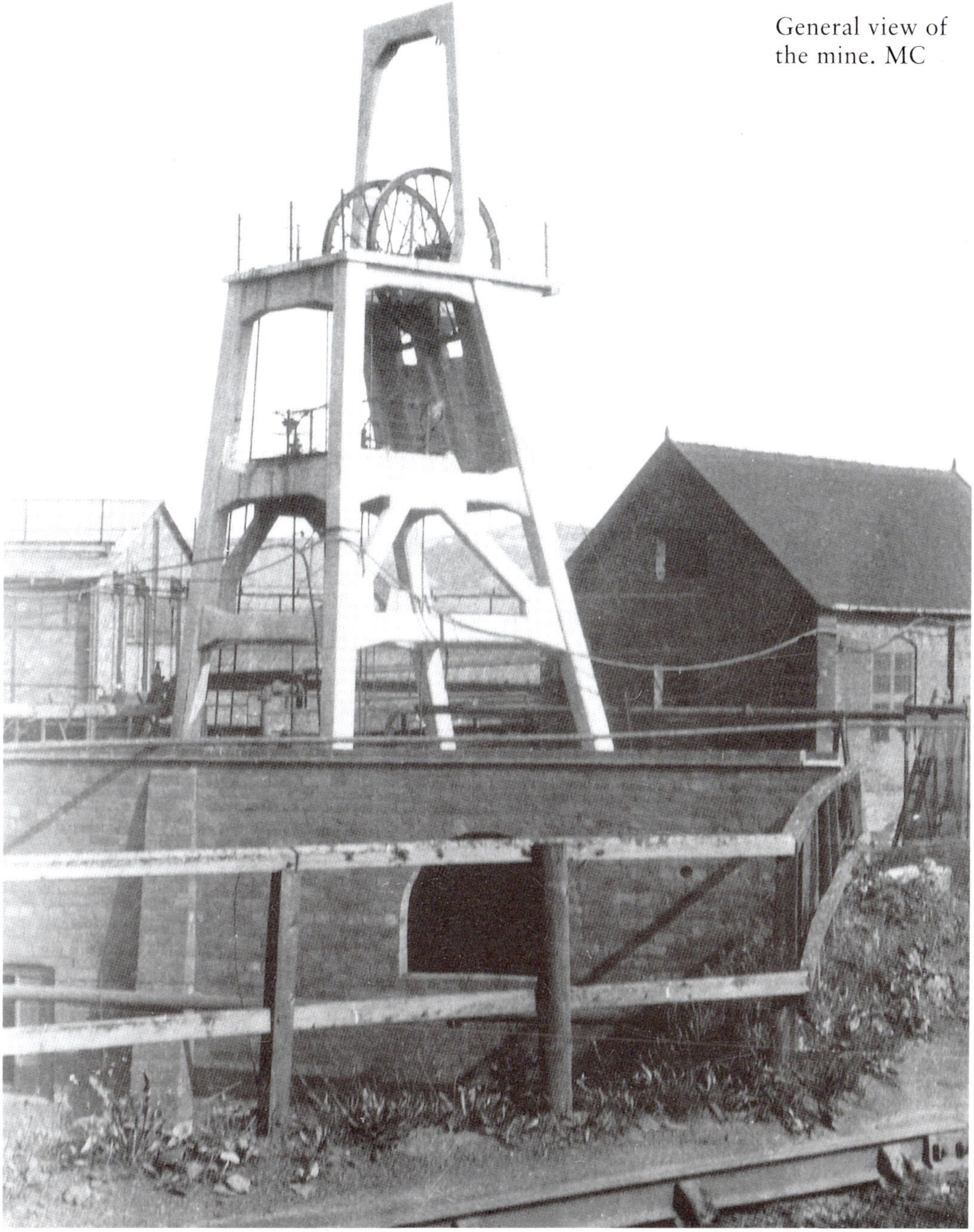

General view of the mine. MC

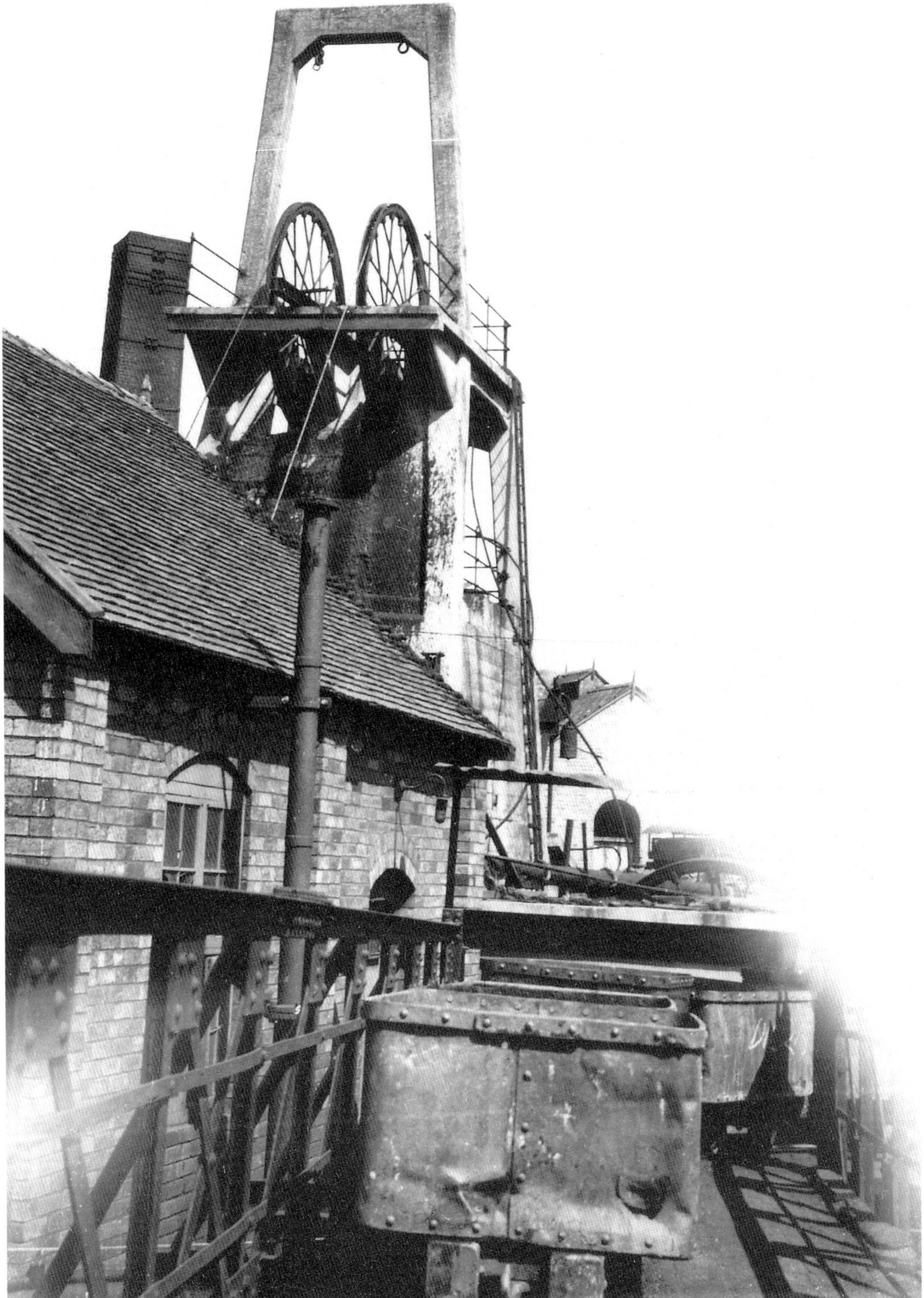

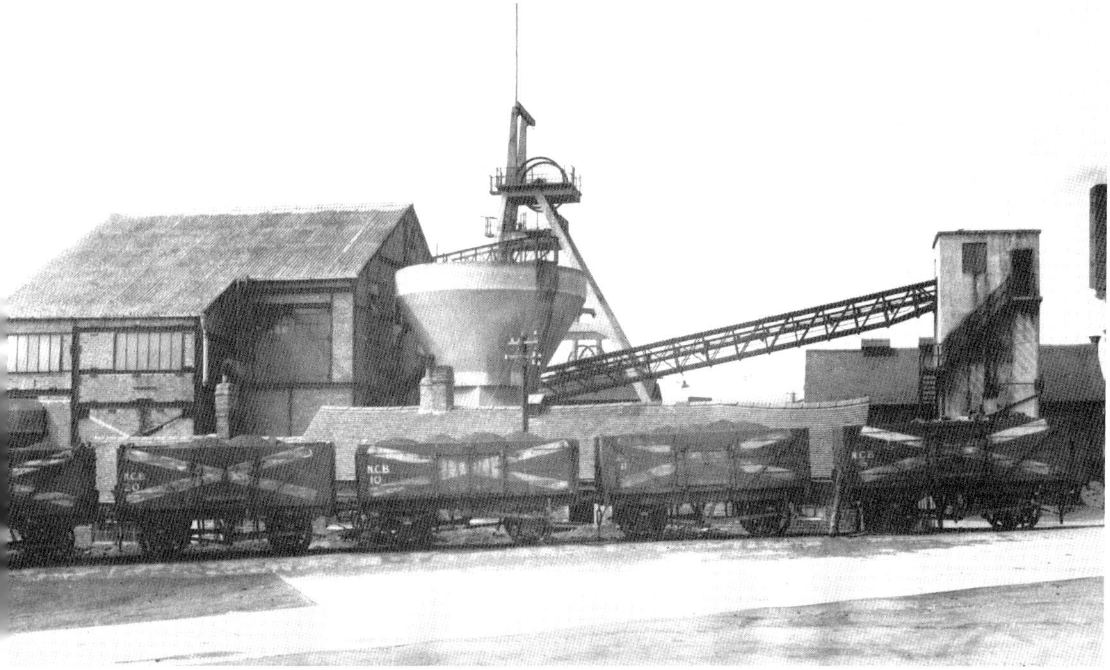

Above: The mine in 1951. MC

Below: General view of the mine. MC

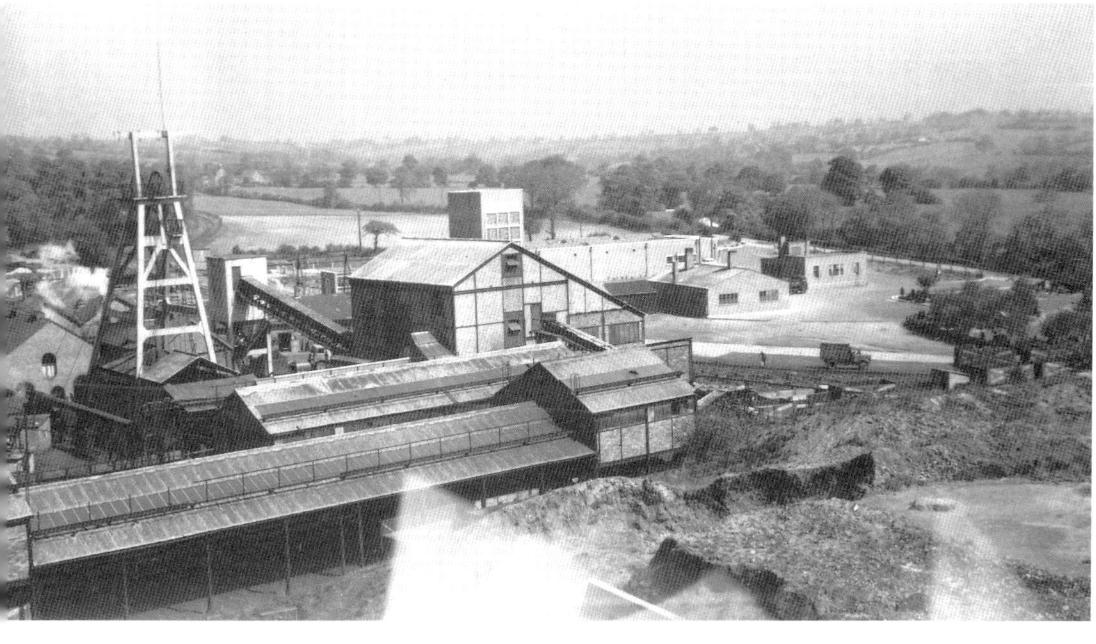

Grange Colliery

Robert Heath & Son (who also operated the Victoria Iron Works)

Situated immediately to the north of Festival Park. A series of shafts were sunk for ironstone as well as coal. It closed in December 1917 after an influx of water from the adjacent Sandbach Colliery to the east, that overpowered the pumps. The remaining reserves were insufficient to justify recovering the mine. A total of nine pits were sunk to work the upper ironstones as well as the Peacock Coal Seam. No. 1 shaft was 214.9m deep and No. 2 shaft was a little deeper at 219.5m deep. MC

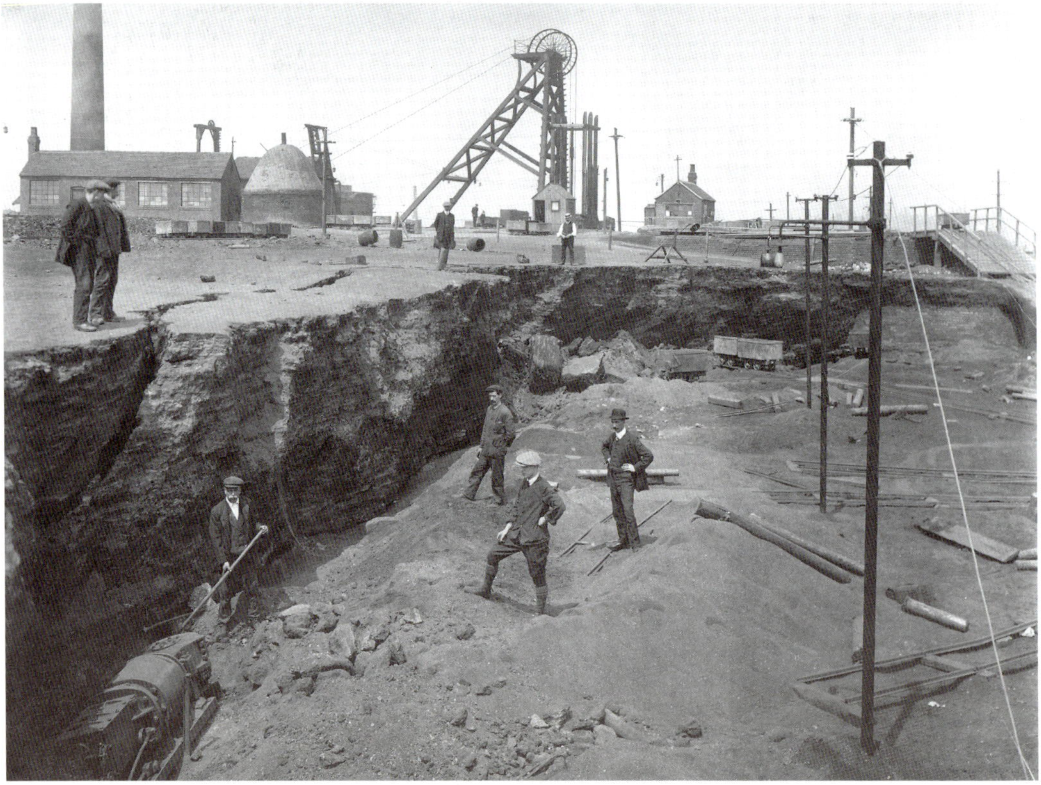

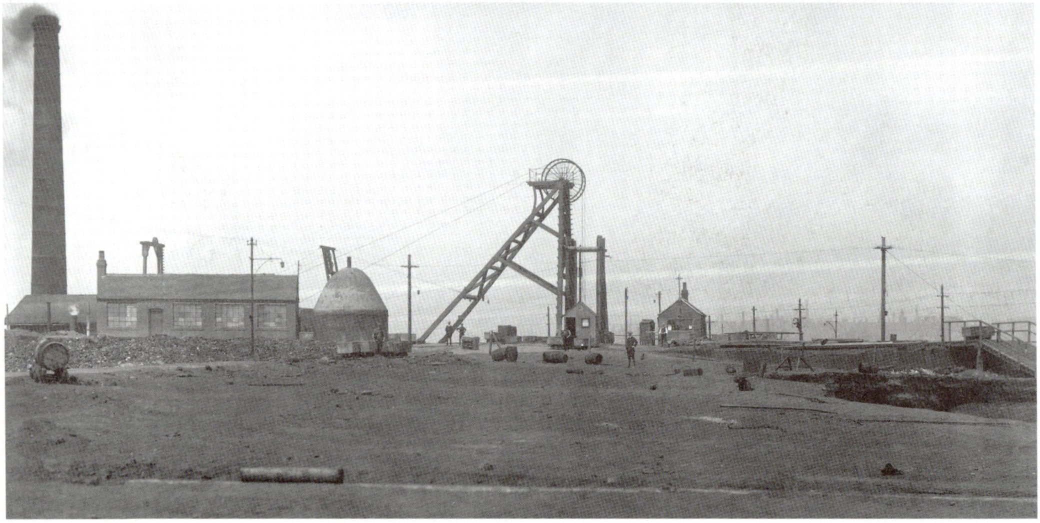

Hanley Deep Pit

Shelton Iron, Steel & Coal Co. Ltd.

Situated on Town Road, Hanley, 500m to the north-east of the Potteries Centre car park. Sunk in c.1850s, the shafts were deepened and widened in 1901. No. 1 shaft was 802.8m deep and No. 2 shaft 811.1m deep. The mine merged with Wolstanton in February 1962. This photograph shows the auxiliary winder at the mine during demolition, on 11 August 1964. The site is now part of Hanley Forest Park. MC

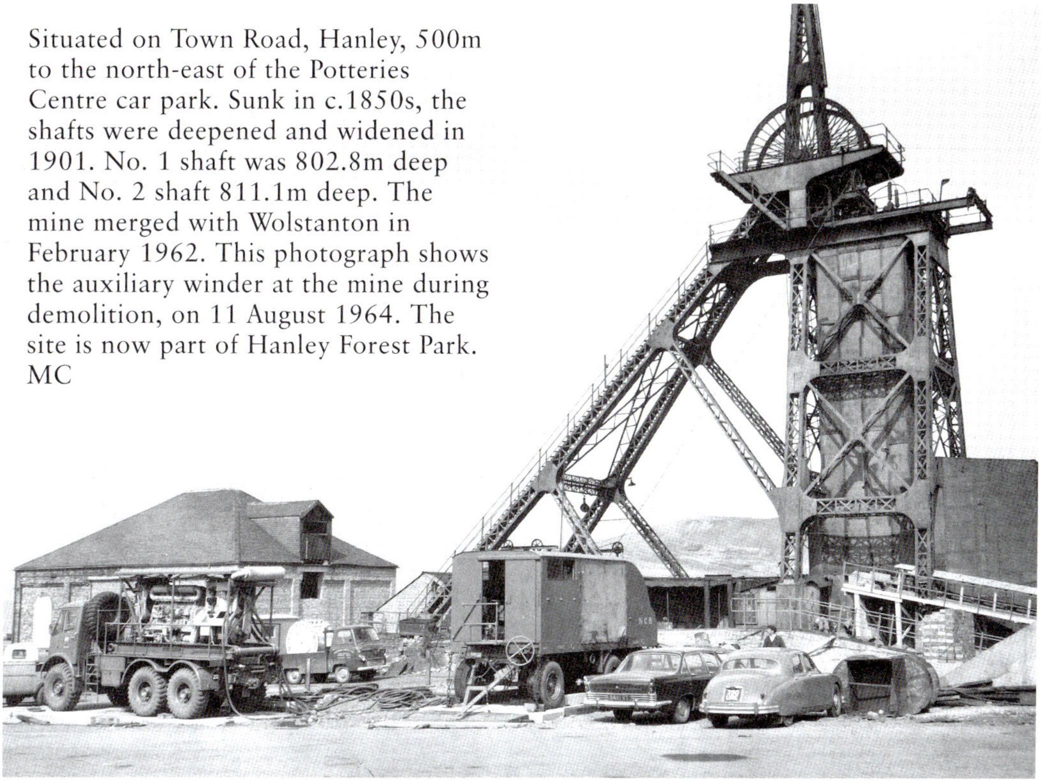

Below: The pit in 1951 MC

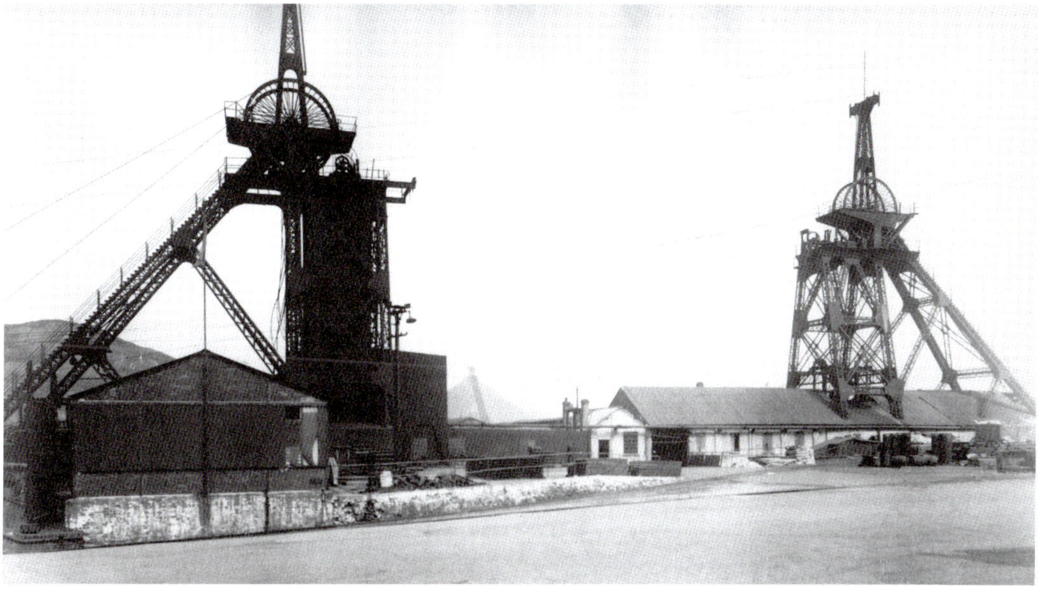

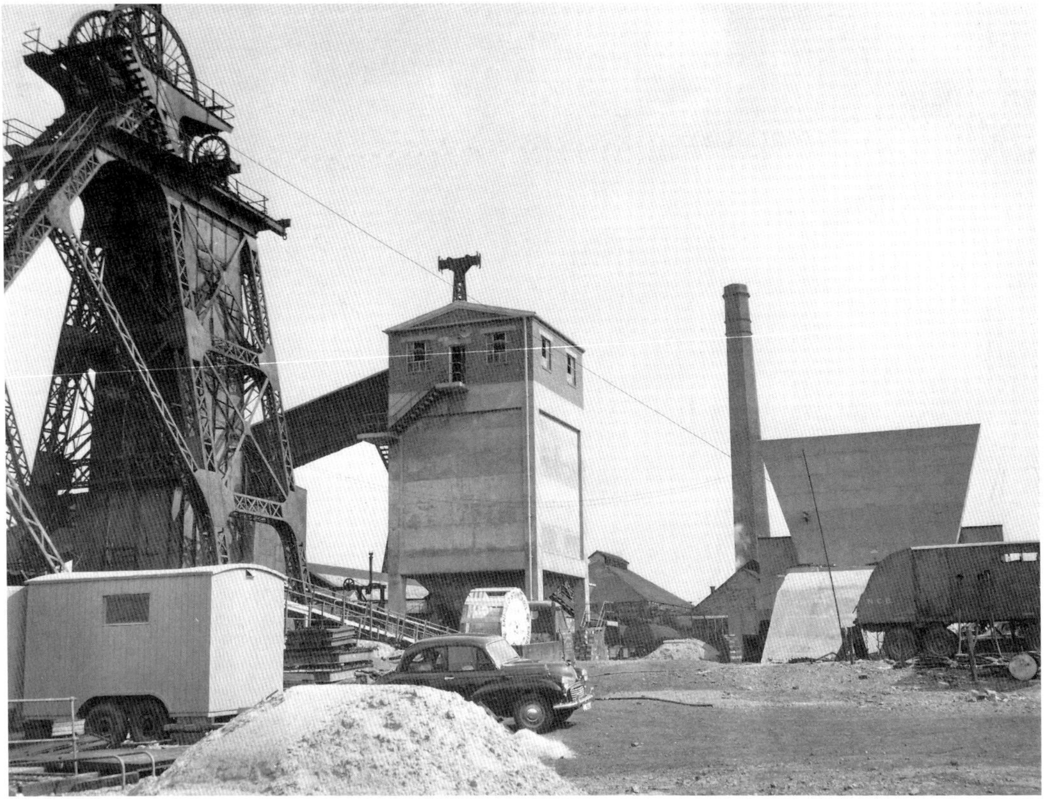

Above: The auxiliary winder in August 1964. MC

Below: Another view of the demolition work in August 1964. MC

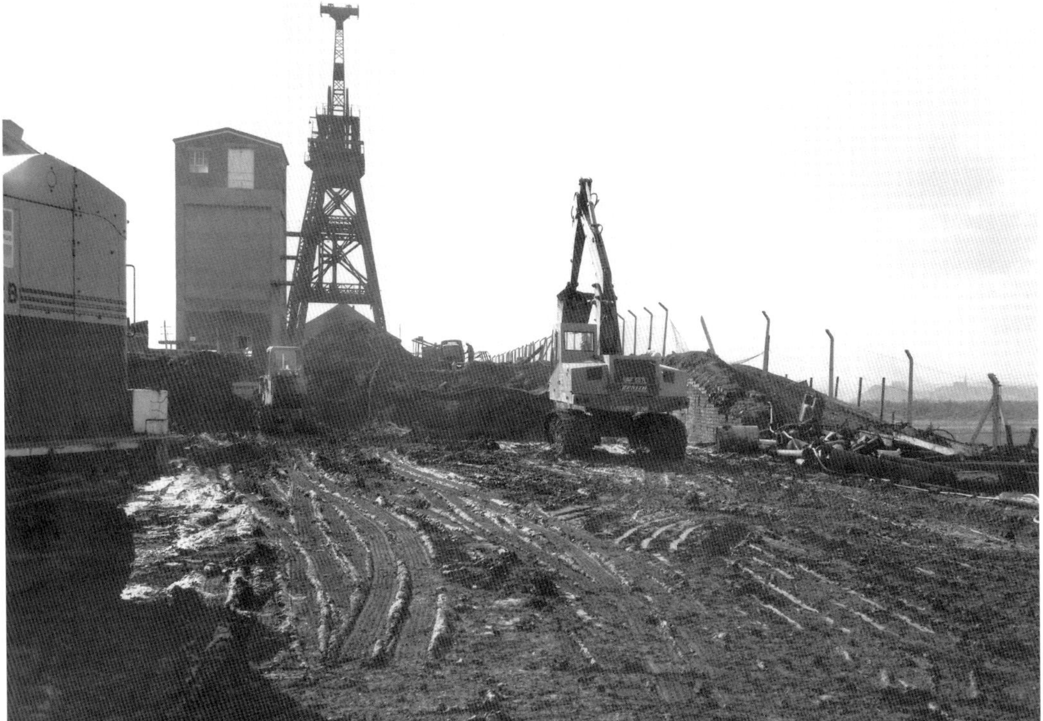

Haying Wood Colliery

J.A. Knight (Coppice) Ltd.

This mine consisted of three footrails: No. 1 working the Winghay Seam, 1971-1984; No. 2, the Rowhurst Seam, 1984-1995; No. 3, the Burnwood Seam, 1994-1997. These were all situated to the west of Haying Wood about 1.5km west of Silverdale church.

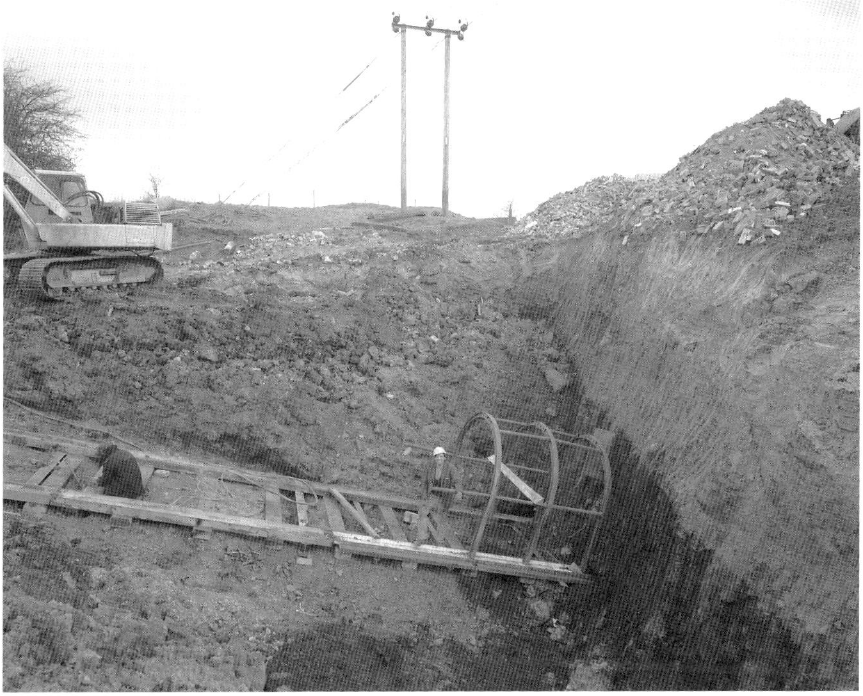

Above: The establishment of Haying Wood No. 2 pit. The excavation was made on the surface and the arches set and covered with steel sheeting before back filling. The roadway was then driven underground. Note the Rowhurst seam outcropping in the right of the excavation. Bernard Ronan on left and Frank Green right, 1984. PD

Below: A general view of No. 2 pit during the winter, January 1992. PD

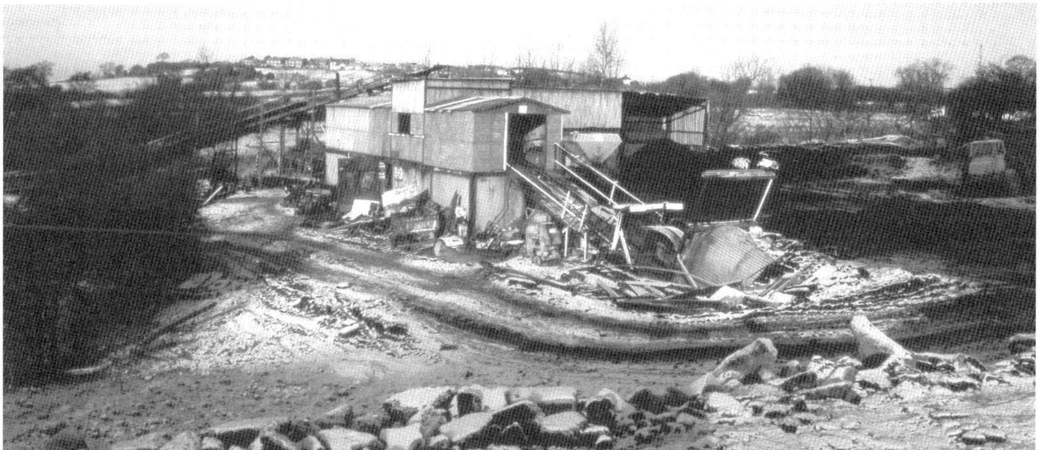

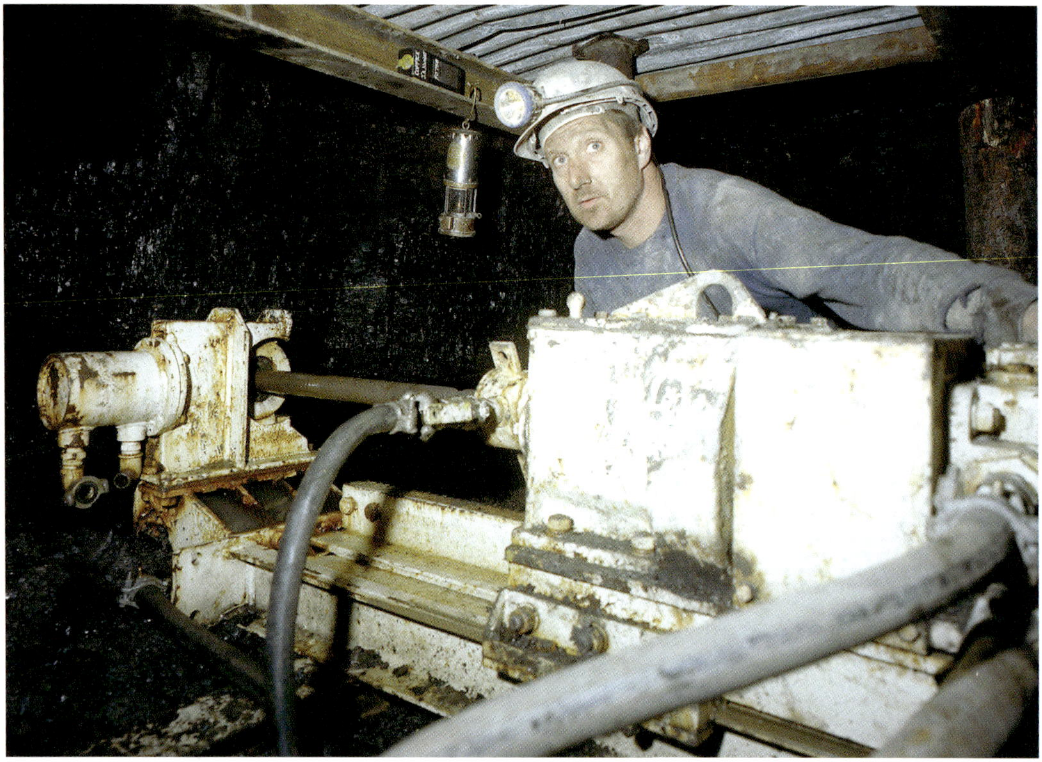

Above: No. 3 Colliery, with Jonnie Harrop undertaking long hole drilling to dewater adjacent abandoned workings, May 1995. PD

Below: Tony Hassell assisting with the long hole drilling, May 1995. PD

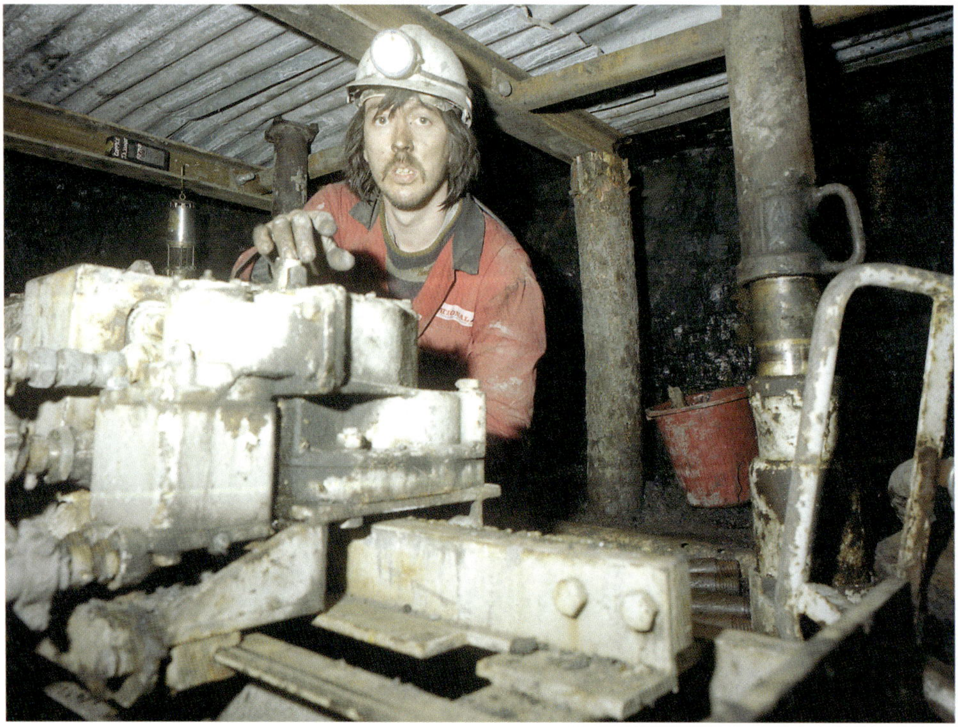

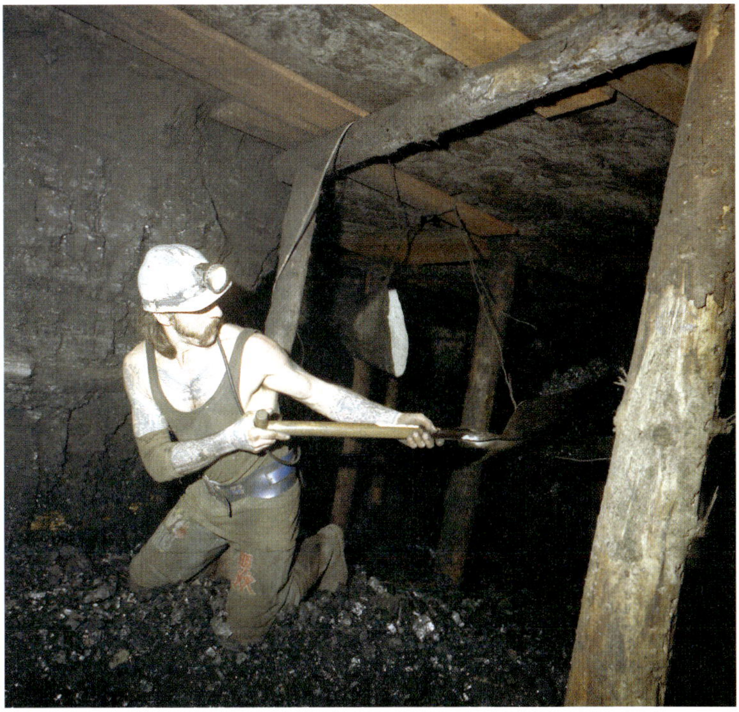

Left: No 2 Colliery. Rees Osoviecki hand loading coal onto dead pans (chutes), that carry the coal down to the tub, February 1993. PD

Below: No 2 Colliery. Gordon Stanier walking along a level in the Rowhurst Seam. The tub track may be seen on the left-hand side of the photograph. Immediately to the right of the track can be seen the electric cable to the drill and air ducting for ventilation in the roof, February 1993. PD

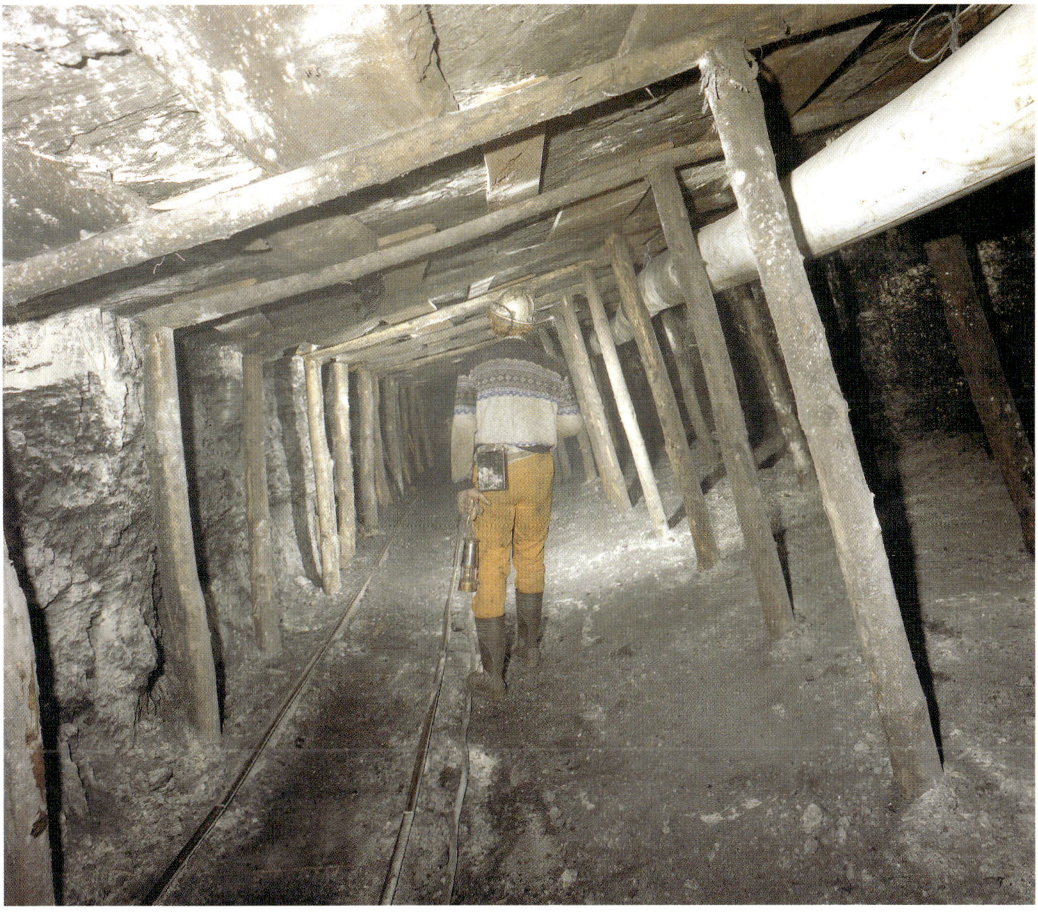

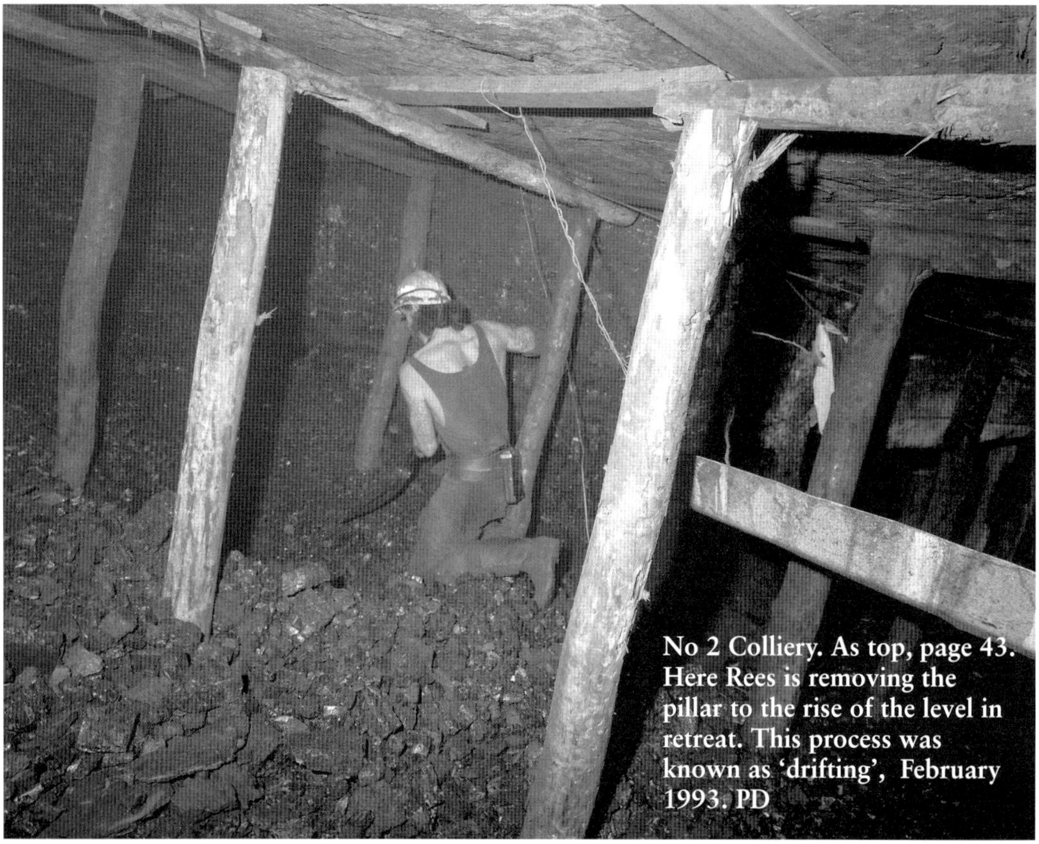

No 2 Colliery. As top, page 43. Here Rees is removing the pillar to the rise of the level in retreat. This process was known as 'drifting', February 1993. PD

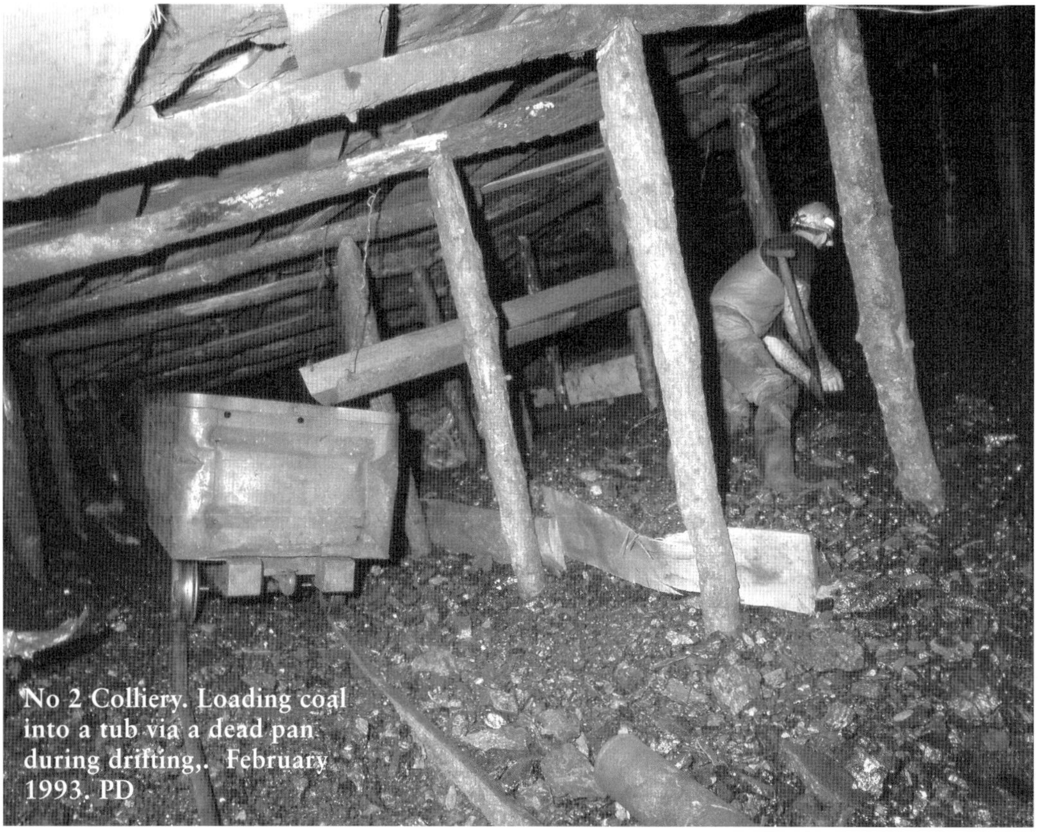

No 2 Colliery. Loading coal into a tub via a dead pan during drifting,. February 1993. PD

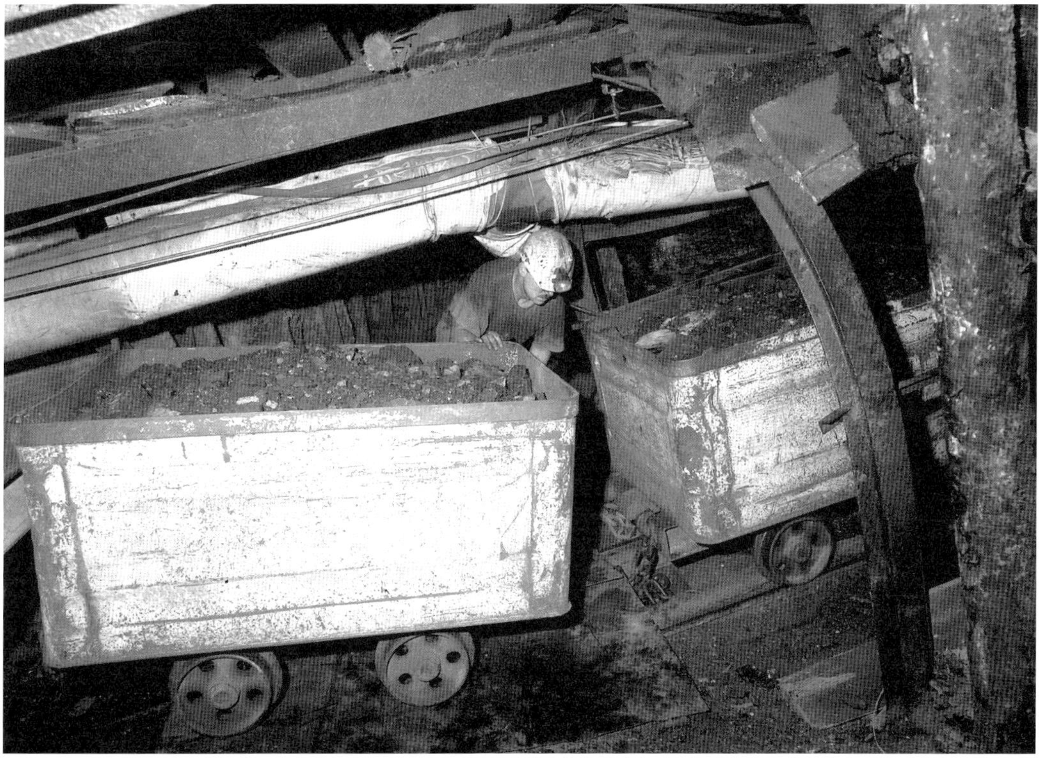

Above: No 2 Colliery. Coupling tubs together ready to be hauled out of the mine, February 1993. PD

Below: No. 2 Colliery. Hand loading into a 0.5 ton tub, February 1993. PD

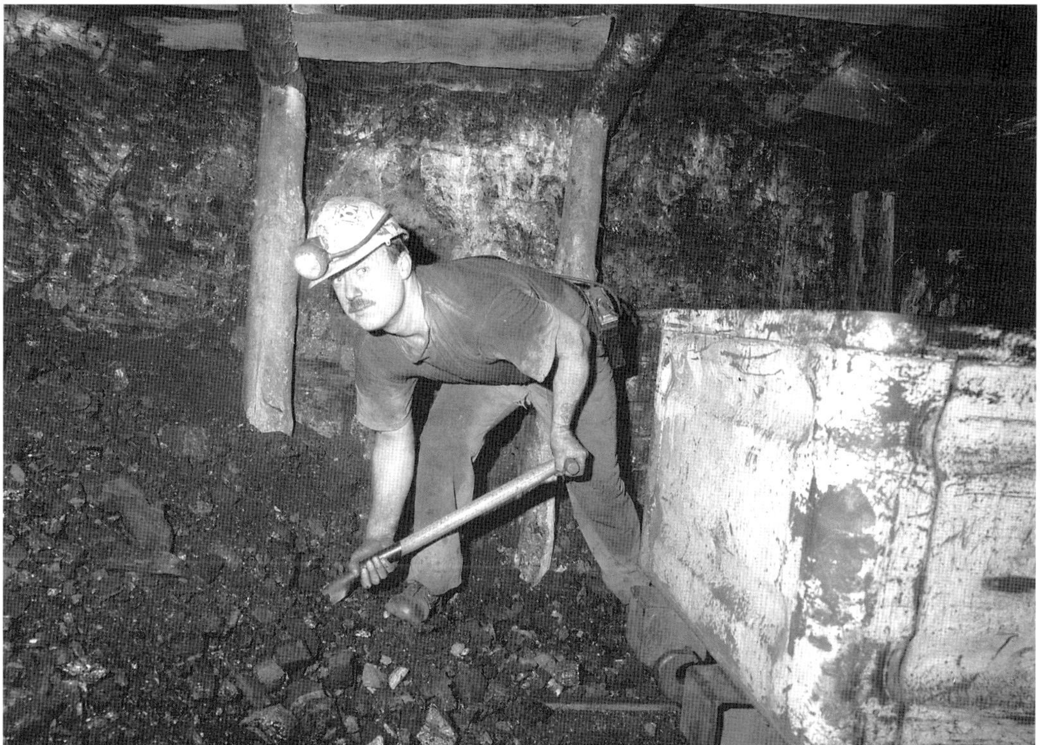

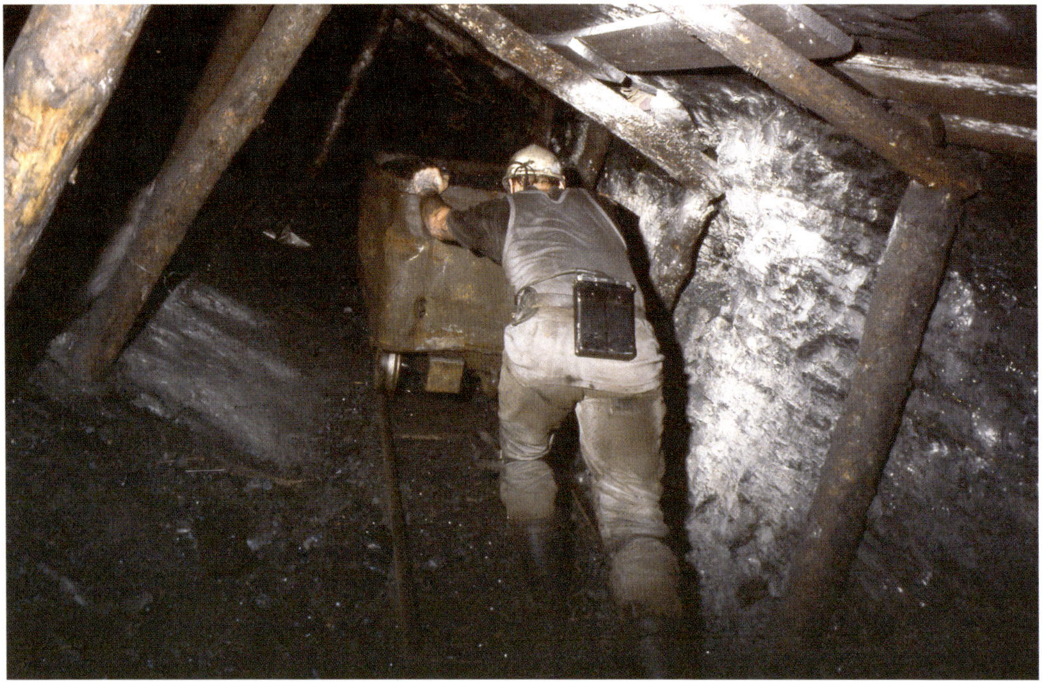

Above: No 2 Colliery. Hand tramming a full tub back to the main drawing dip with Rees Osoviecki, February 1993. PD

Below: No 2 Colliery. Ray Shelley unloading timber from a tub, January 1992. PD

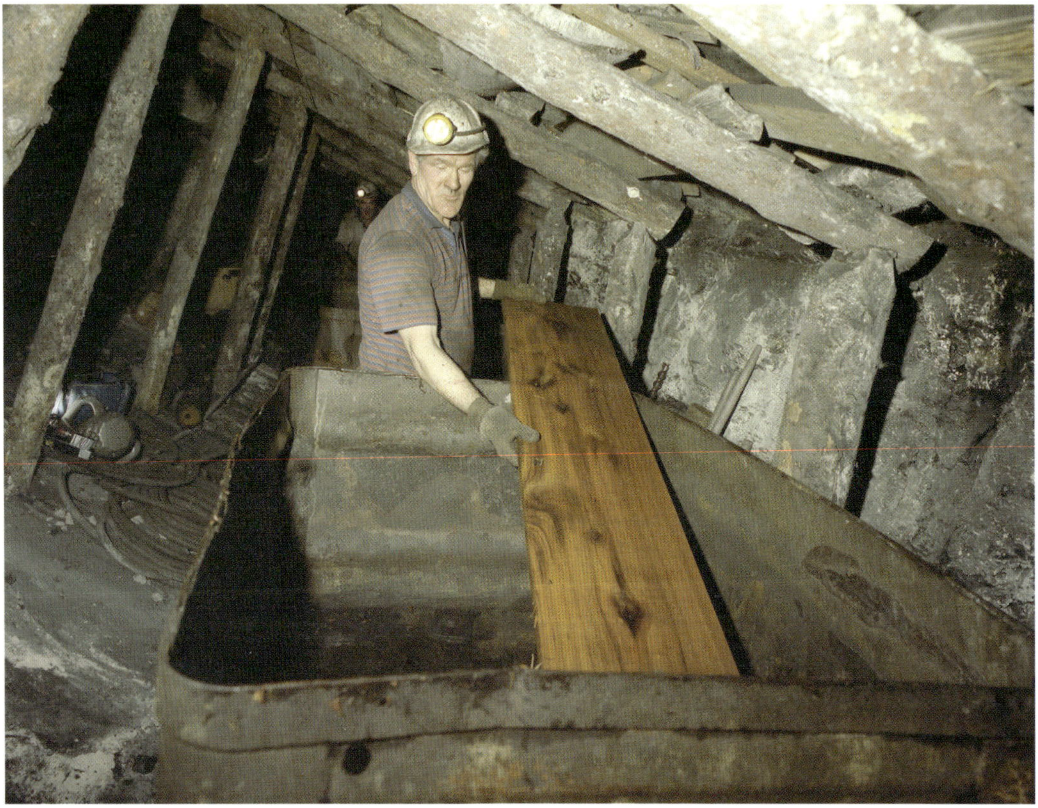

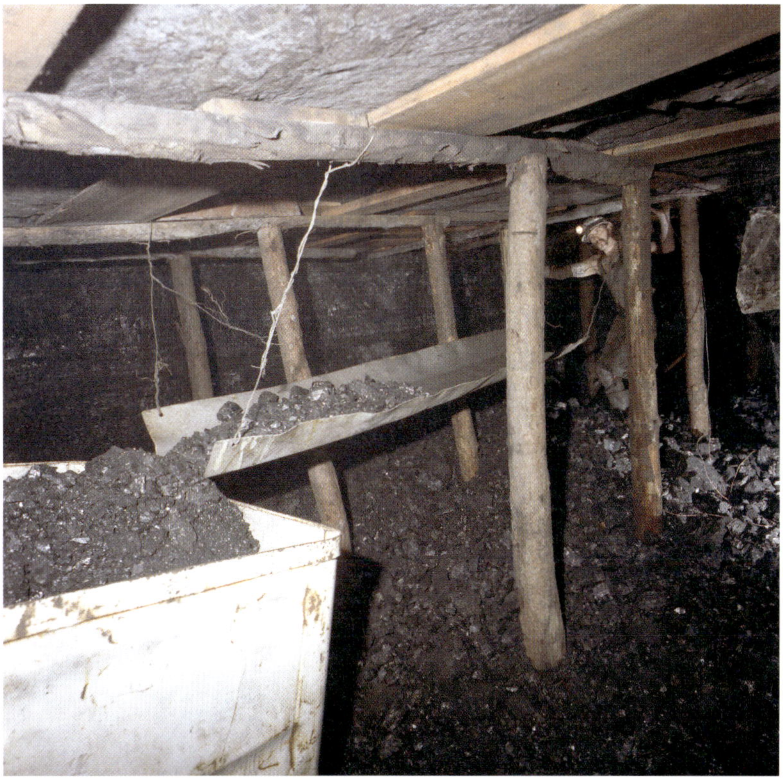

Right: No 2 Colliery. Deadpans delivering coal to a tub during drifting, February 1993. PD

Below: Archie Maiden loading explosives into shot holes at No. 3 Colliery, September 1996. PD

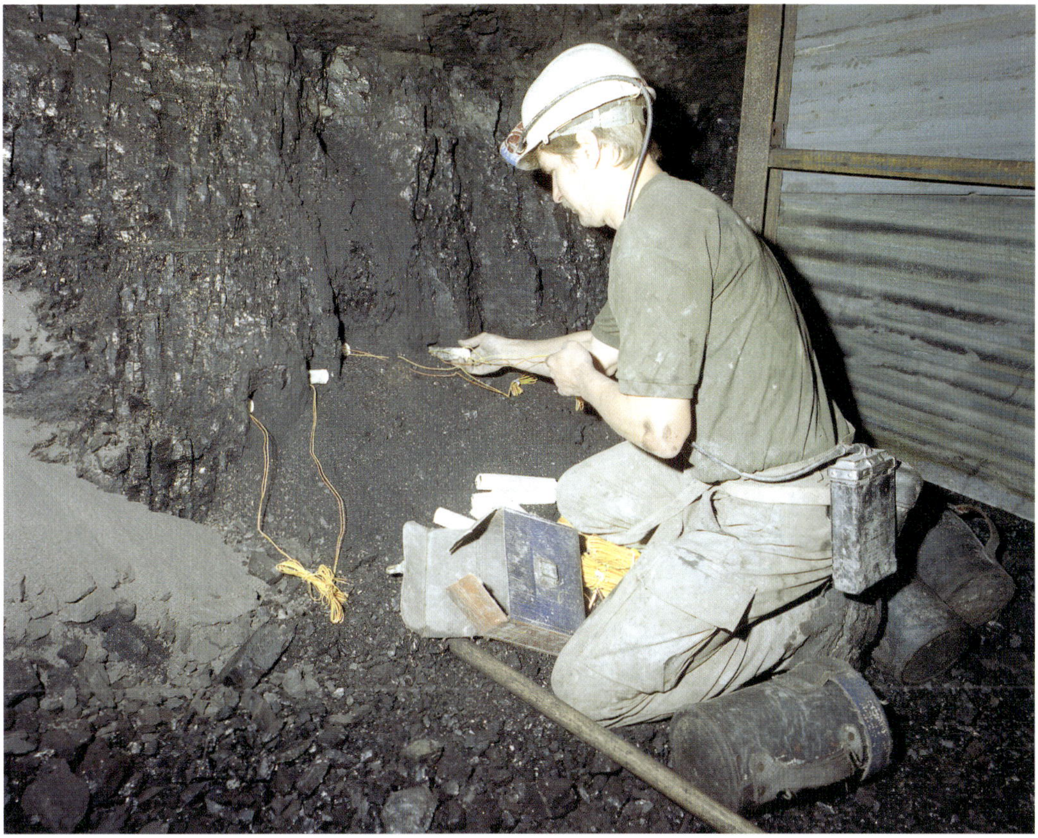

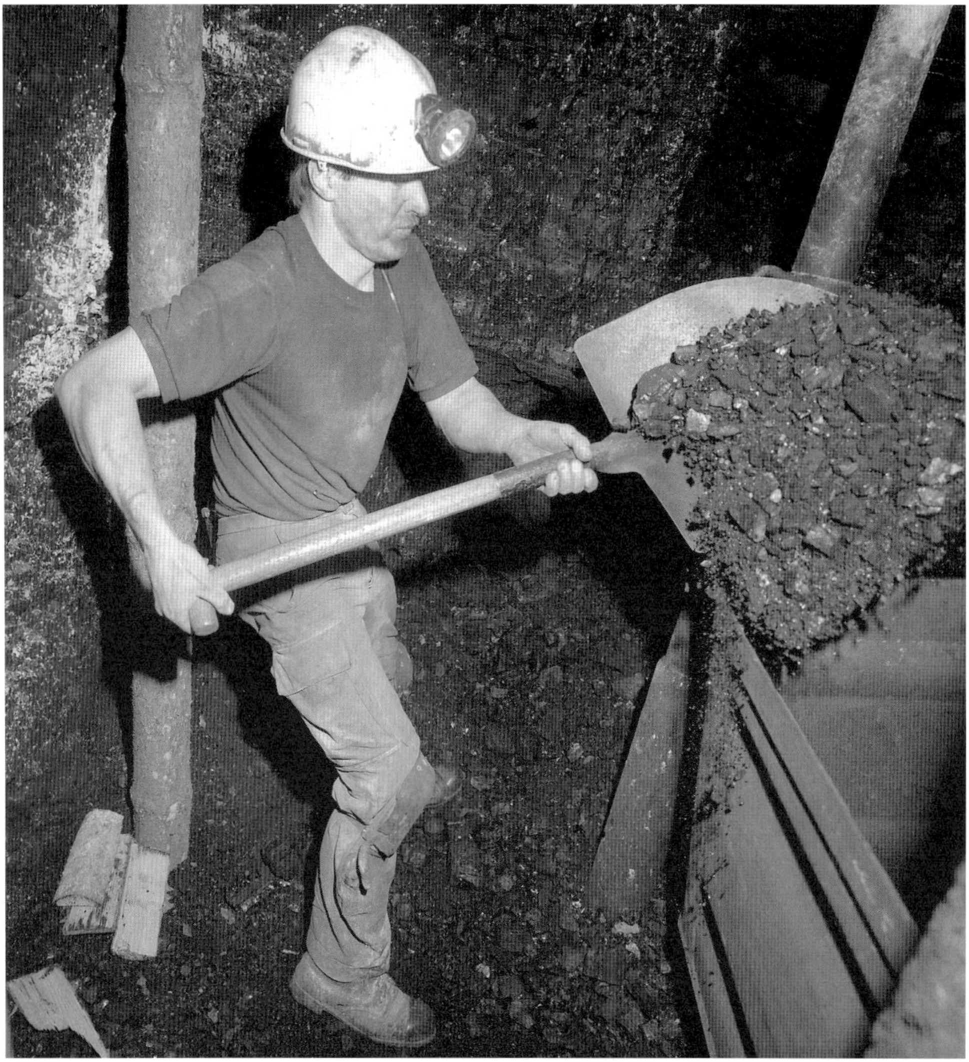

Above: No 2 Colliery. Archie Maiden loading coal into a tub, April 1994. PD

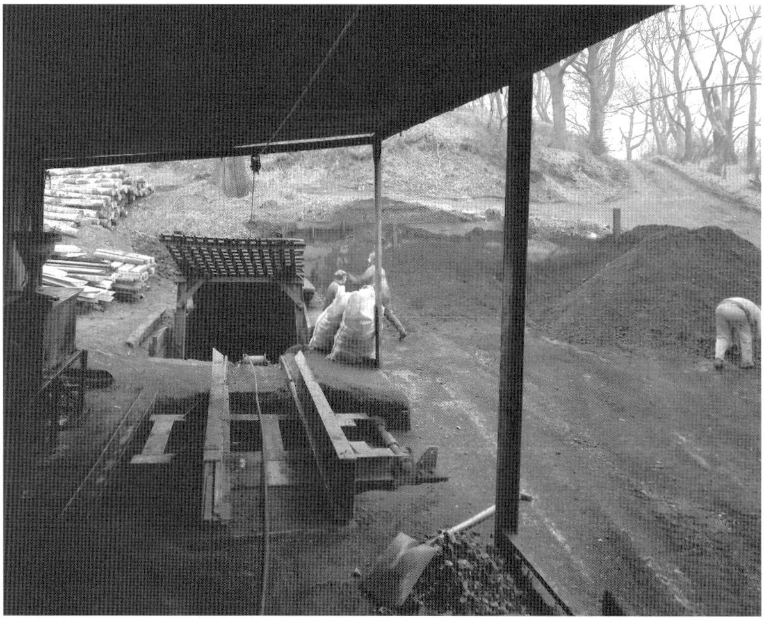

Right: No. 1 Colliery was a small footrail immediately to the west of the Parksite housing estate at Silverdale. The sites of all three footrails have now been restored to agriculture and nature conservation.

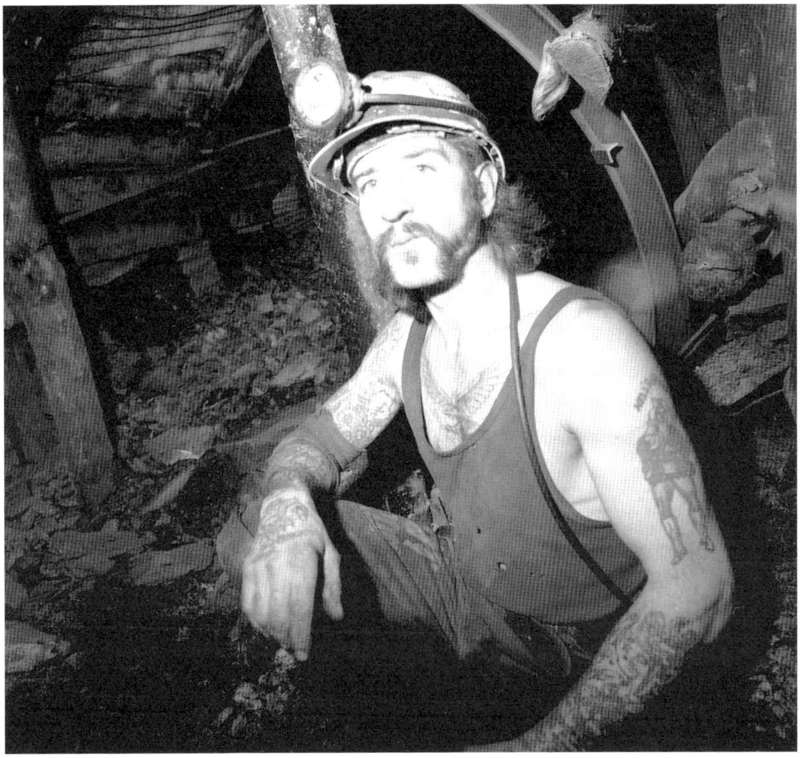

Right: Rees Osoviecki taking a break, February 1993. PD

Below: Rees Osoviecki standing by the deputy with a round of shots ready to be fired, February 1993. PD

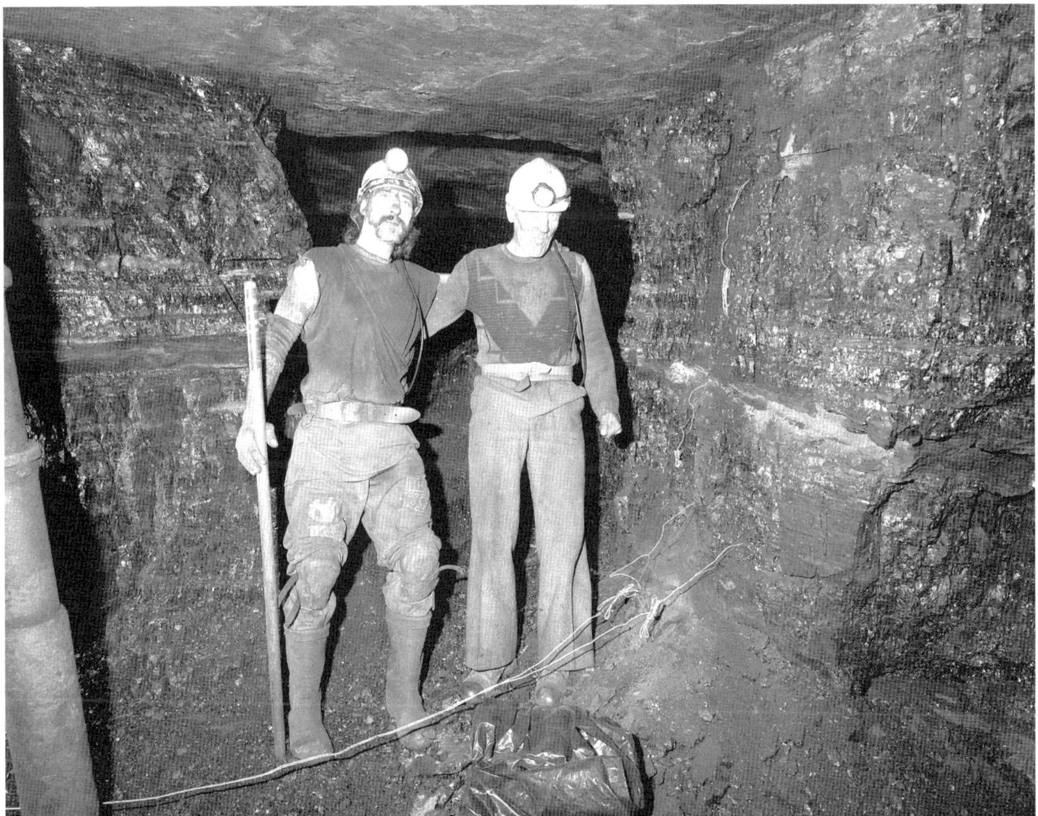

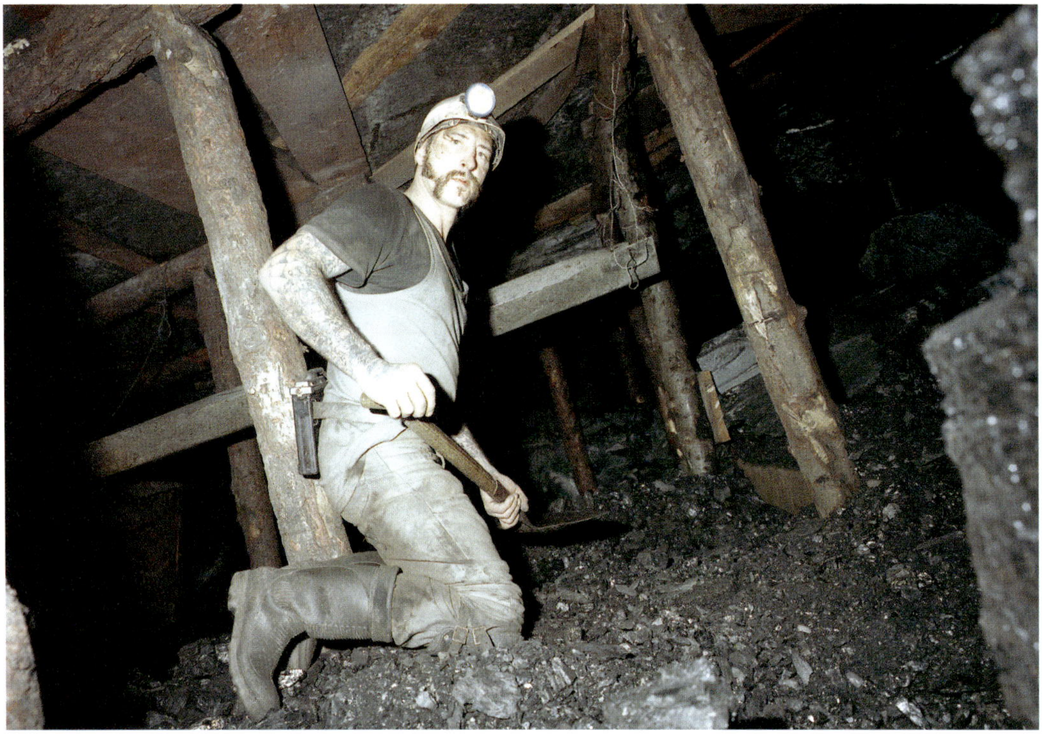

Above: No 2 Colliery. Rees Osoviecki loading coal in a drift, February 1993. PD

Below: Selwyn Johnson and Bernard Ronan (on right) loading a tub during the drivage of the intake dip, Haying Wood No. 2 Colliery, 1984. PD

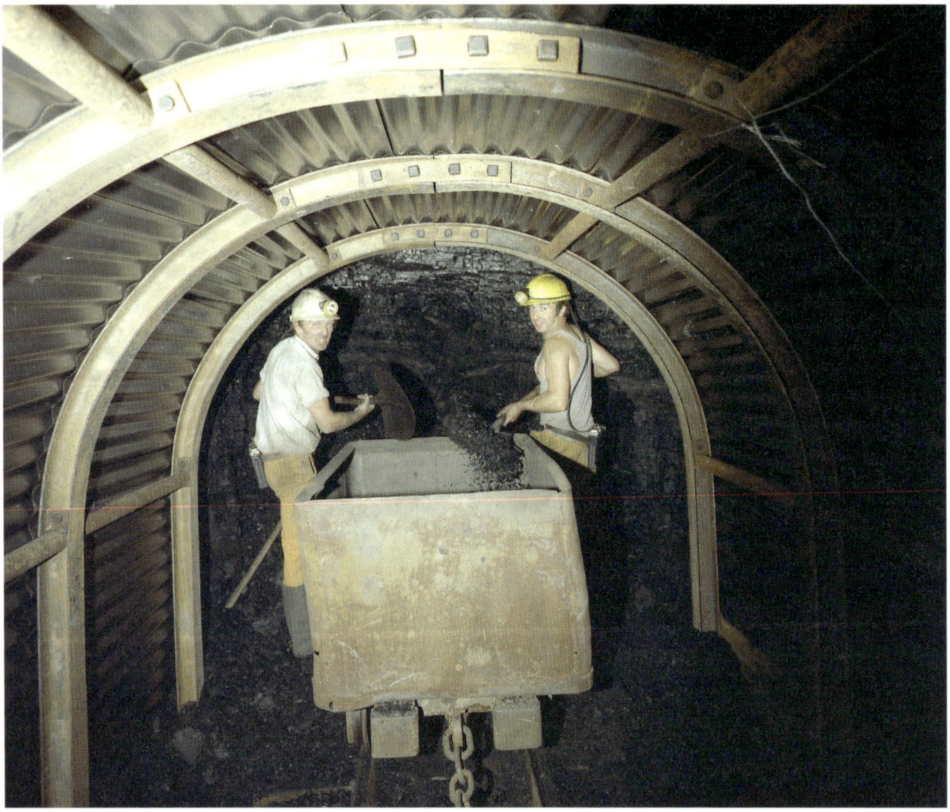

Hem Heath Colliery

Stafford Coal & Iron Co. Ltd.

Situated at Trentham immediately to the east of the railway and north of the A5035 main road.

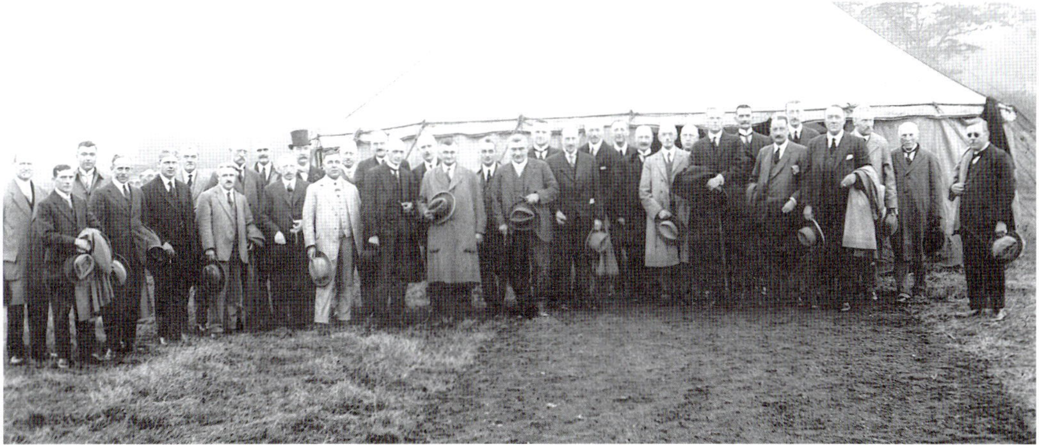

Above: Cutting the first sod for No. 1 shaft in 1924. MC

Below: No. 1 shaft, 1930s. MC

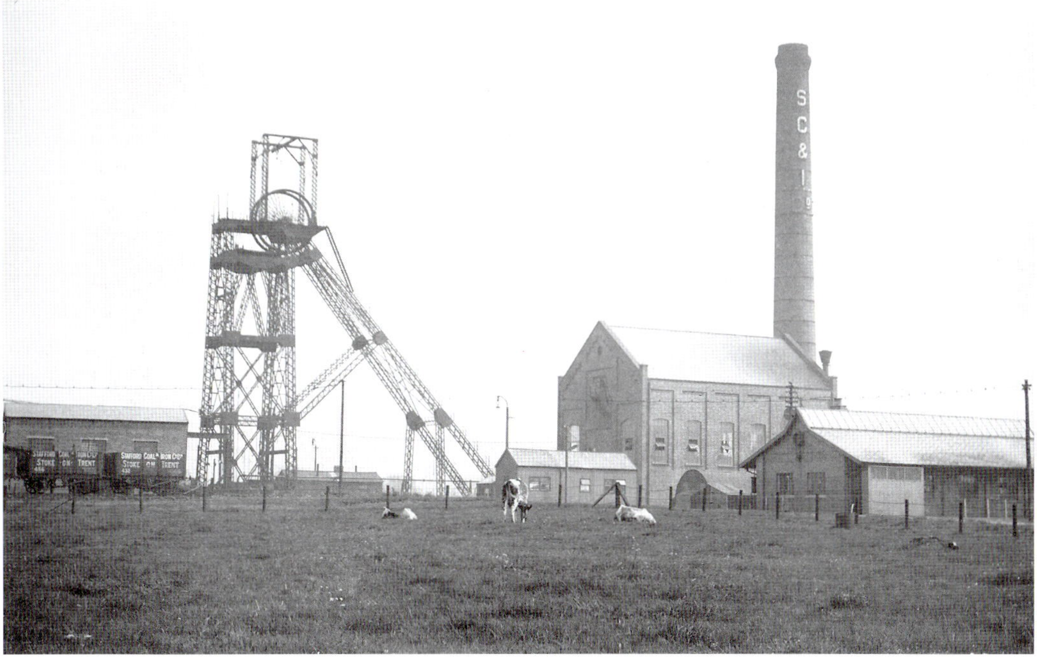

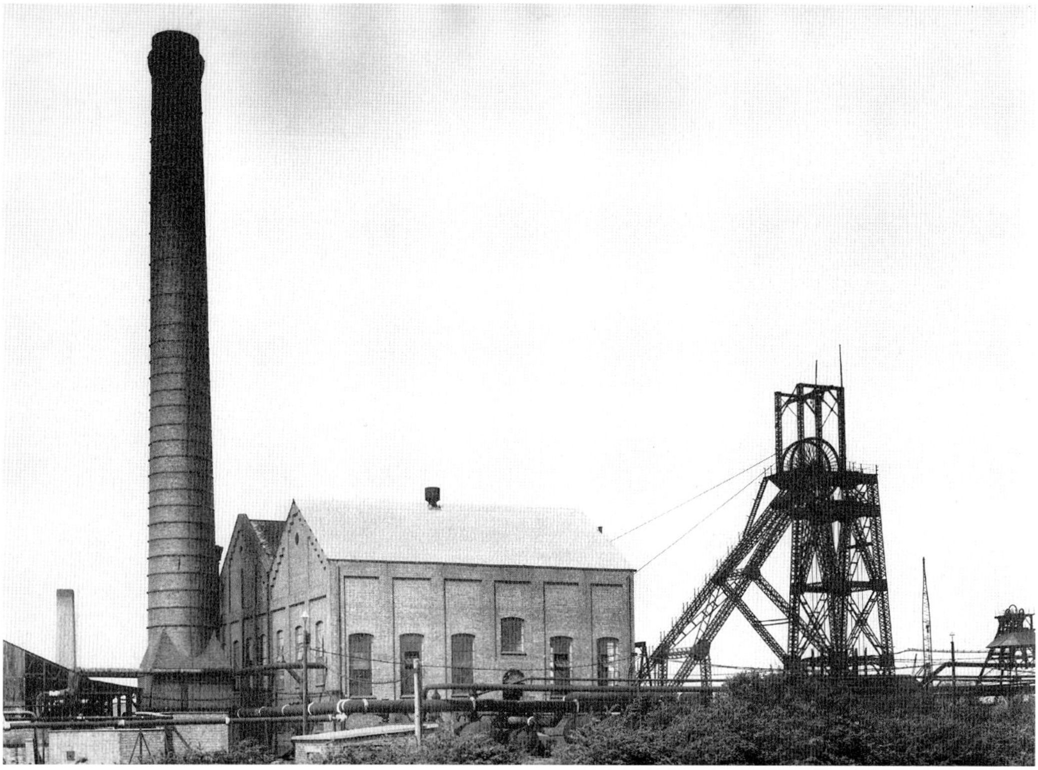

Above: No. 1 shaft in 1951. MC

Below: A view of No. 1 shaft in 1951. The winding engine house was subsequently used as the electrical equipment stores. MC

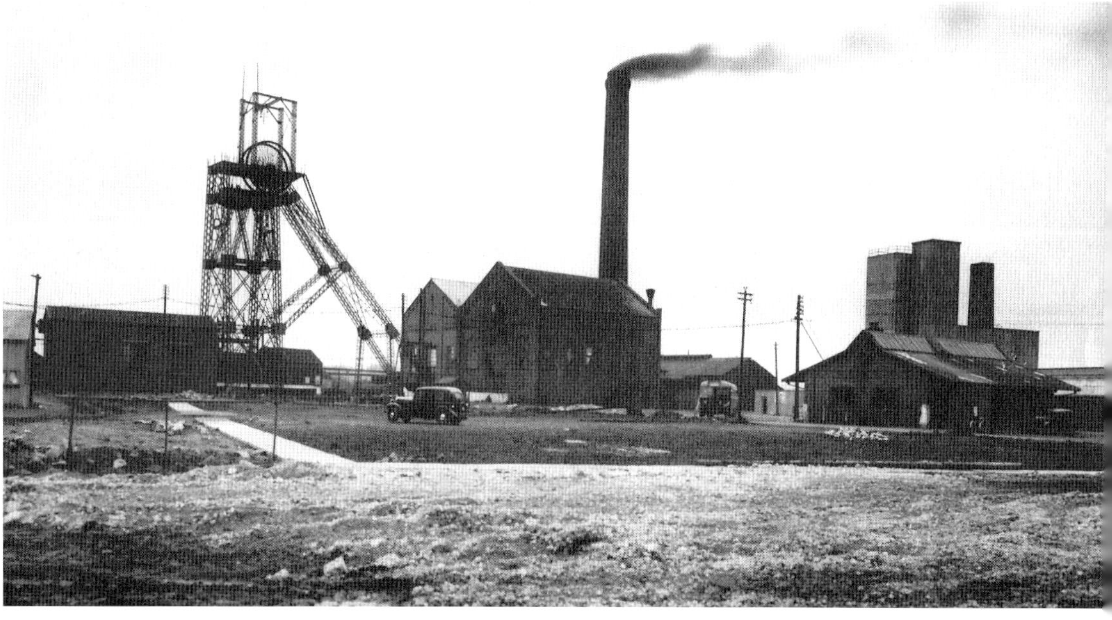

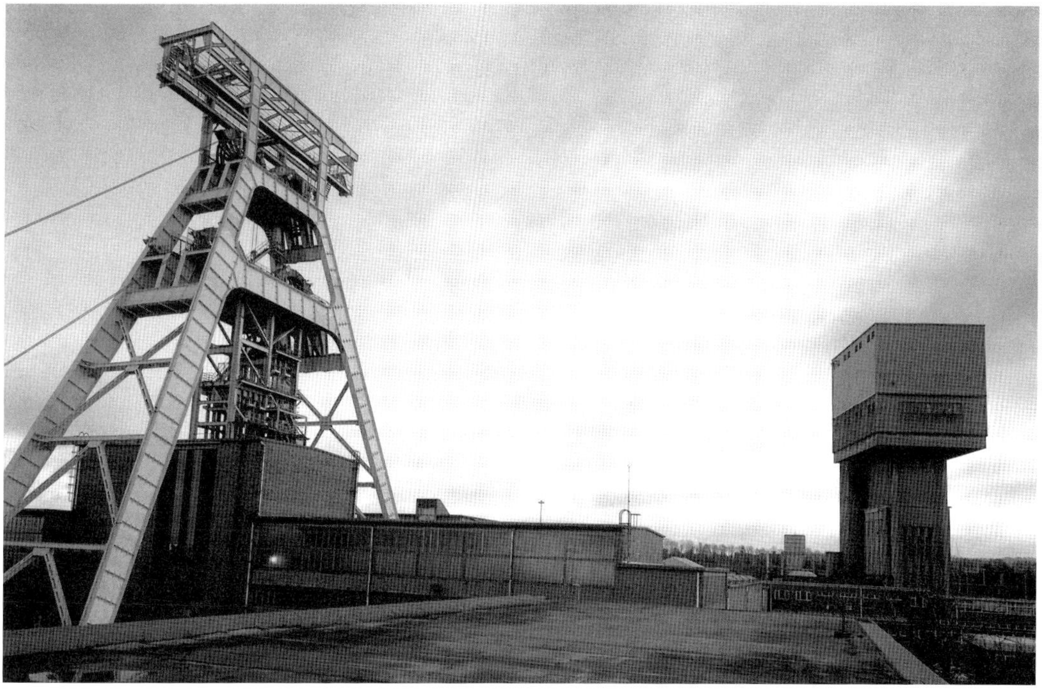

Above: No. 2 shaft on the left and No. 1 shaft winding tower on the right, 15/2/1997. PD

Below: A view of the No. 2 shaft headstocks from the top of No. 1 shaft. No. 2 shaft was sunk in the 1950s-60s and was 7.3m in diameter and 1,020.6m deep. 15/2/97, PD

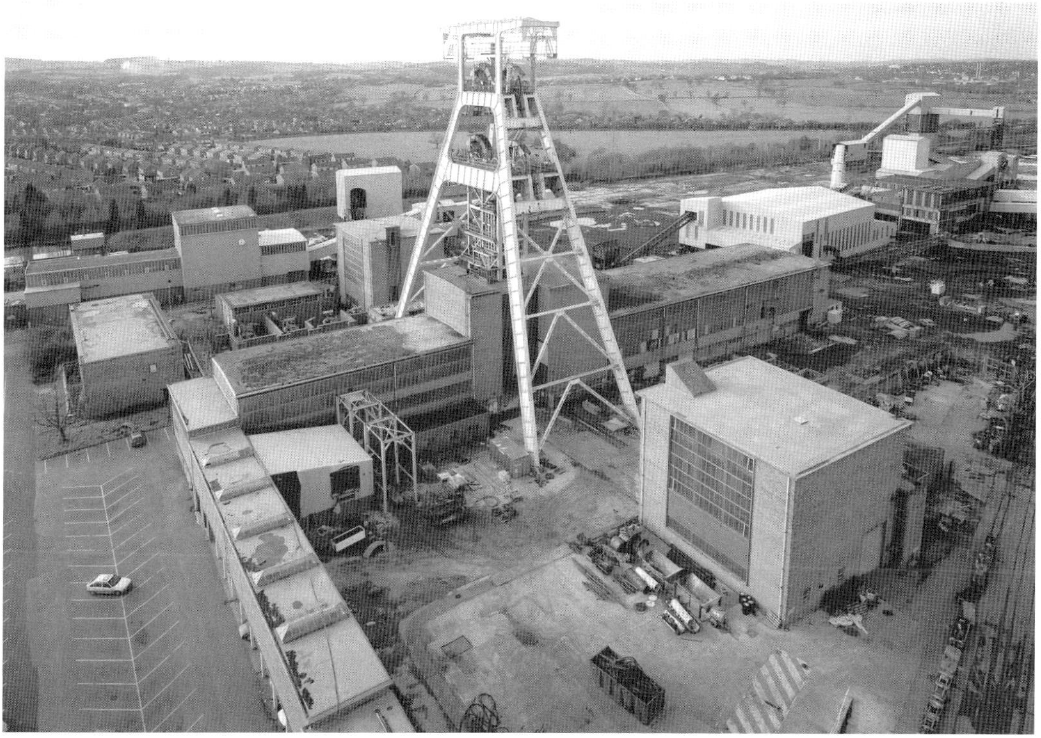

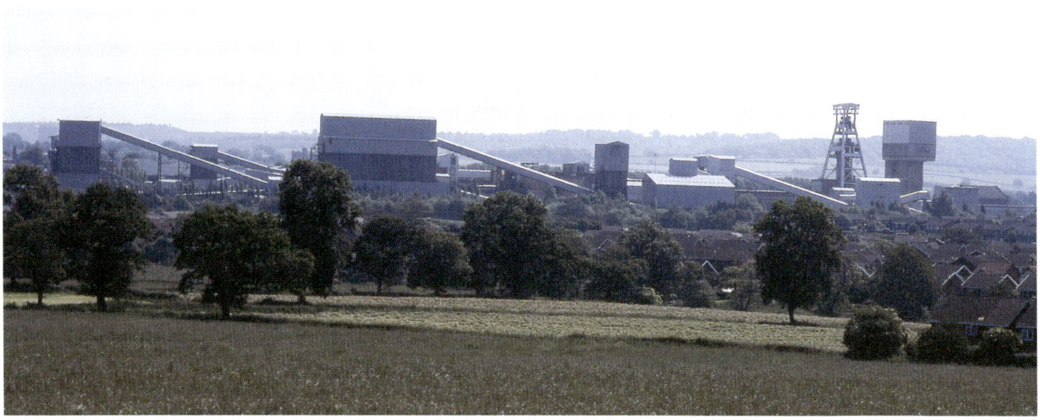

The colliery from New Inn Lane, Trentham, in 1994. The mine was abandoned in June 1997 and is now a housing estate. PD

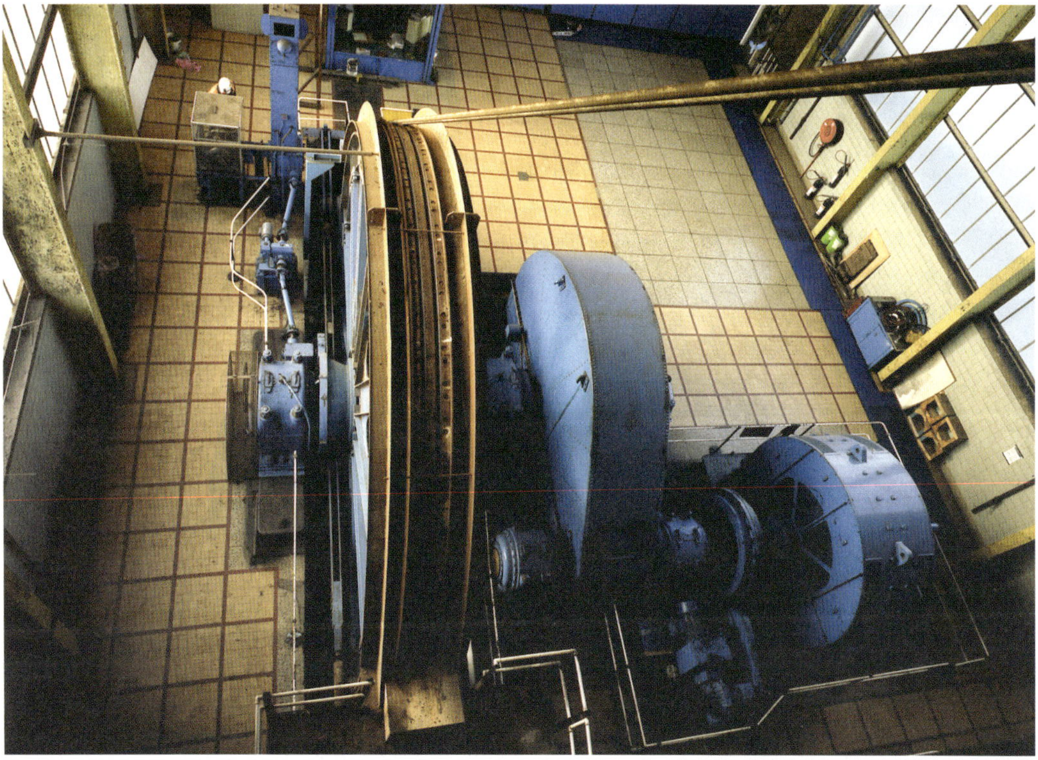

The Keope Winding Engine on the No. 2 shaft. 15/2/97, PD.

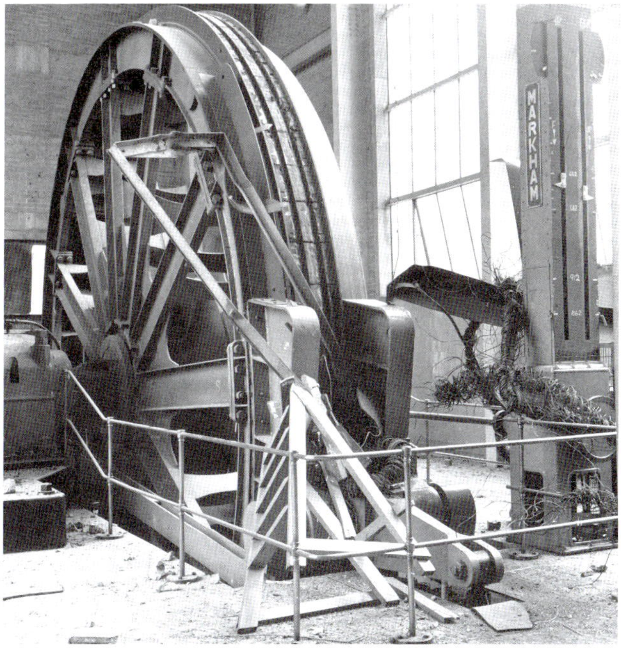

Right: No. 2 shaft Keope Winder after a winding accident in the 1960s. All of the rope ended up at the bottom of the shaft. No 2 shaft was sunk in the 1950s-60s and had a final depth of 1,020.6m. MC

Below: The 3,300 bhp 4-rope friction winder on the No. 1 shaft. The shaft was initially sunk to a depth of c. 600m and sinking commenced in 1924. It was deepened in the 1950s/60s, to a final depth of 1,039.8m. The shaft is 5.8m in diameter and was lined with brick in the upper portion and with reinforced concrete in the lower portion. There was a single cage with a balance weight. It had four decks, each with a capacity of 24 men, allowing 96 men to be carried at a time. No. 1 shaft was sunk as a southerly extension to Kemball Colliery and operated with one shaft, with a connection to Kemball until the 1960s. 15 Feb 1997, PD

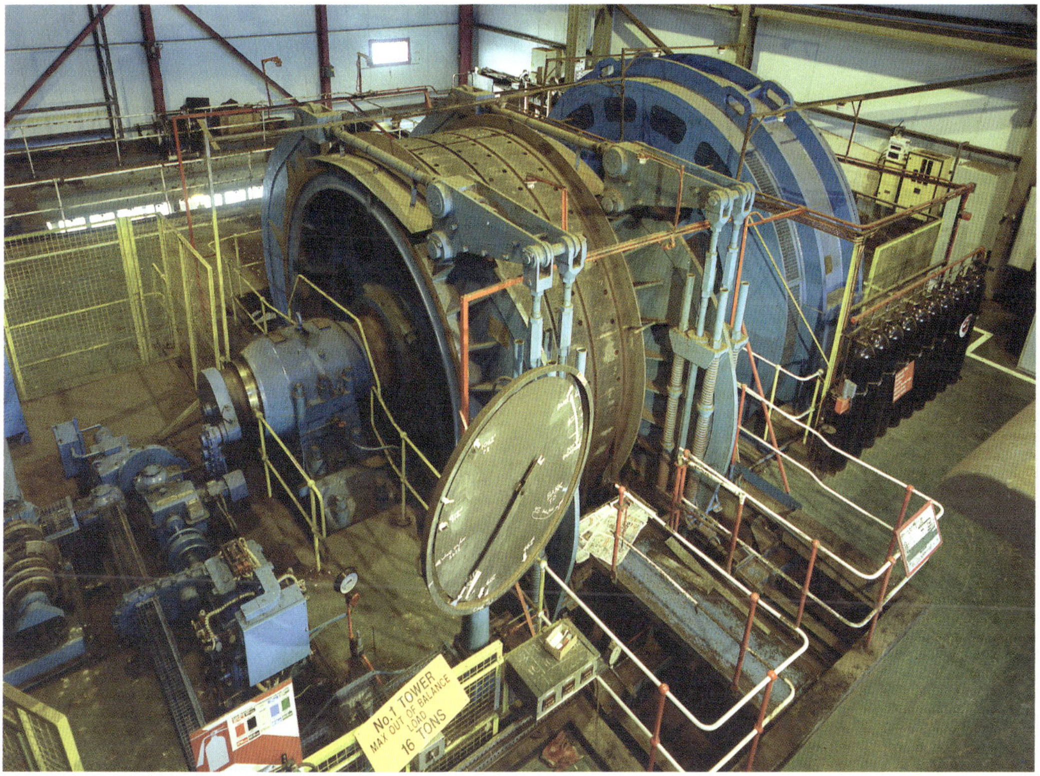

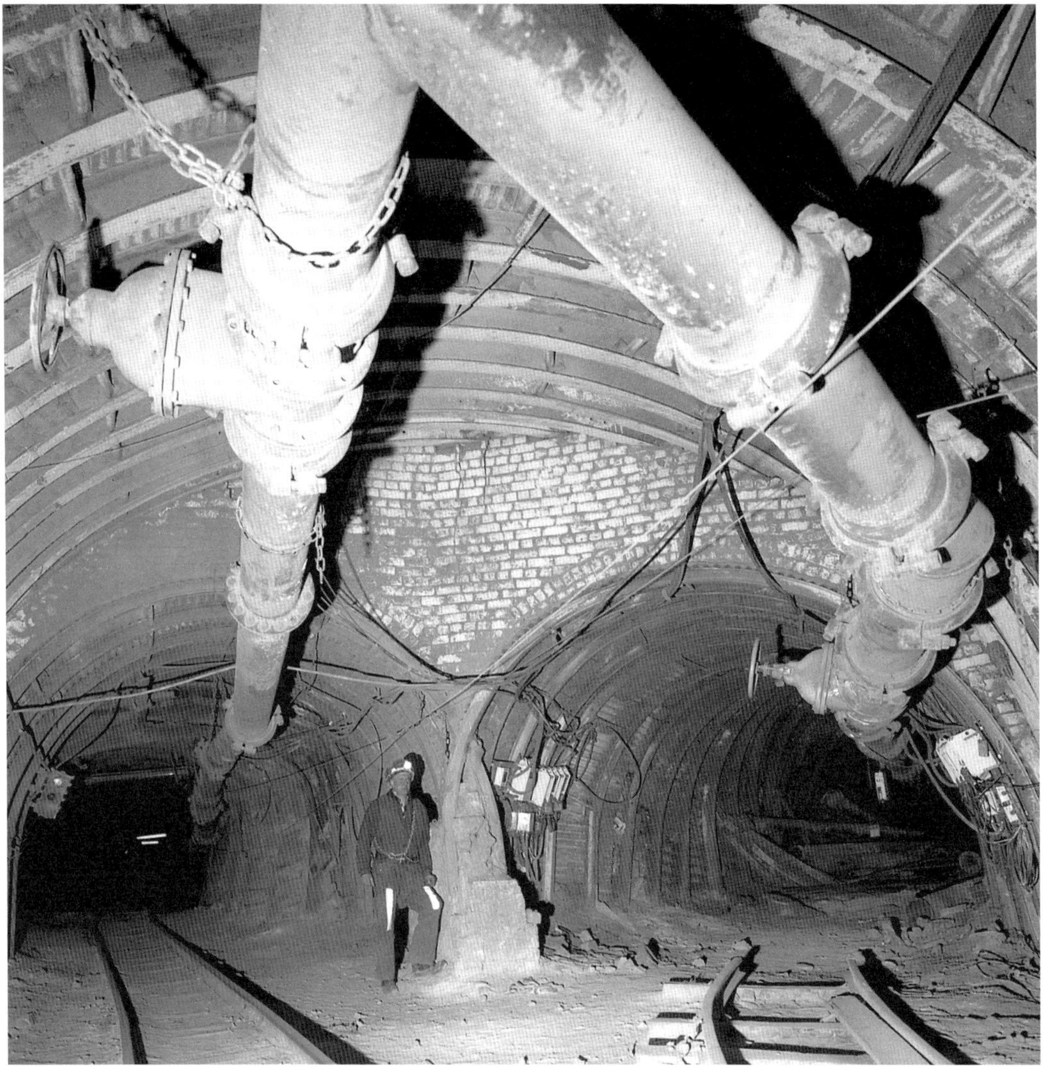

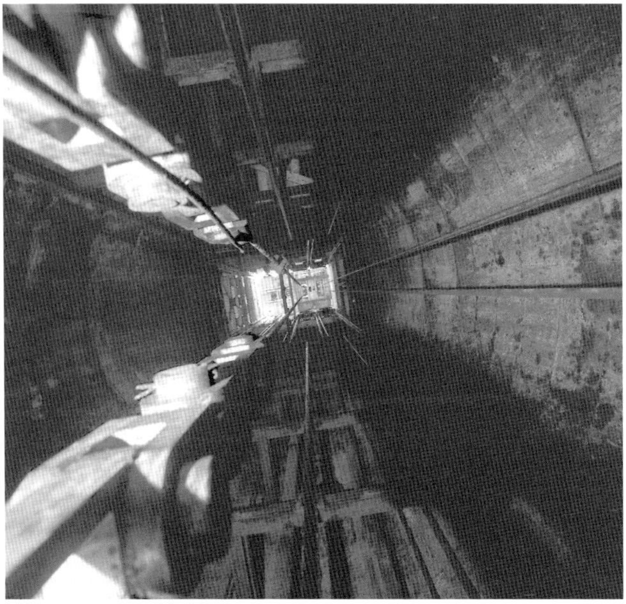

Above: The 1,062 Yard pit bottom looking south. These roads led to the west and east 1,062 Yard horizons. The pipes are for methane drainage, 15/2/1997. PD

Left: The view up No. 2 shaft, from the top of the cage. On the left may be seen the rope capels, 15/2/97. PD

Right: The balance weight in No. 1 shaft. The block and tackle were used to adjust the tension of the ropes to ensure that they were all applying equal friction on the winding drum, 15/2/97. PD

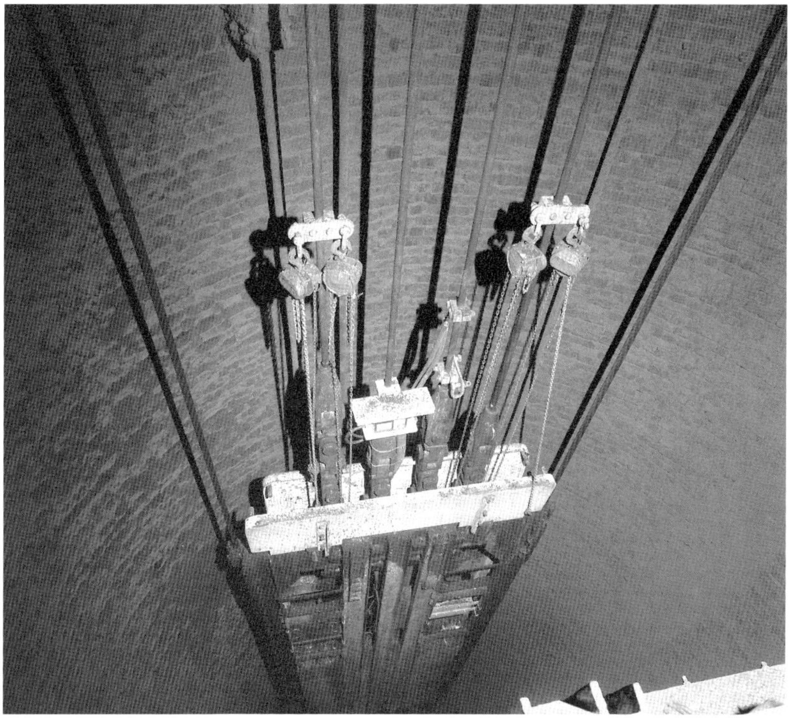

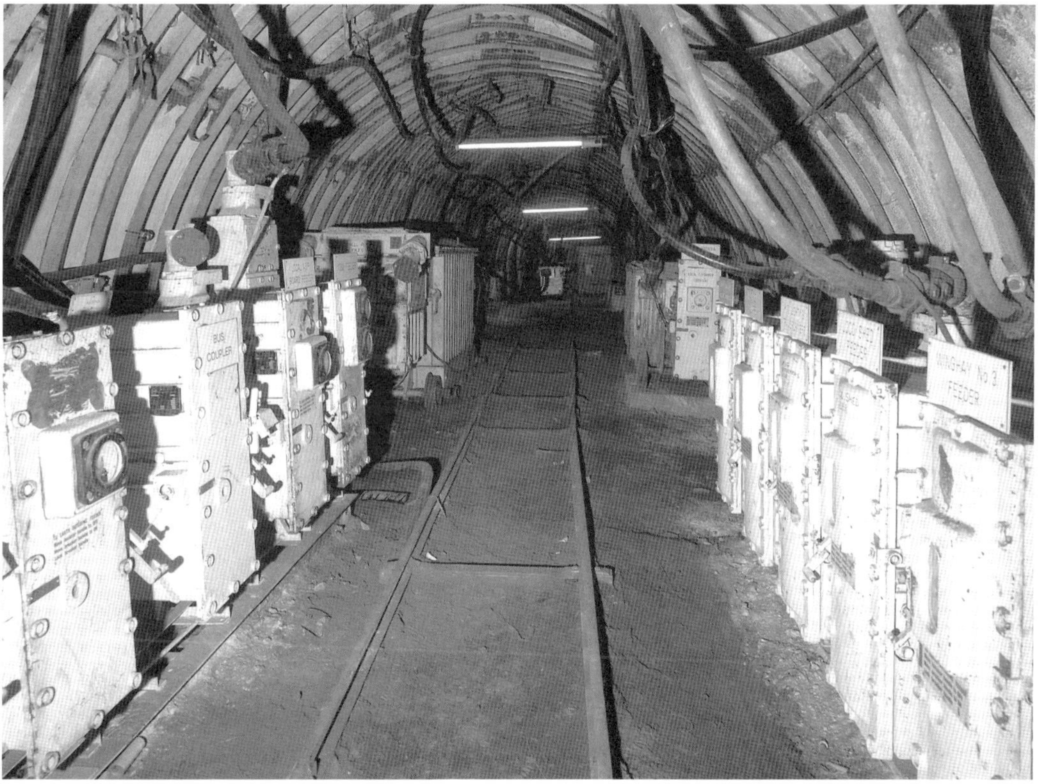

Above: The 612 yard pit bottom electric sub-station, 15/2/97. PD

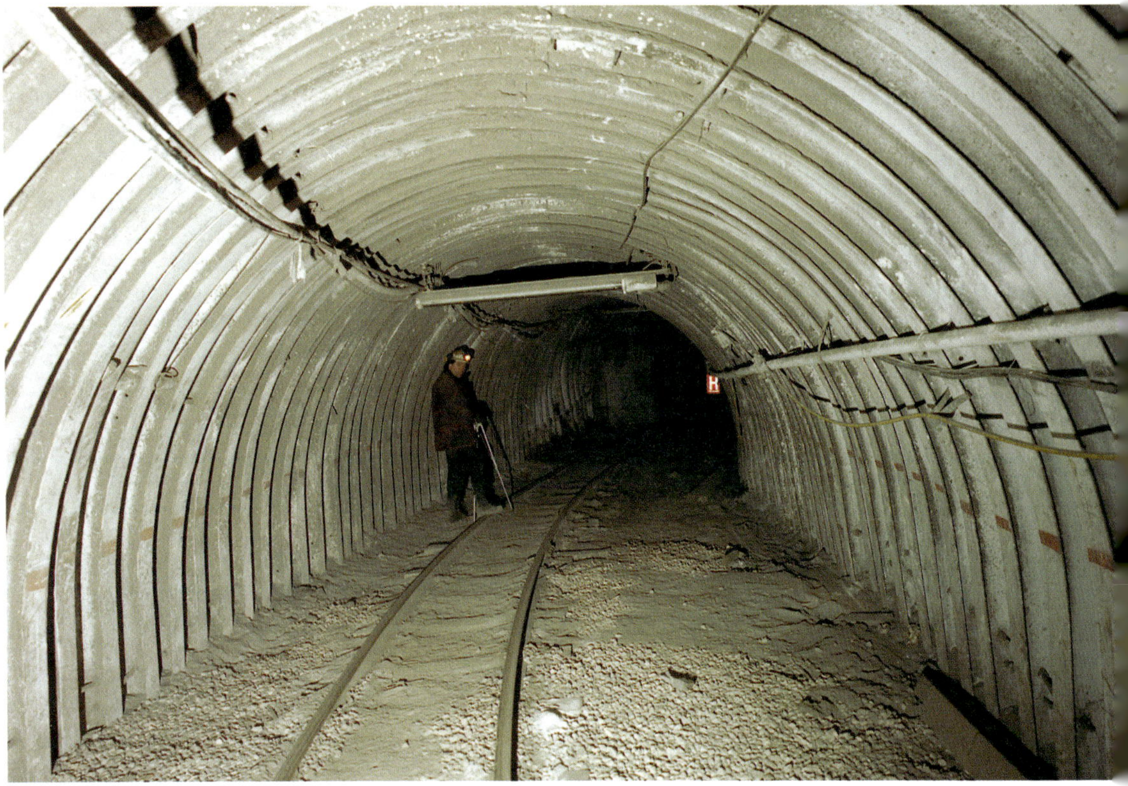

Above: Connecting roadway between the shafts, 15/2/97. PD
Below: The inset at the 612 yard level of the No. 1 shaft, 15/2/97. PD

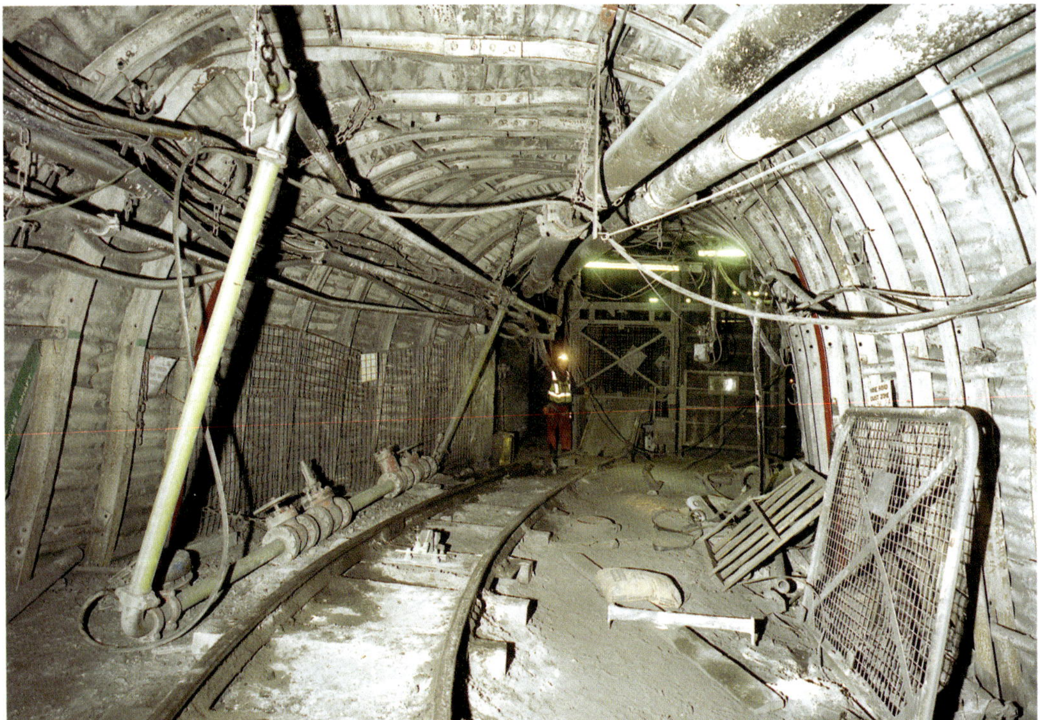

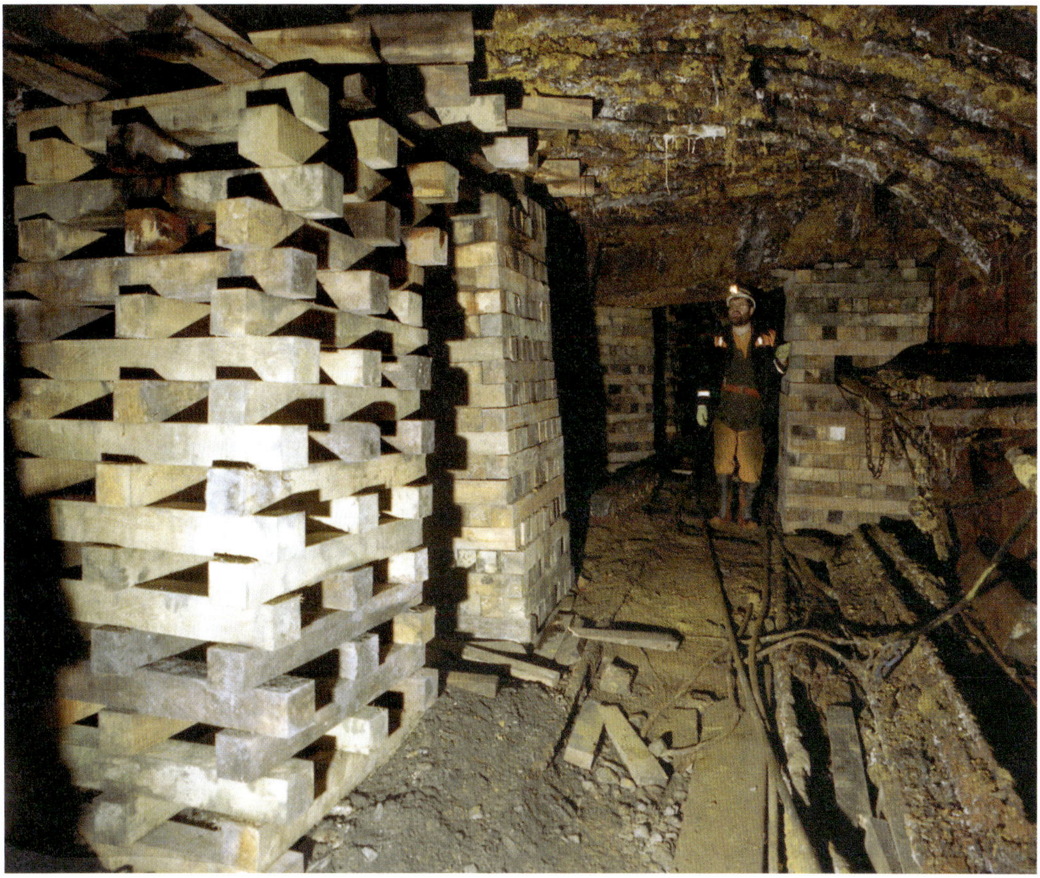

Both views: The 612 yard pump lodge,15/2/97. PD

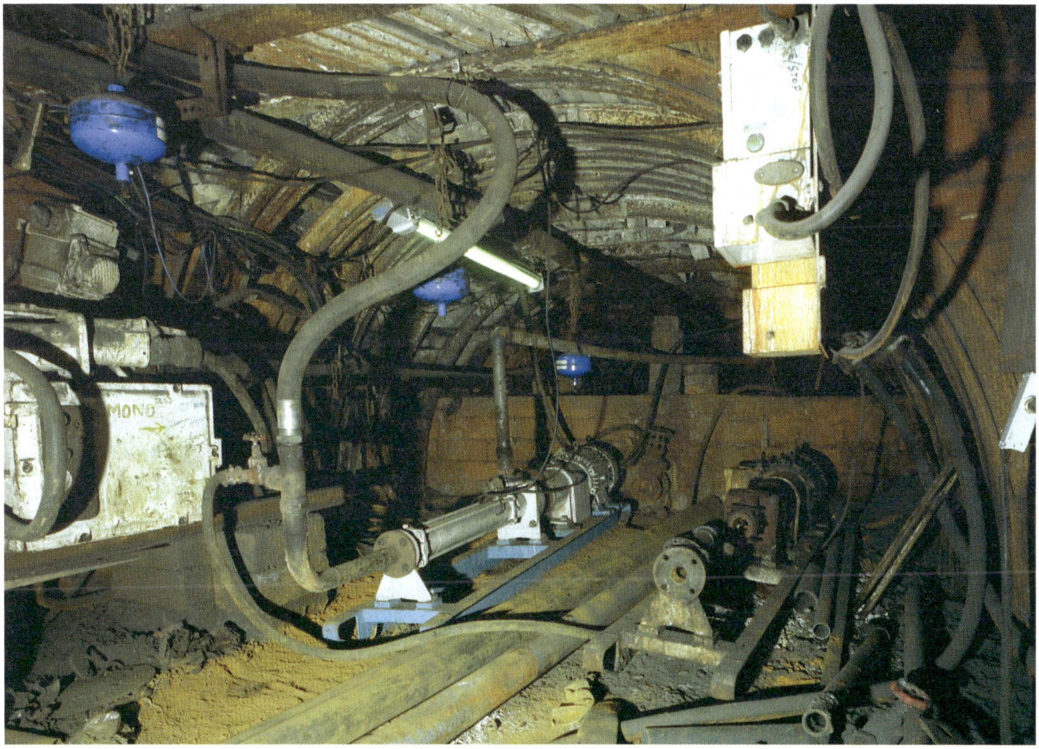

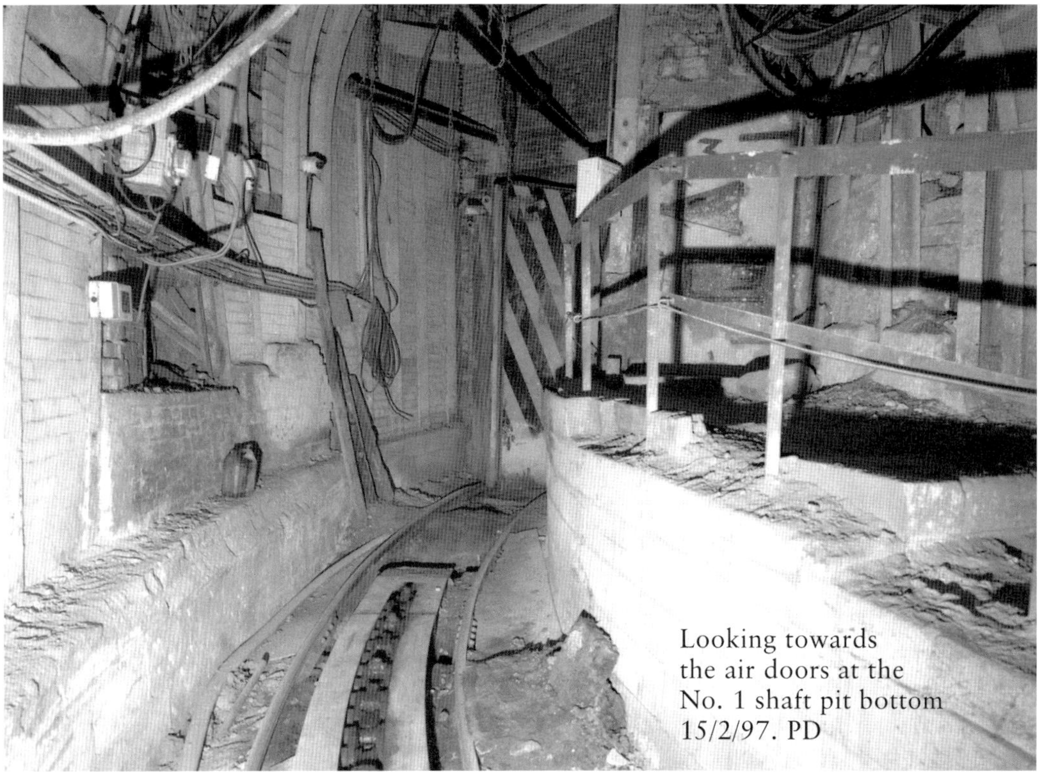

Looking towards the air doors at the No. 1 shaft pit bottom 15/2/97. PD

Below: The surface drift, driven almost due south at a gradient of 1:4. It is 2.49km long. Electric cables may be seen above the conveyor belt on the right of the photograph. The conveyor is a cable belt, where the belt rests on two steel ropes, which transmit the power. On the left is the track for transporting materials, 15/2/97. PD

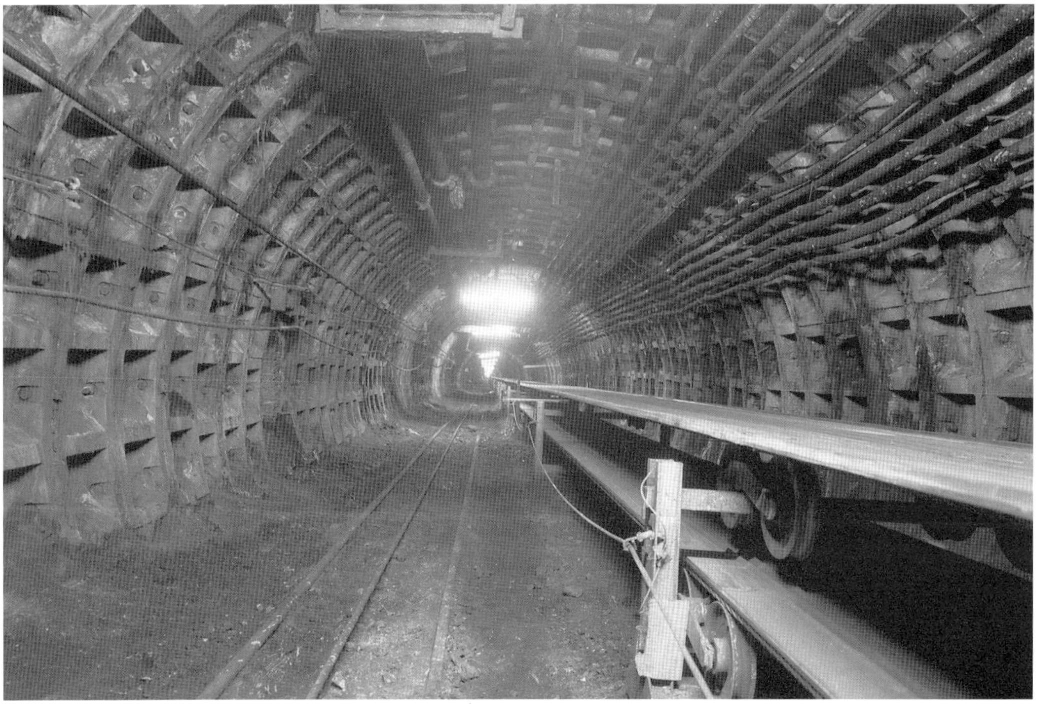

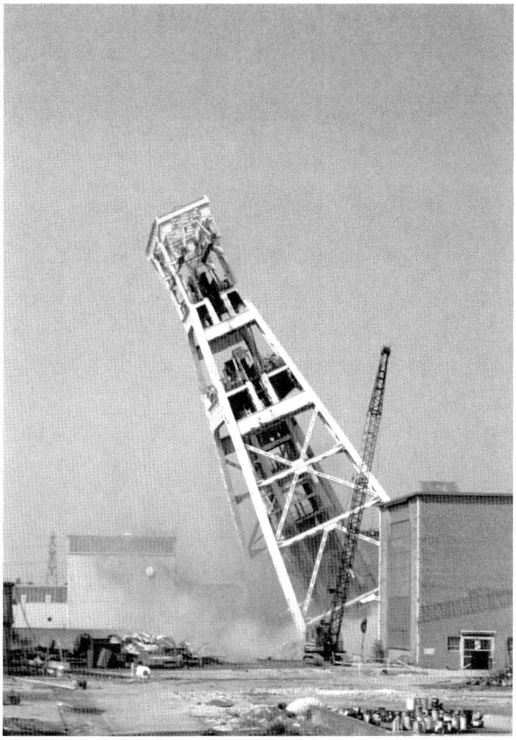

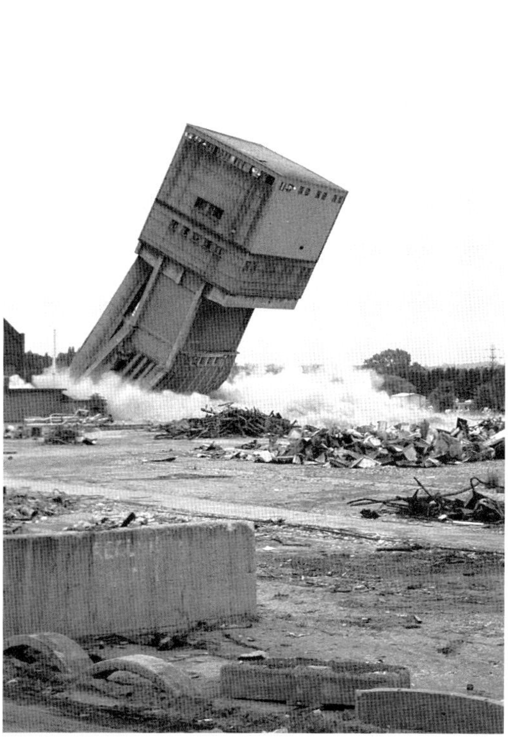

Above left: The A-frame going down on No 2 shaft. Above right: The winding tower on No 1 shaft during demolition, 13th August 1997. Photos: Albert Baines Davies

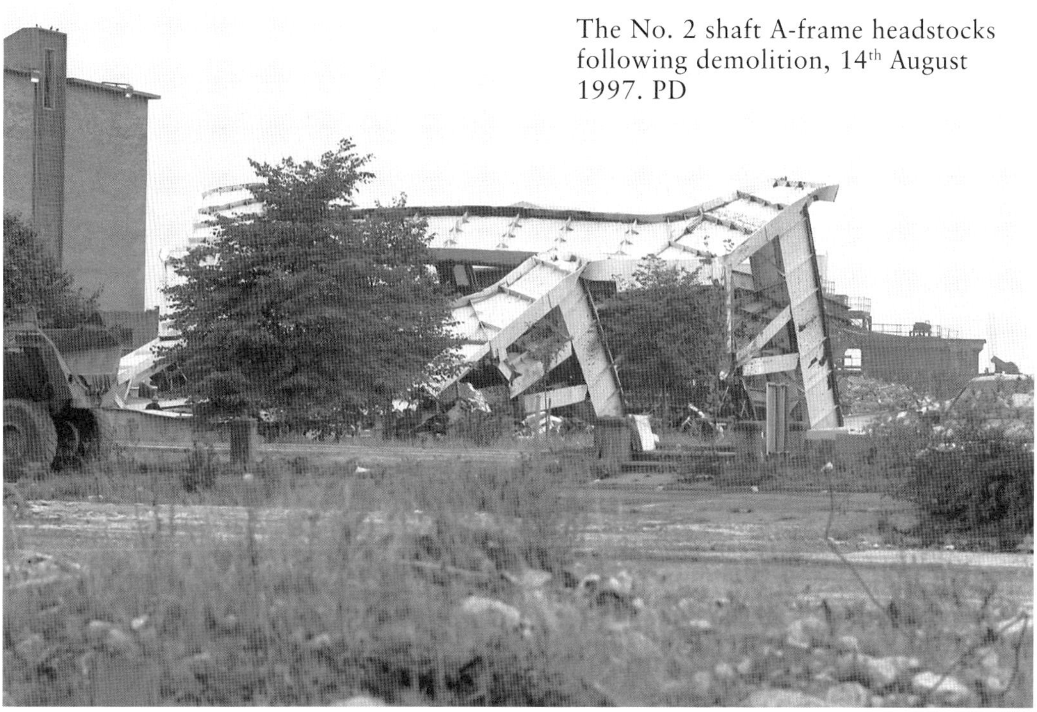

The No. 2 shaft A-frame headstocks following demolition, 14th August 1997. PD

High Lane Opencast Site

NCB Opencast Executive contractors – Crouch

General view of the High Lane site showing an O & K hydraulic shovel loading a 90-ton payload Cat 777B dump truck. It took three passes to fill the dumper with overburden. Immediately to the left, a loader and a 25-ton articulated lorry are dwarfed by the Cat 777B. The site worked the steeply dipping seams on the east side of the Apedale anticline. It eventually worked c. 4.5 million tons of coal and handled an average of 100,000 tons of overburden per day over its seven-year operational life, May 1992. PD

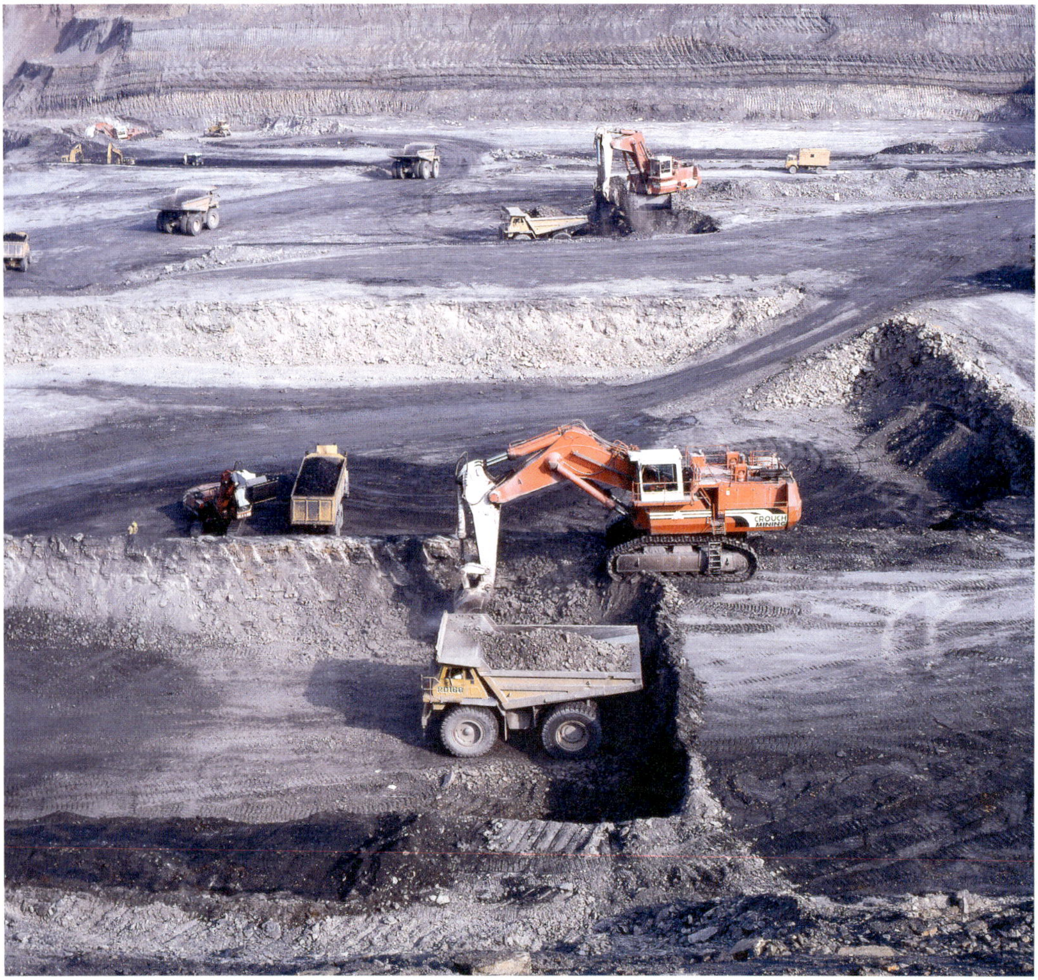

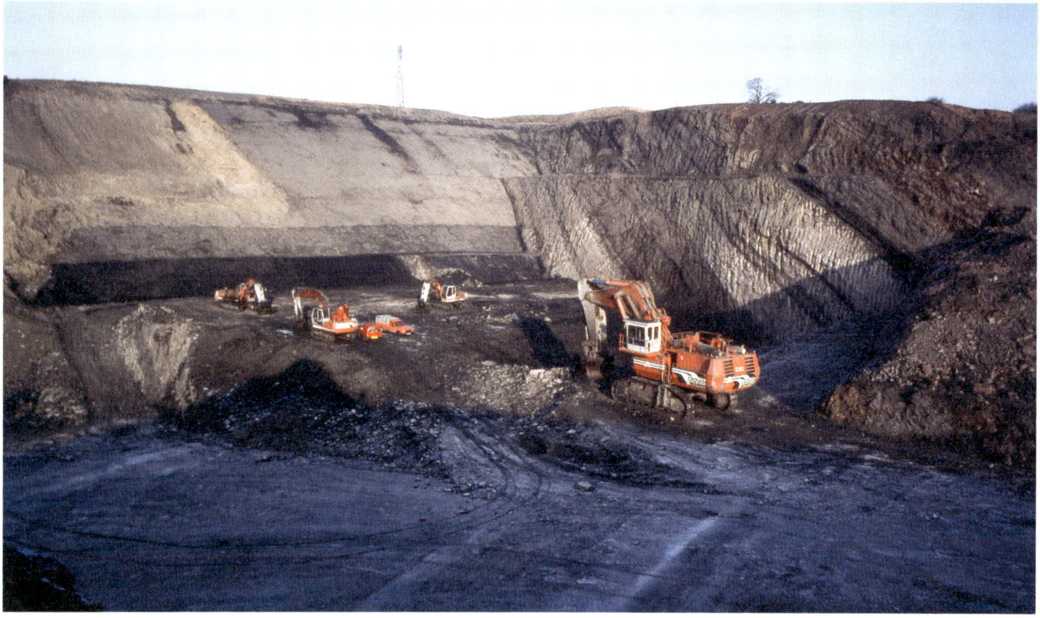

Above: A view of the site to the west of Black Bank Road. The coal seams can be seen as black lines in the high wall on the right side of the photograph. The background shows the base of the seam with the coal extracted. The black strip towards the base immediately above the plant shows a strip of coal in the Chalkey Seam waiting to be lifted, December 1991. PD

Below: A Cat 777B dump truck crossing Black Bank Road. The car waiting at the traffic lights shows the scale, May 1992. PD

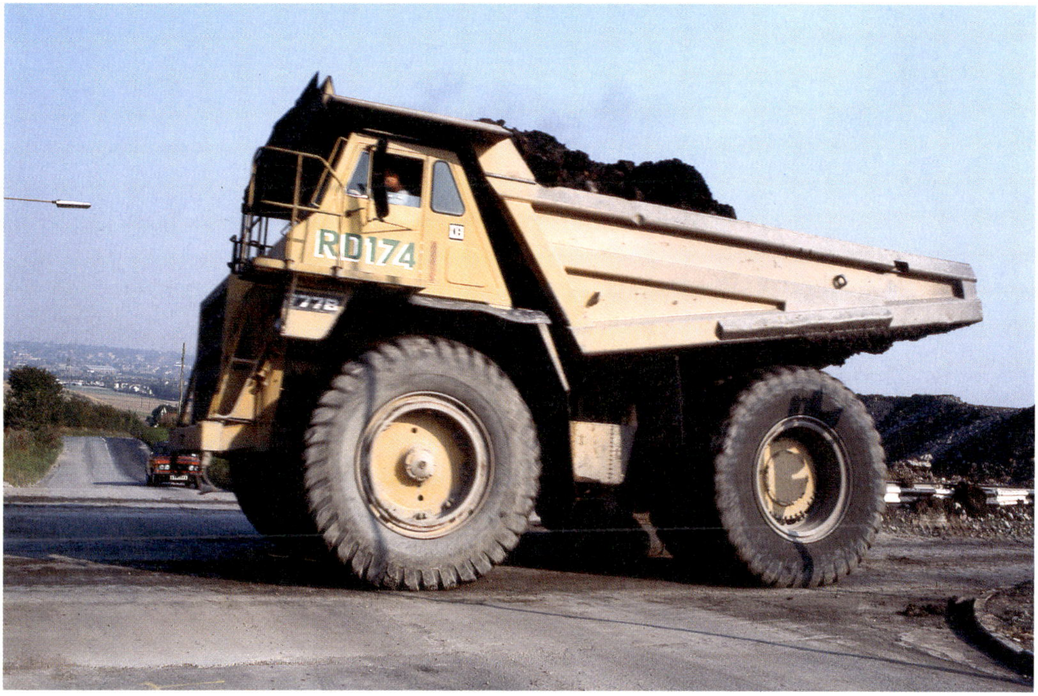

HOLDITCH COLLIERY

Holditch Mines Ltd.

This mine was sunk in c.1918 by the Brymbo Company as an iron mine. No. 1 shaft is 624.6m deep and No. 2, 625m deep. The mine initially worked the Black Band, Red Mine and Red Shagg ironstones and the Great Row coal. The shafts did not reach the coal seams and the pit bottoms were in the Black Band ironstone. The mine subsequently drove cruts (cross-cuts) through the 700m Apedale Fault to exploit the Moss and deeper seams lying between the Apedale Fault, the Millbank Fault and old workings from Apedale Colliery (Burley Pit). The mine closed in August 1989. Situated 1.5km south of Chesterton Church.

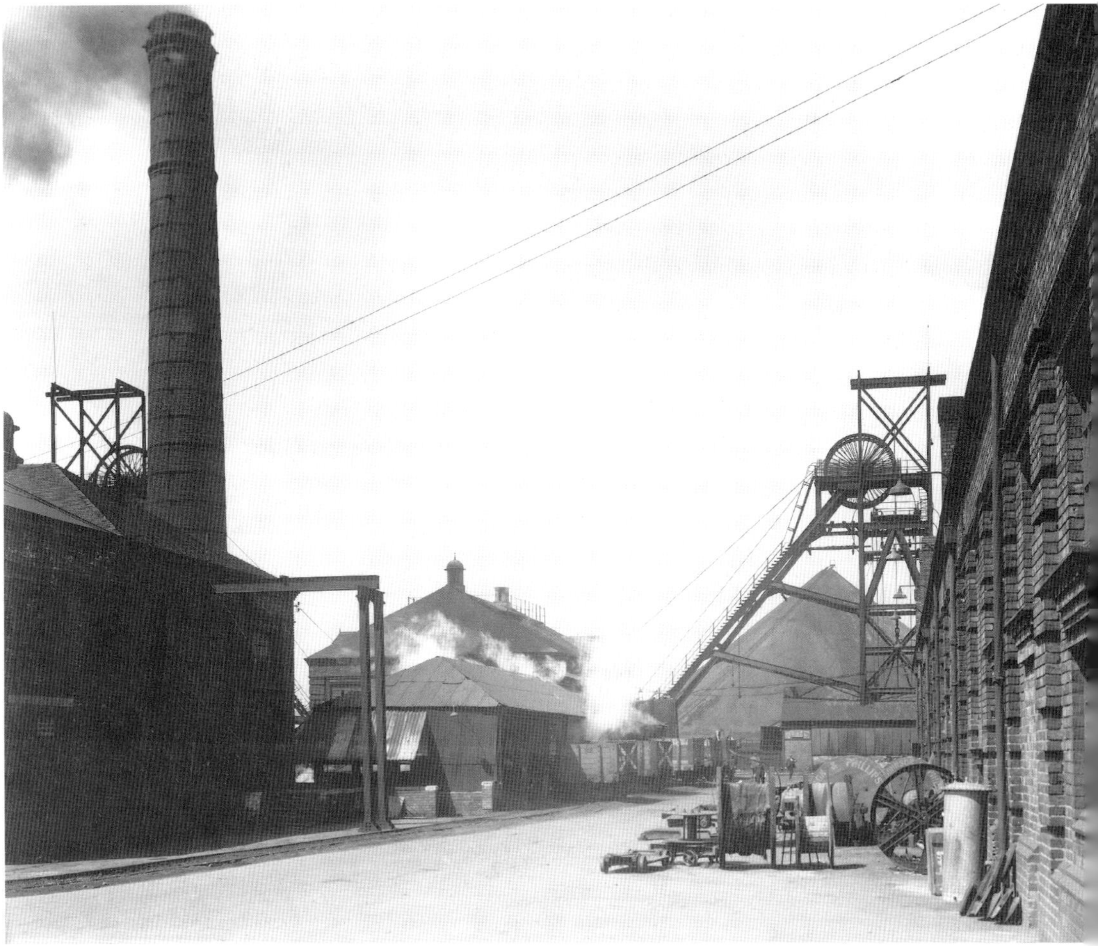

View of the colliery in 1951. The site is now an industrial estate. MC

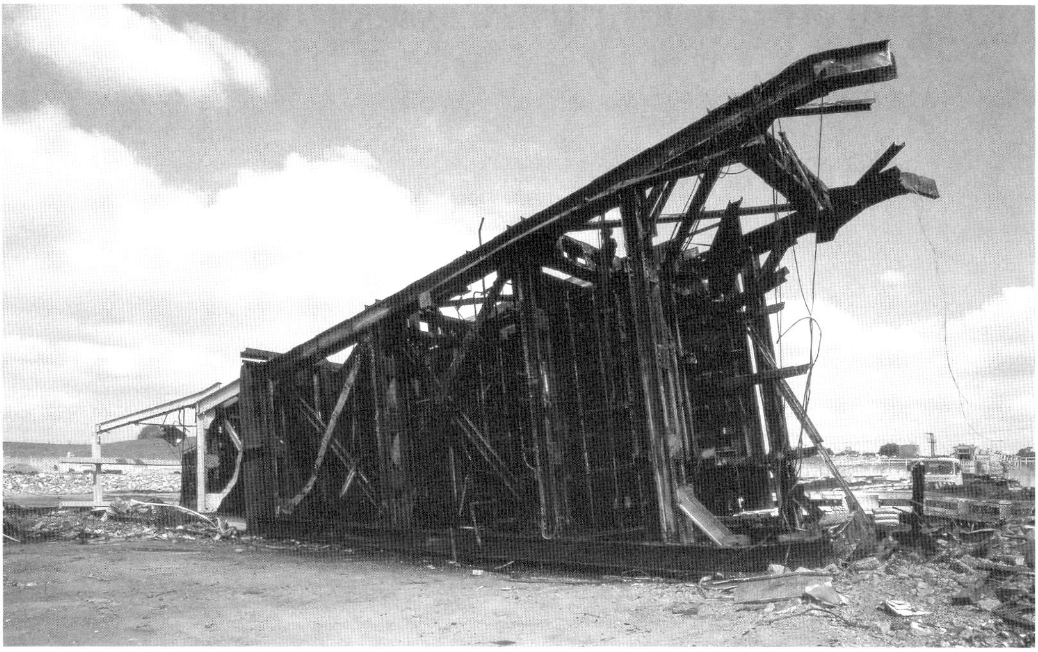

Above: The headstocks lying on their side, c. 1991. PD

Below: The remains of the headstocks following demolition in c. 1991, with Barry McCready. PD

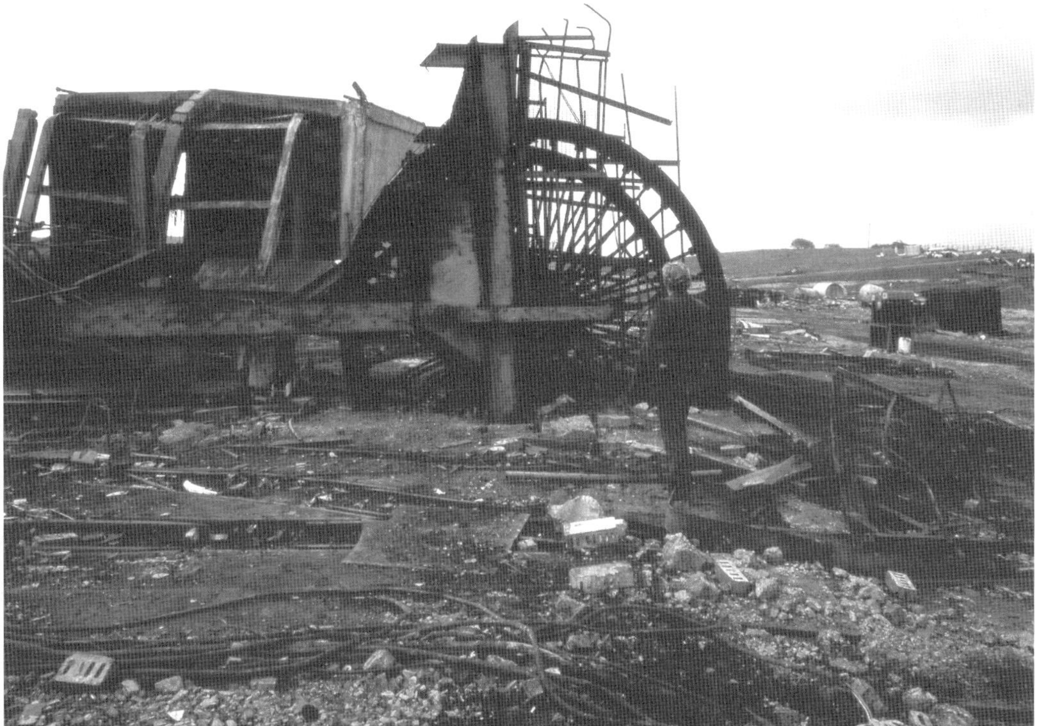

KEMBALL COLLIERY (OR STAFFORD NO. 1)

Stafford Coal & Iron Co. Ltd.

This mine had four shafts: Pender – 310.3m deep; Bourne – 525.5m deep; Kemball 310.4m deep and Hem Heath No. 1. The mine was sunk in 1876 and ceased coal production in November 1963, with the completion of the No. 2 shaft at Hem Heath. The surface facilities were retained as a training centre. It ws situated 2.5km west of Longton and 200m south of the A50. The site is now the Holiday Inn near the Britannia Stadium.

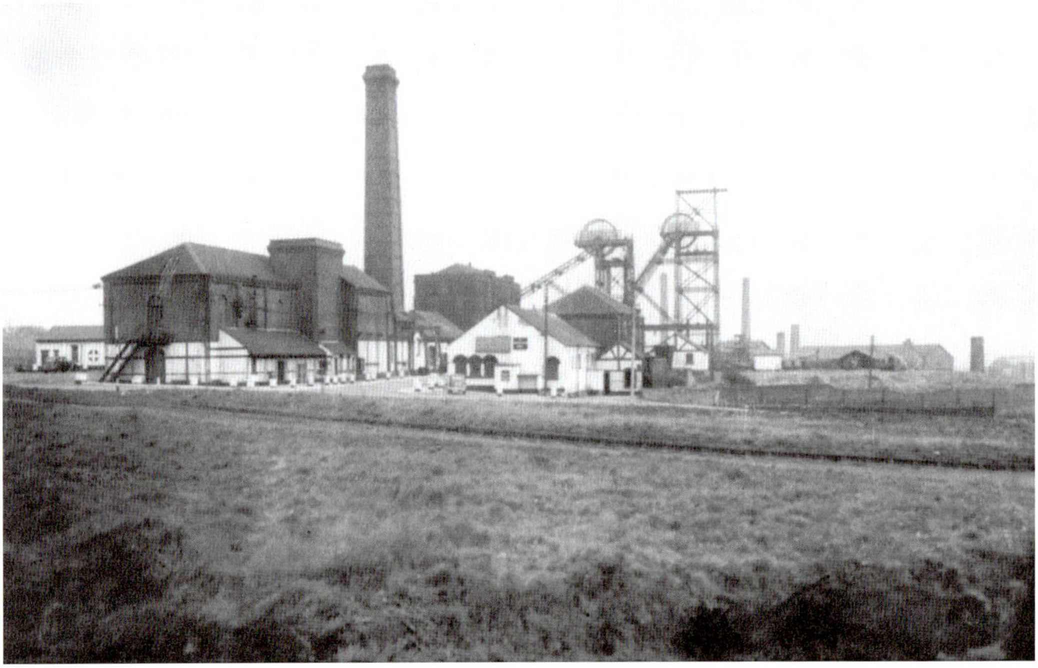

View of the colliery taken in 1951. MC

Little Sherriff Colliery

Apedale Hall Colliery Co. Ltd.

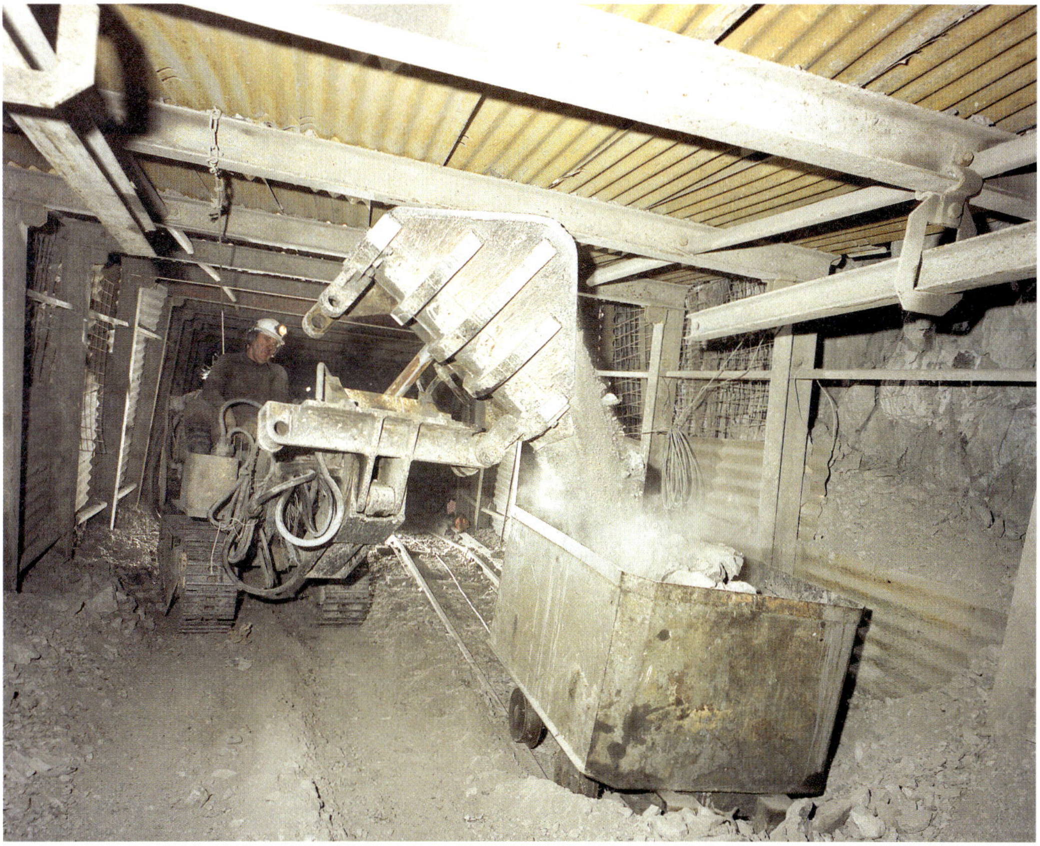

Little Sherriff was a small mine situated in the curtilage of Silverdale Colliery. It was sunk in the 1980s to work a strip of the Winghay Seam lying between the Watermills footrails and Silverdale Colliery. It operated until 1996.

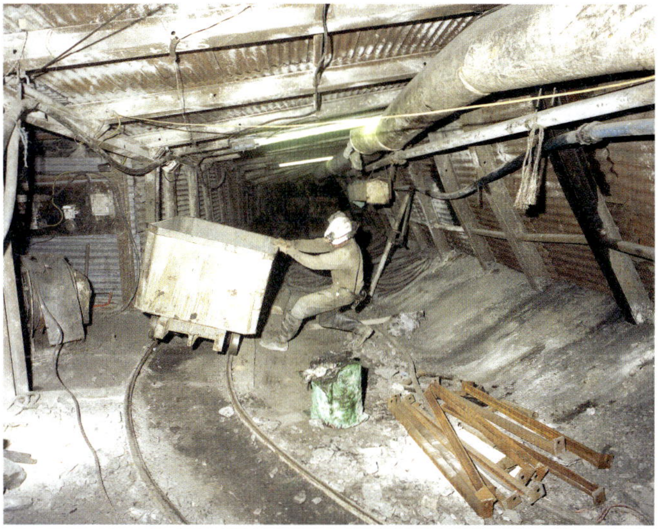

Above: Loading a tub with a Mindev 90 loader during the driving of a crut, May 1994. PD

Right: A collier putting a tub back onto the rails, May 1994. PD

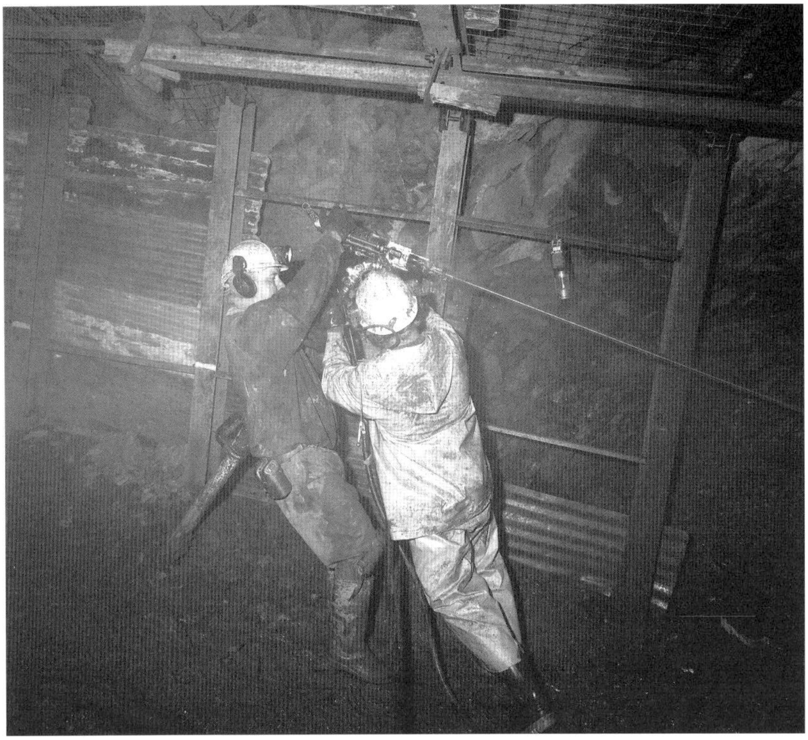

Left: Two colliers removing a drill steel from a shot hole, May 1994. PD
Below: Dressing down the roof after firing and prior to loading, May 1994. PD

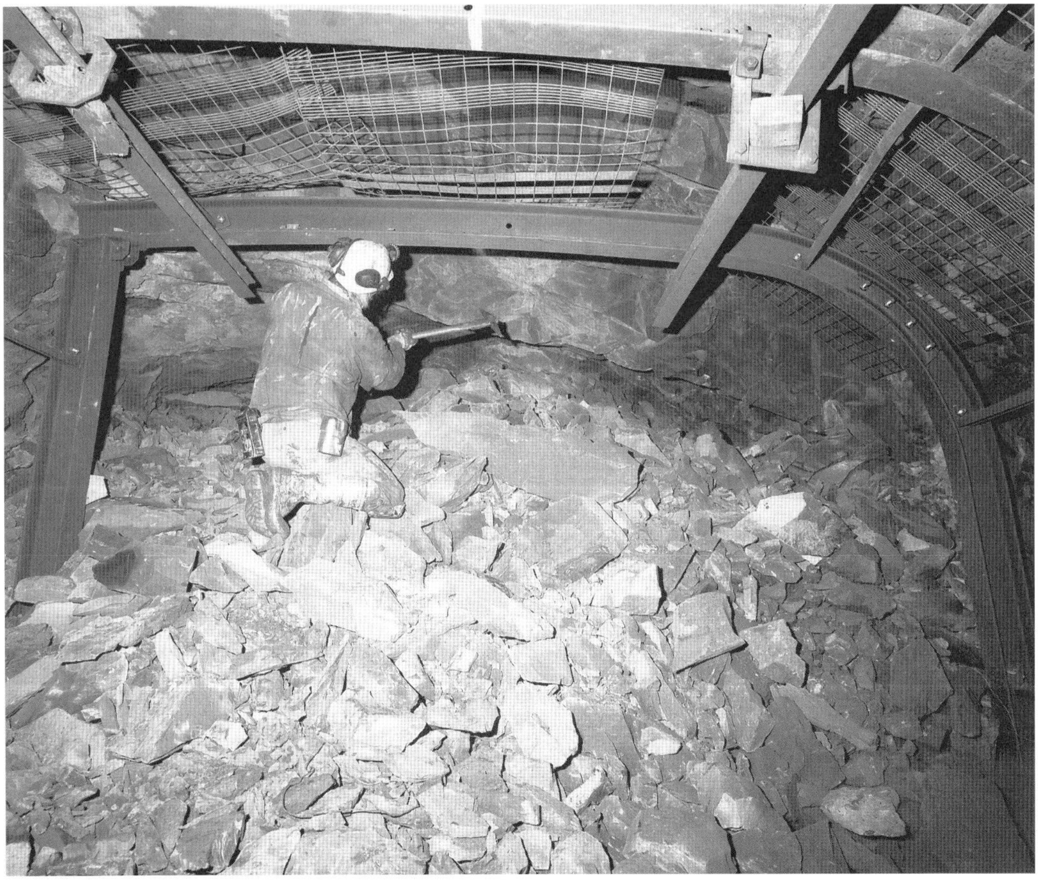

These photographs show a miner using a compressed air drill and air leg whilst driving a crut from the Winghay Seam to the underlying Burnwood Seam, May 1994. PD

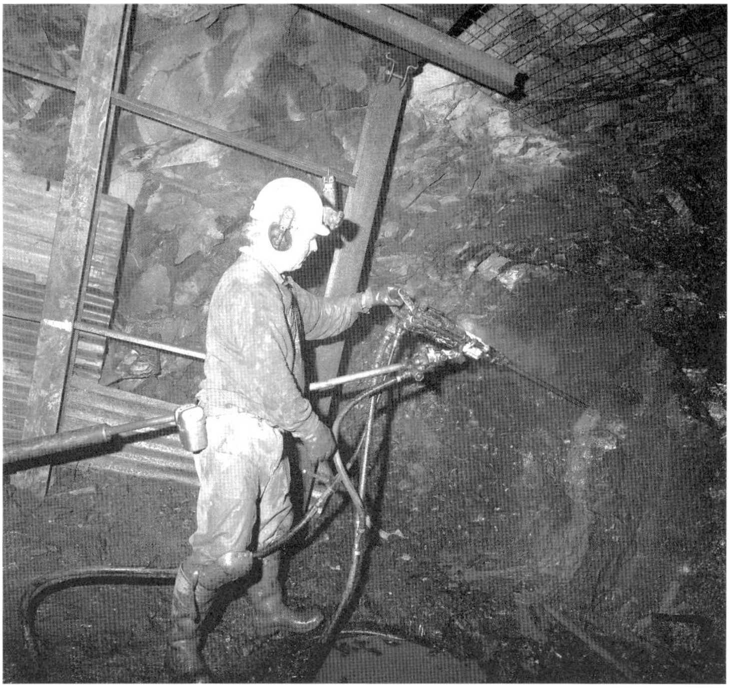

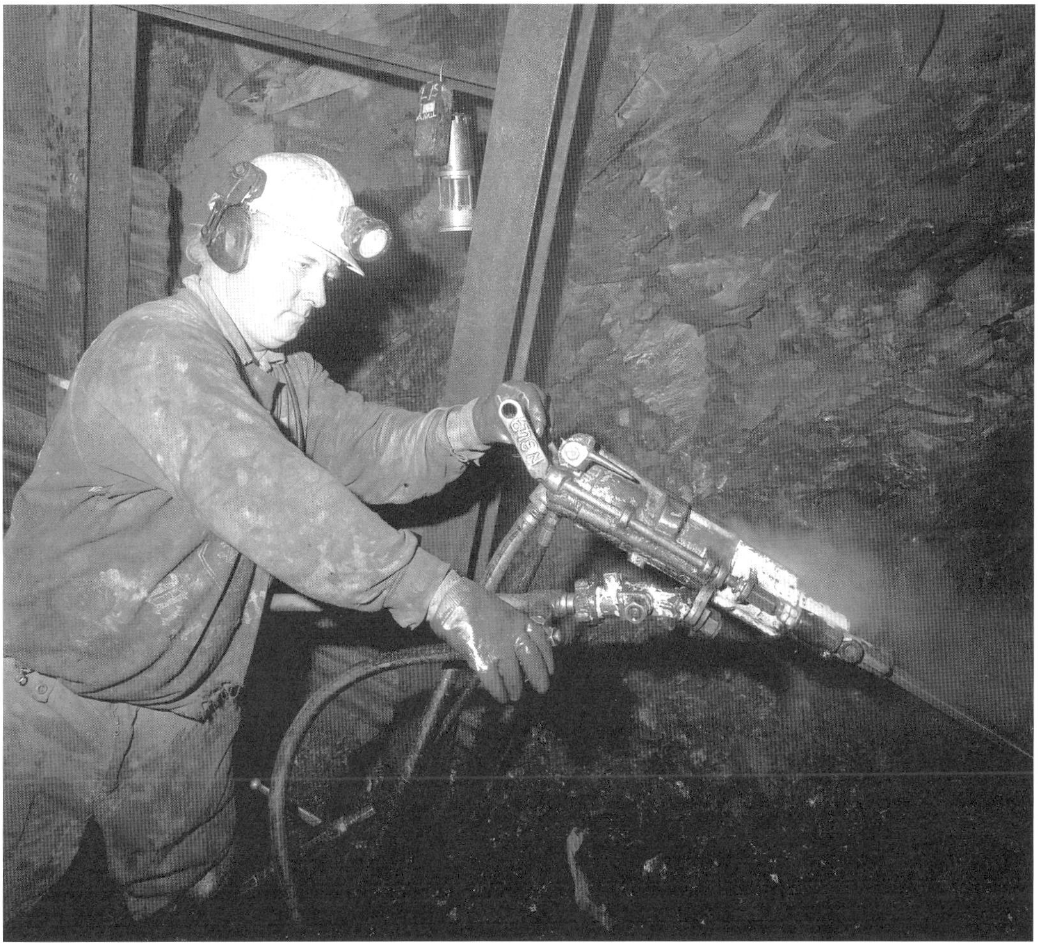

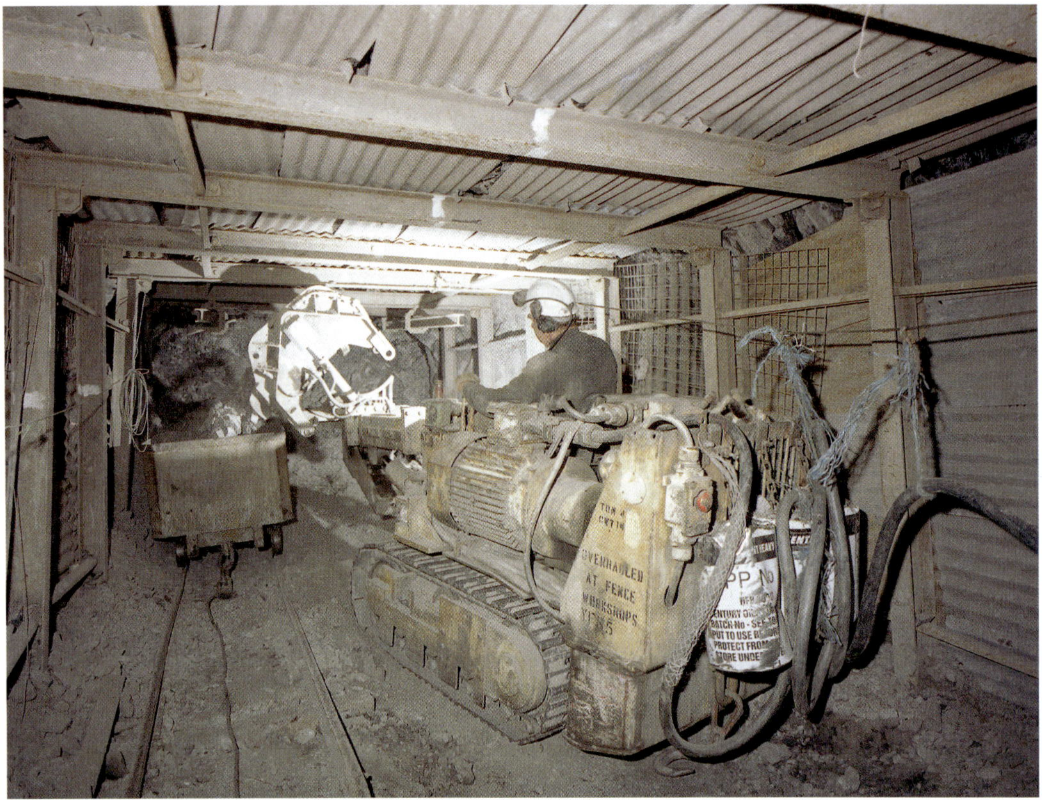

Above: Loading stone into tubs whilst driving a crut using a track-mounted Mindev 90 loader, May 1994. PD

Below: Mindev 90 loader, loading stone at the face, May 1994. PD

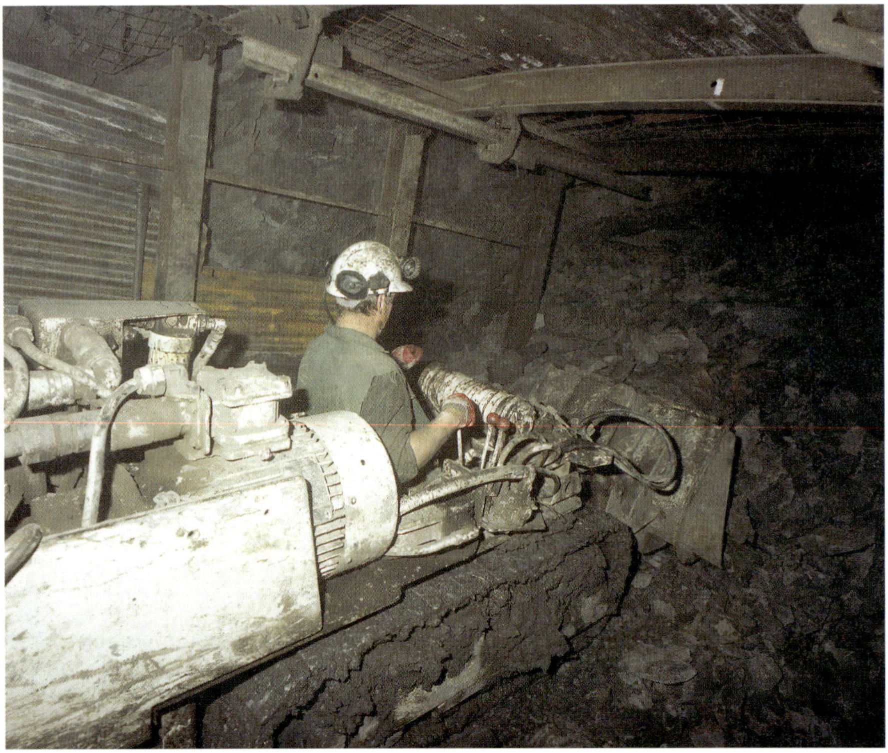

Madeley Colliery
(also known as Leycett Colliery)
Madeley Collieries Ltd.

Situated 1km north of Middle Madeley.

The shafts here were sunk in 1880, and the mine closed in September 1957. It had five shafts: Fair Lady (519.4m); Bangup (448.3m); Clarkes (332.5m); Harrison (403.8m); and Woodburn (295.2m). Clarkes was the Upcast shaft and Harrison and Woodburn were both pumping shafts.

A view of Bangup Shaft. Today the site has been restored to agriculture. MC

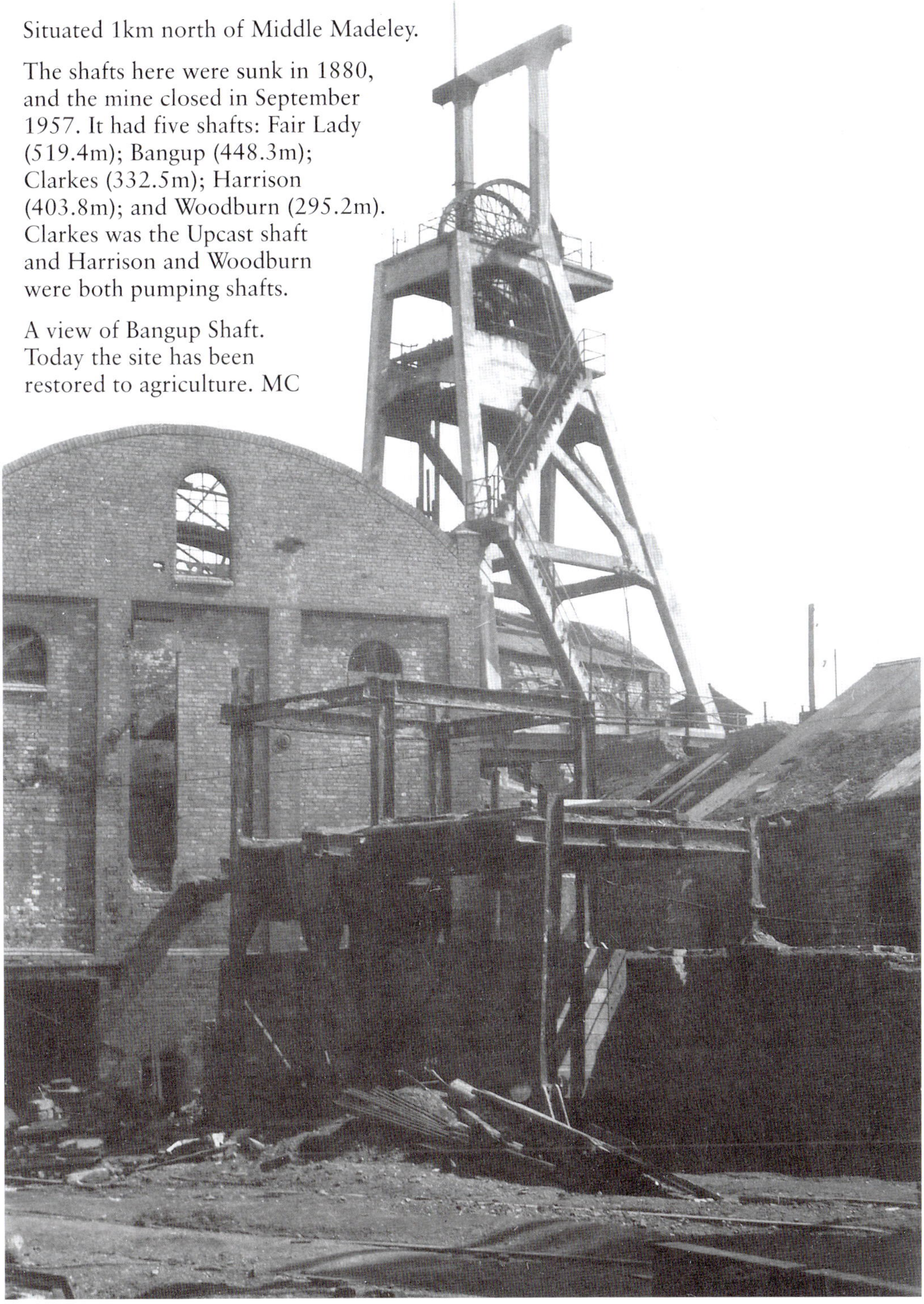

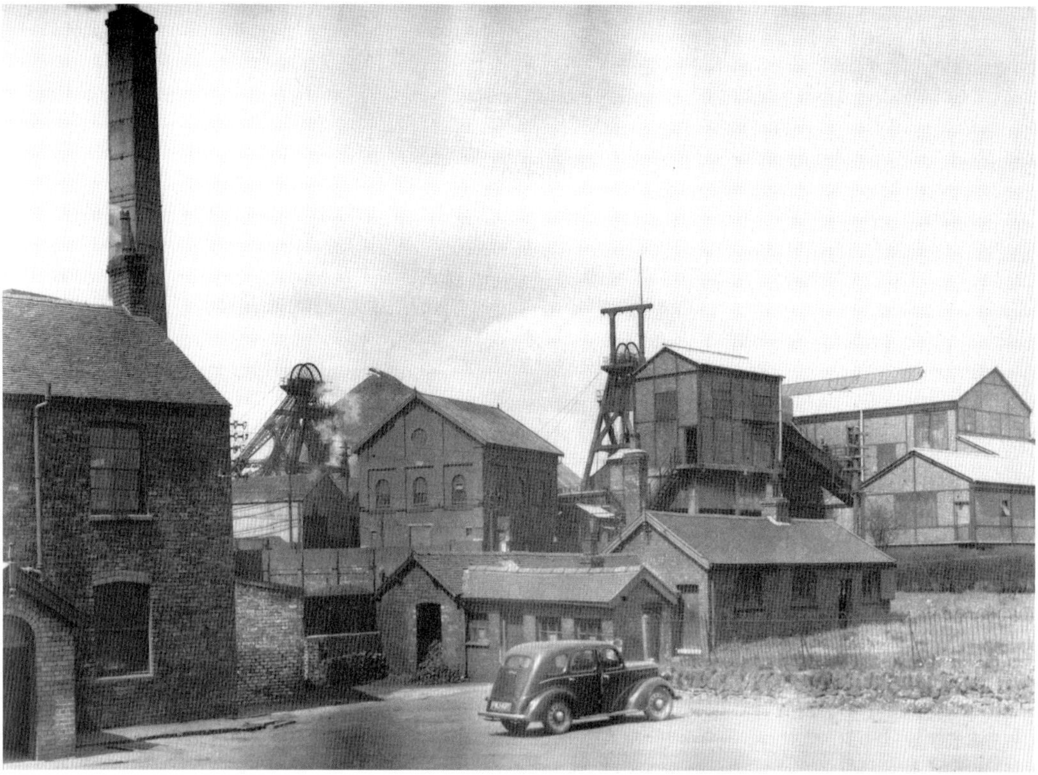

Above: A view taken in 1951. Fair Lady Shaft, with the lattice steel headstocks is on the left and Bangup Shaft with the concrete headstocks is on the right. Both of these shafts were downcast and Fair Lady Shaft was used for coal winding. MC
Below: The colliery from the west, with Bangup Shaft on the left and Fair Lady Shaft in the centre. MC

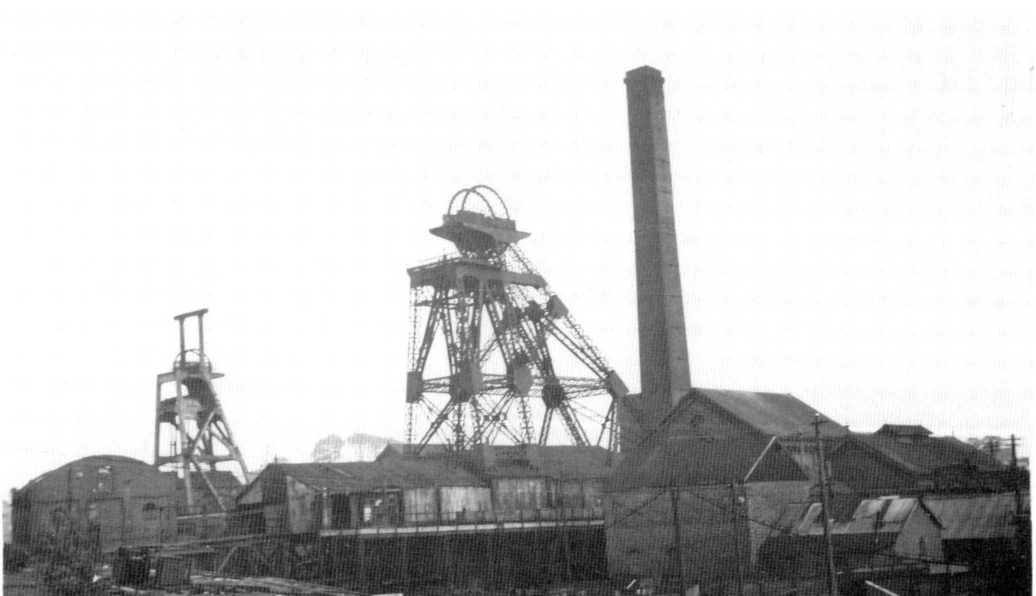

MINNIE PIT
Midland Coal, Coke & Iron Co., Ltd.

Situated immediately to the west of Halmer End.

The mine was part of Podmore Hall Colliery and the scene of a disaster on 12 January 1918. See pages 152-53 for photographs of the monument to all of the 156 men and boys who died in the disaster. This scene clearly records happier times. Note the men on top of the headstocks. Today's Health and Safety officials would no doubt take a dim view of that! MC

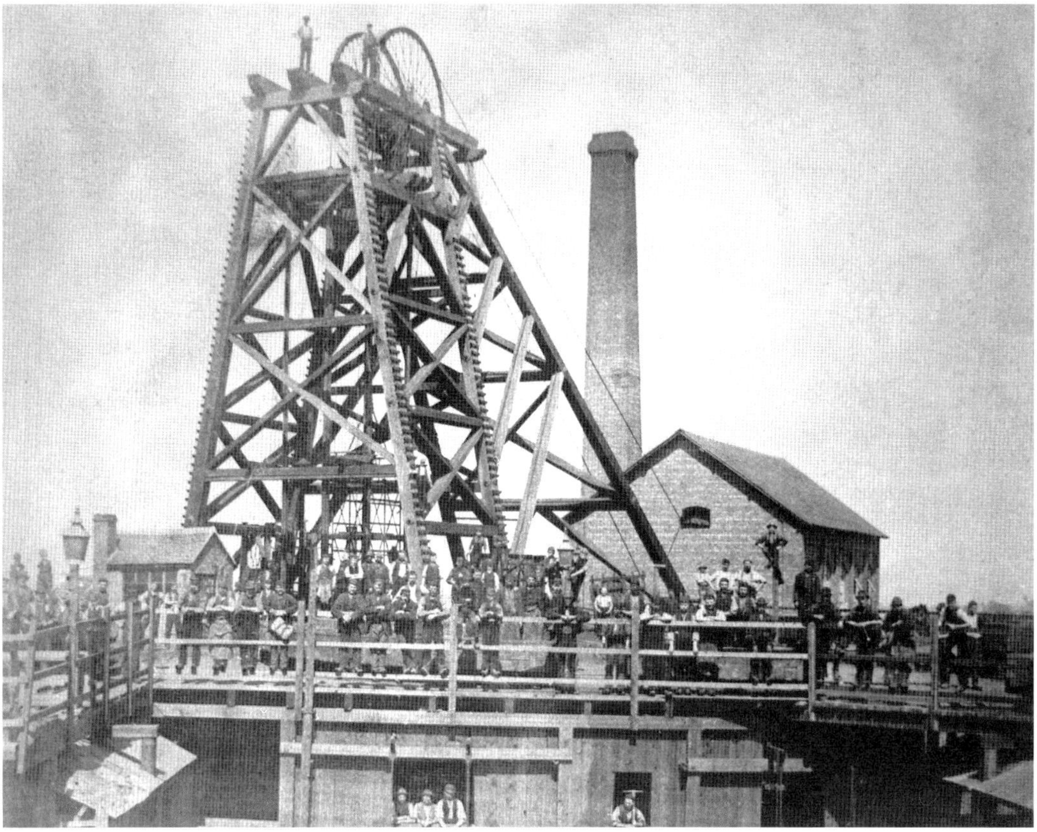

Mossfield Colliery

Mossfield Colliery Ltd.

Situated 1.7km north of Longton town centre.

There were four shafts: No. 1, (99.7m); No. 2, (123m); No. 3, (405.9m) and No. 4, (417.1m). It was connected to Adderley Green No. 9 pit (see page 75) for ventilation and the second means of egress. The colliery closed in May 1963. The site is now open fields. The views were taken in 1951. MC

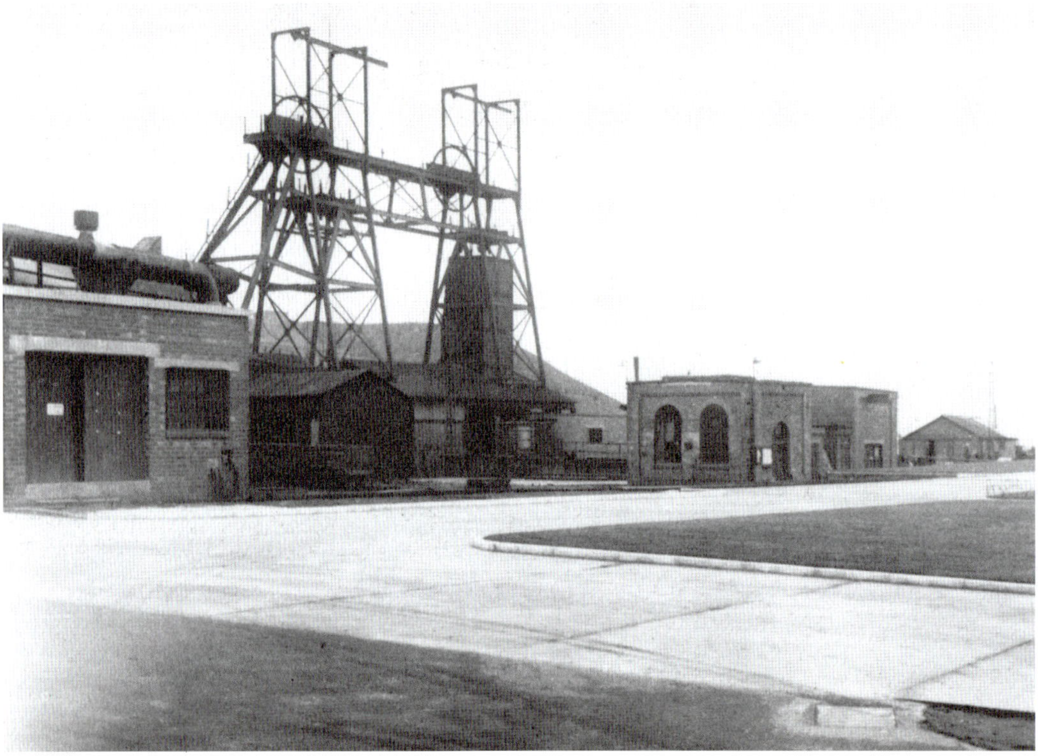

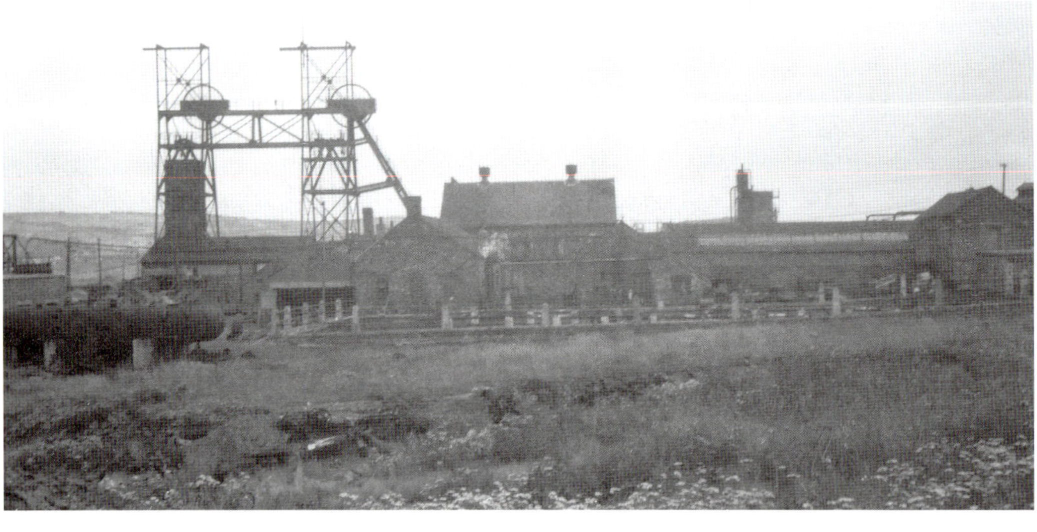

ADDERLEY GREEN COLLIERY

Mossfield Colliery Ltd.

Situated 1.5km north-east of Longton town centre.

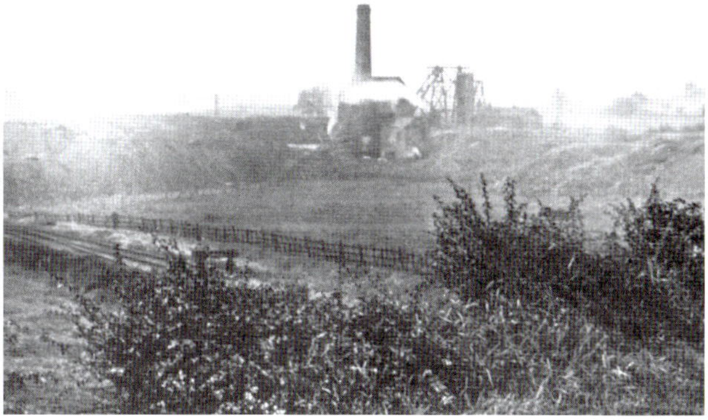

Right: Meadow Pit, Adderley Green. Meadow Pit No. 2 Shaft was 246.9m deep and No. 3 was 164.6m deep. MC

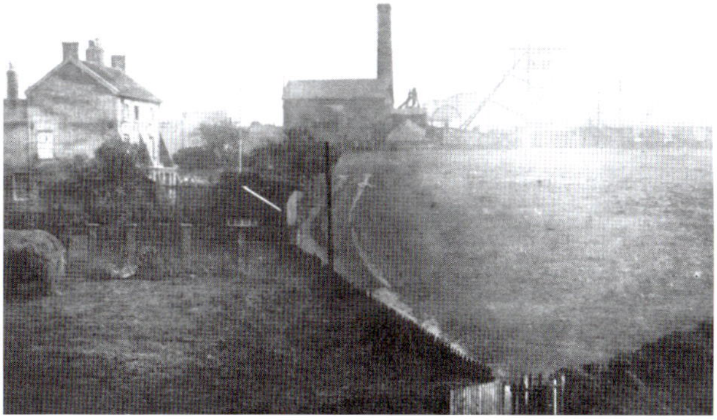

Below: Adderley Green Nos. 8 and 9 pits, which ceased production in September 1940, having merged with Mossfield Colliery in April 1940. No. 9 shaft was subsequently used for ventilation and the second means of egress for Mossfield until the latter closed in 1963. No. 8 shaft was used for pumping and ventilation for Parkhall Colliery. No. 8 shaft was 256.2m deep and No. 9 shaft was 381m deep. MC

NEW HADEN OPENCAST SITE

Contractors: Agate Ltd.

Situated 500km south of Brookhouses, Cheadle, Staffs

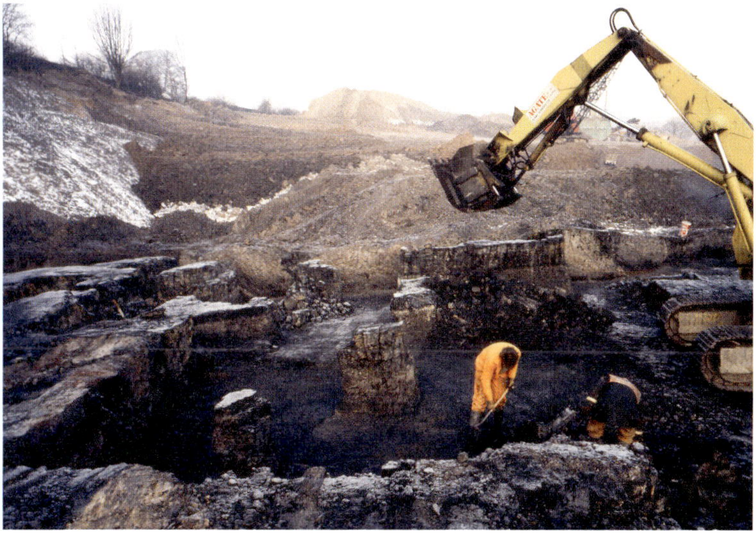

Clearing up old workings in the Two Yard Seam at New Haden opencast site. Remnant pillars left by old miners may be seen as the area was cleared with a brush and shovel prior to the coal being lifted. 1987. PD

NORTON COLLIERY

Norton and Biddulph Collieries Ltd.

Situated at Norton in the Moors, between the A53 and B5051 roads.

The mine's tandem winder, working two shafts. The mine was sunk in 1860 and closed in 1977 after 117 years at work. The site is now open ground and being developed for housing. MC

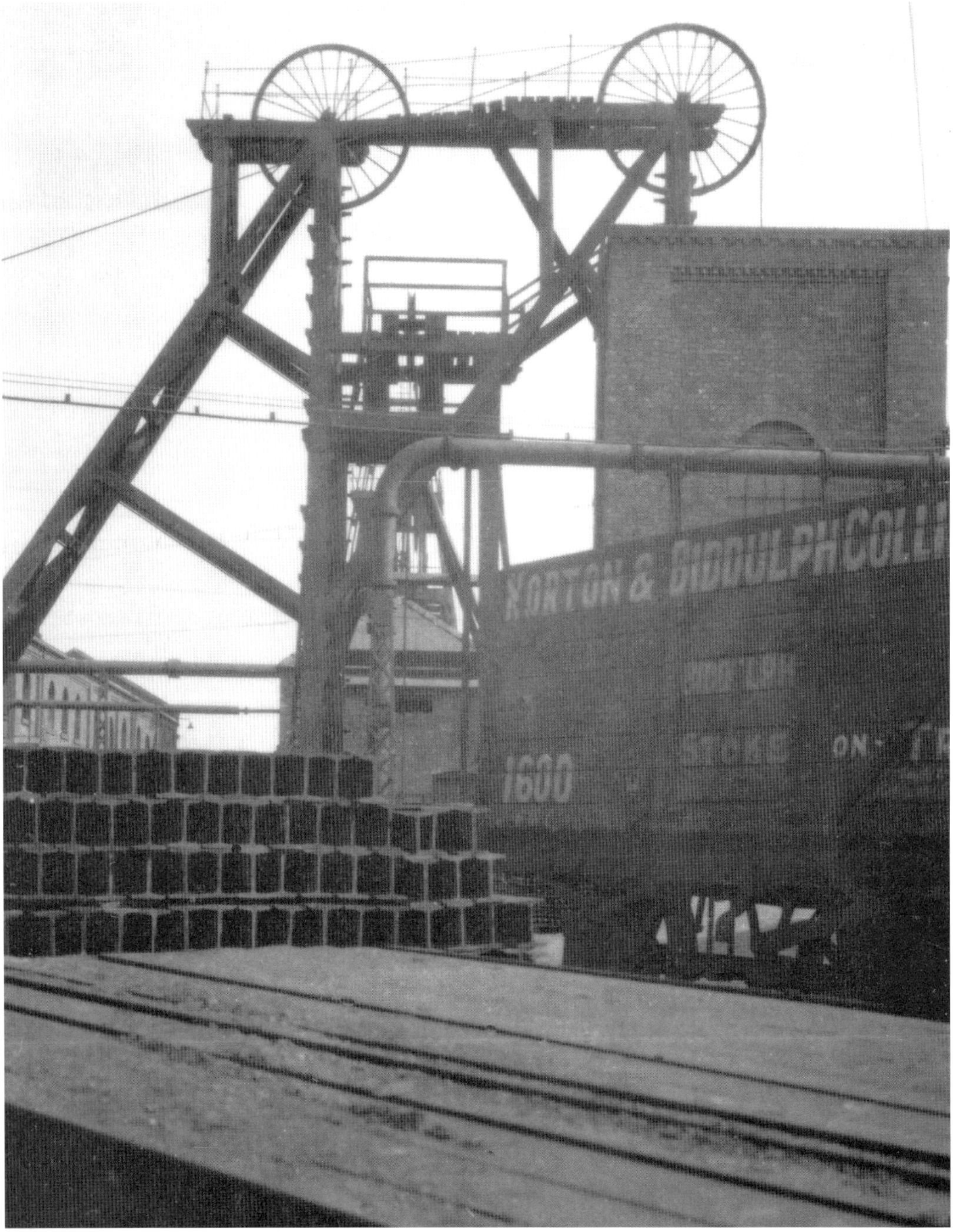

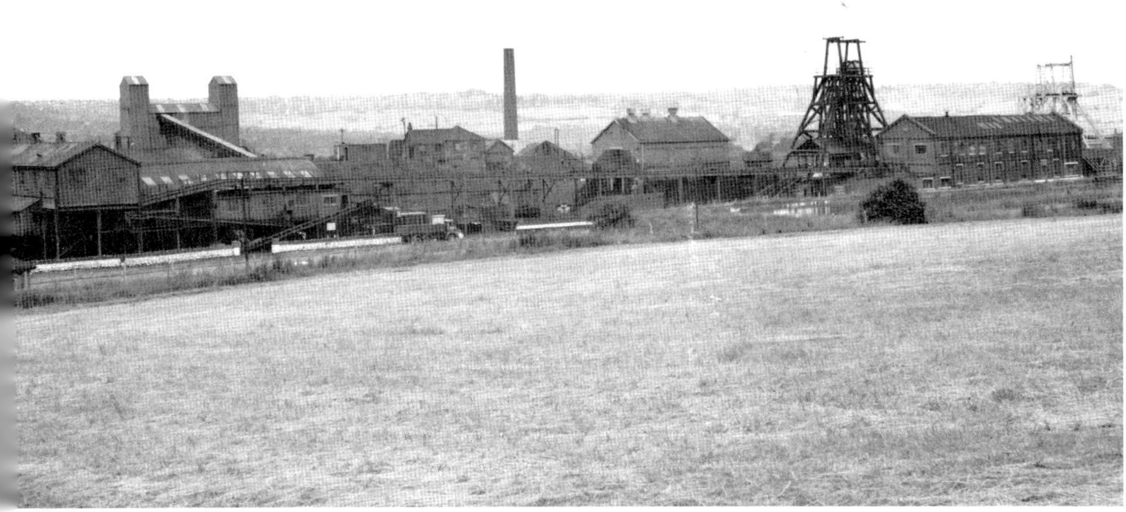

Two views taken in 1951. There were four shafts: No. 1, 358.2m; No. 2, 352.3m; No. 3, 449.3m; and No. 4 (depth not known). MC

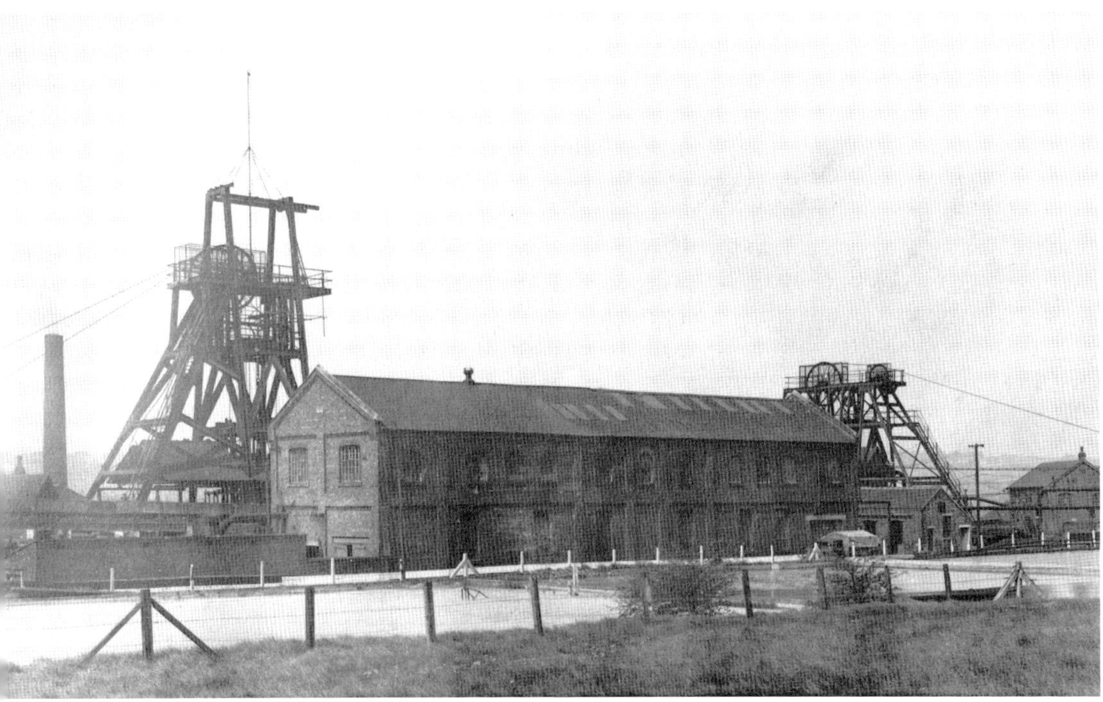

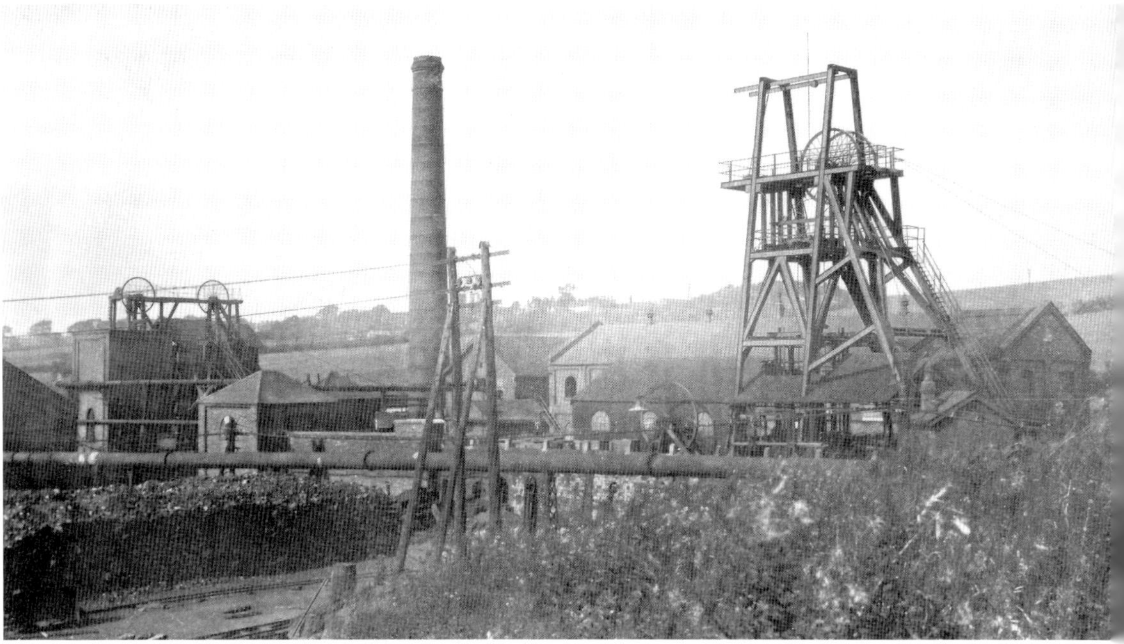

Two other views taken in 1951.MC

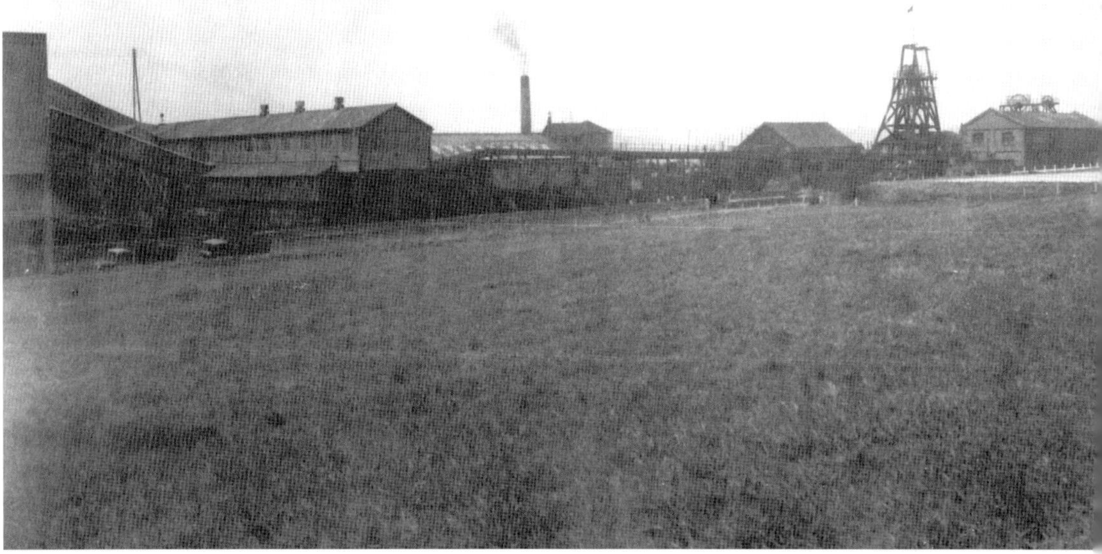

PARK HALL COLLIERY

Mossfield Colliery Ltd.

Situated about 1.5km east of Longton town centre and immediately to the north of Weston Coyney road.

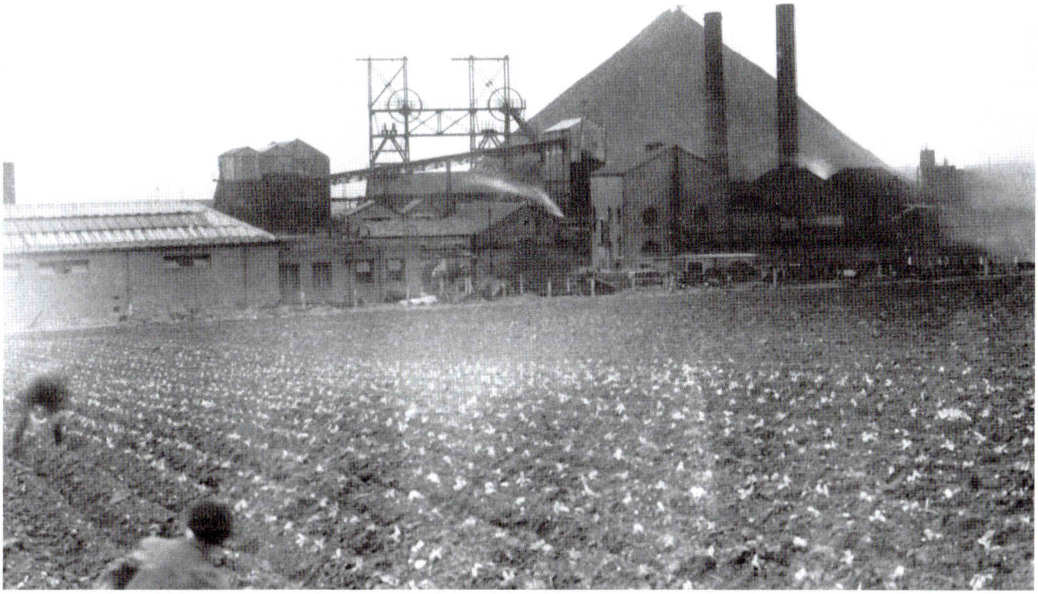

Two views of Nos. 5 and 6 shafts. No. 5 was 365.8m deep and No. 6 was 366.1m deep. Note the conical tip. These were a common site around the city but none remain in this form. The mine was sunk in 1860 and closed in December 1962. The site is now Cinderhill Industrial Estate. MC

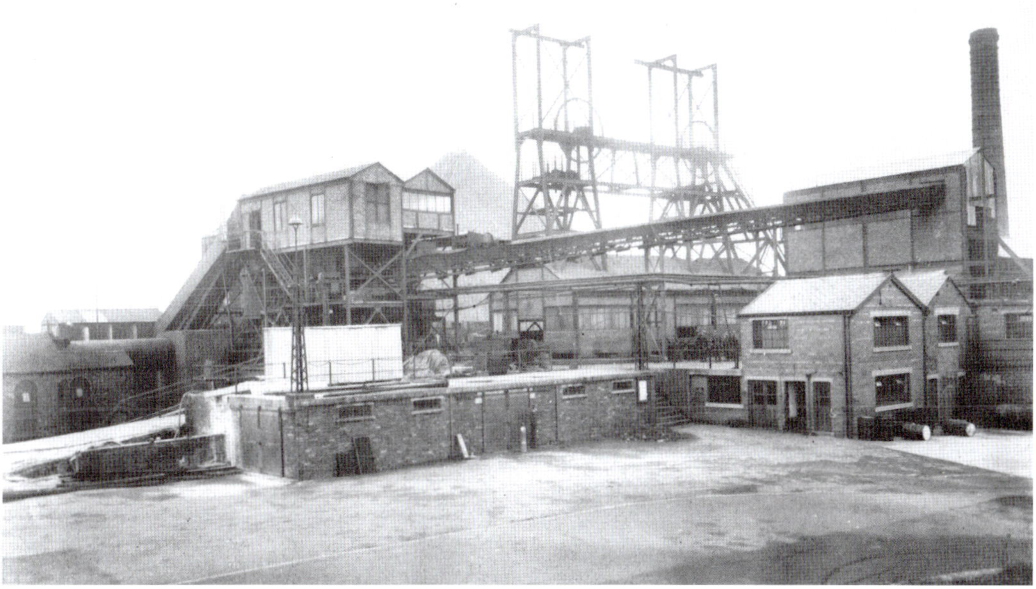

Parkhouse Colliery, Chesterton

Parkhouse Colliery Ltd.

This mine was sunk in 1874 with No. 3 shaft being sunk in 1917. It was situated 1km north-east of Chesterton Church.

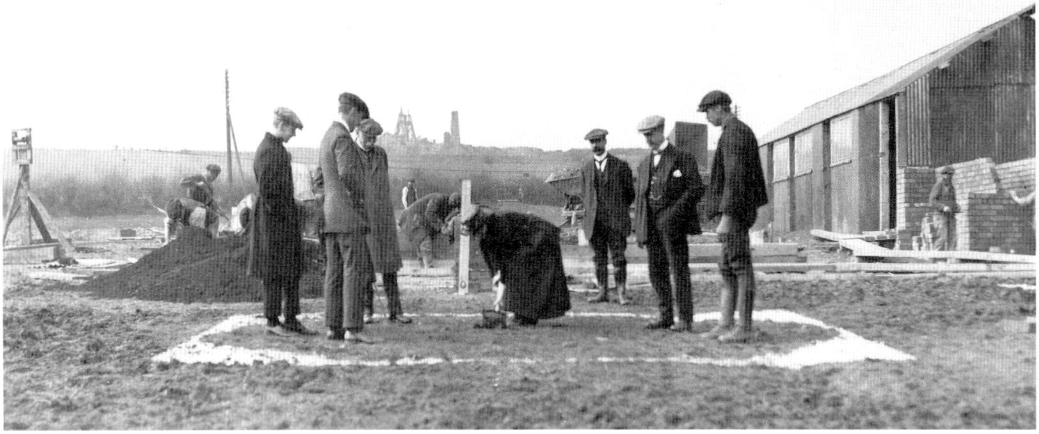

Above: Cutting the first sod of No. 3 shaft on 26 May 1917. On the horizon may be seen the New Hem Heath Colliery (not to be confused with Hem Heath Colliery at Trentham). MC

Below: The form work at the top of No. 3 shaft during the sinking, 1917. MC

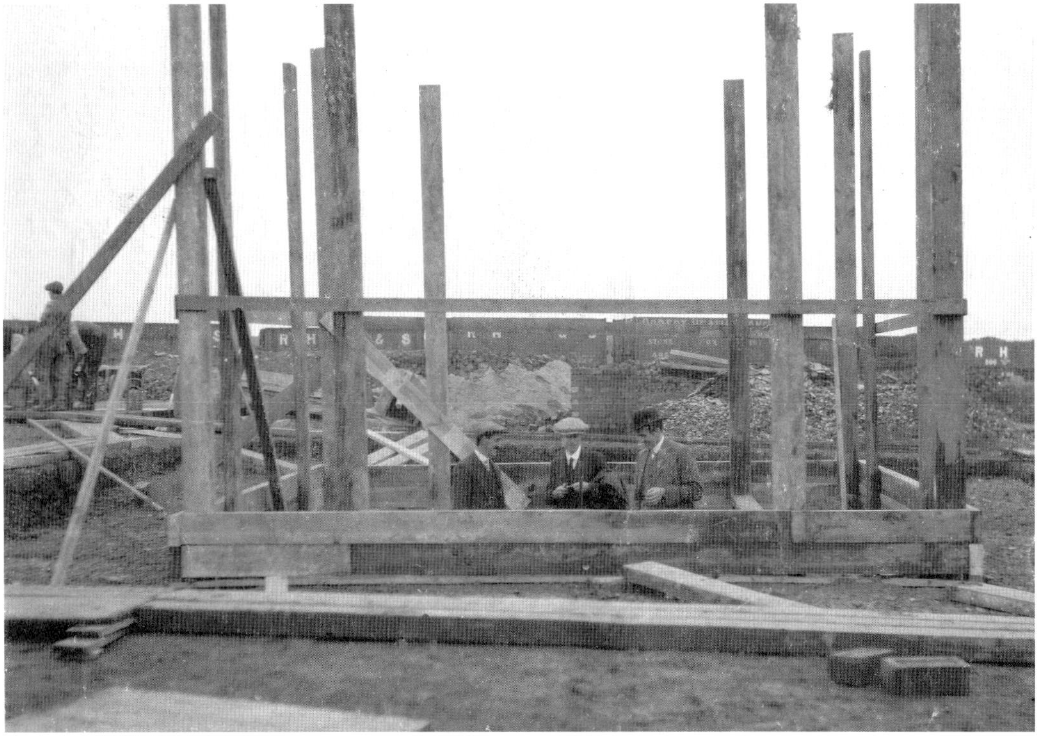

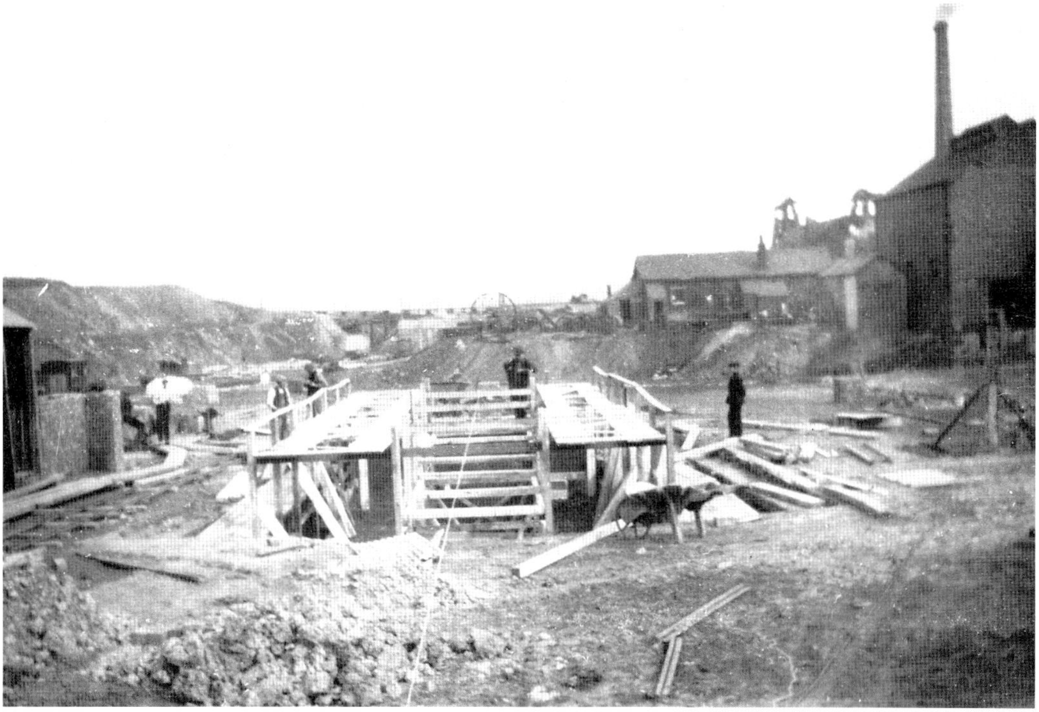

Above: The view of the top of No. 3 shaft with Nos. 1 and 2 Pits to the right. No. 1 was 299.9m deep and No. 2, 474.2m, 1917. MC
Below: The temporary headstocks and winding house on No. 3 shaft, 18 September 1917 . MC

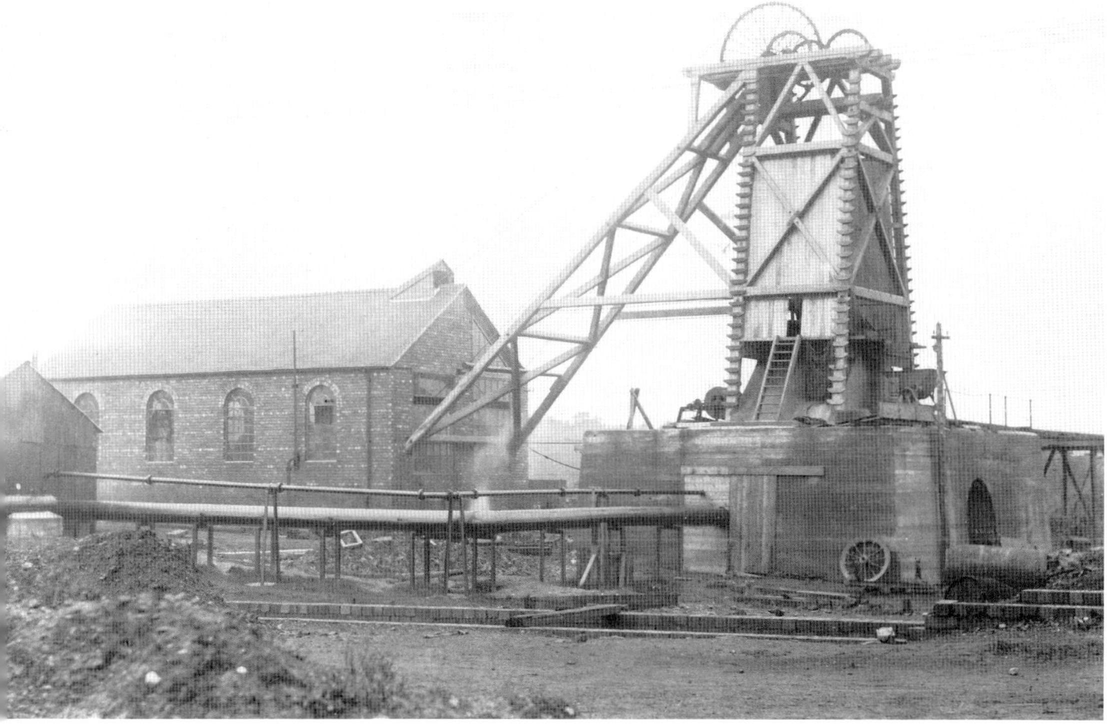

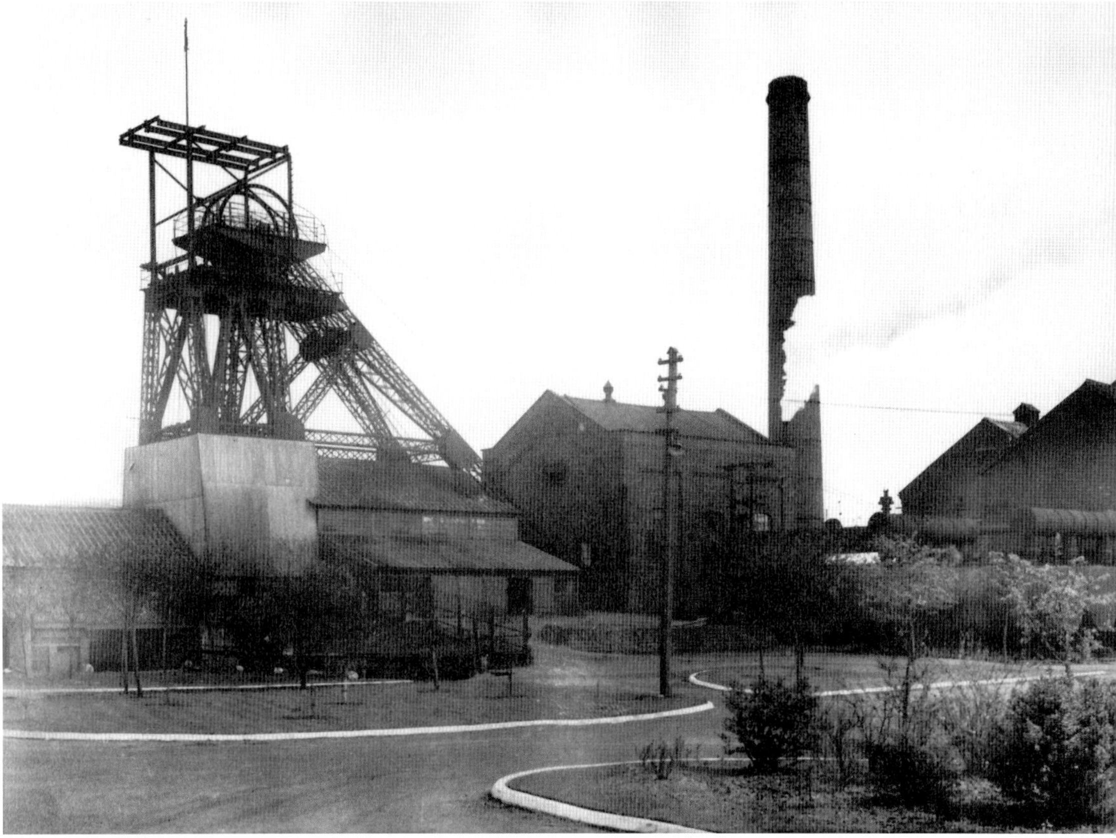

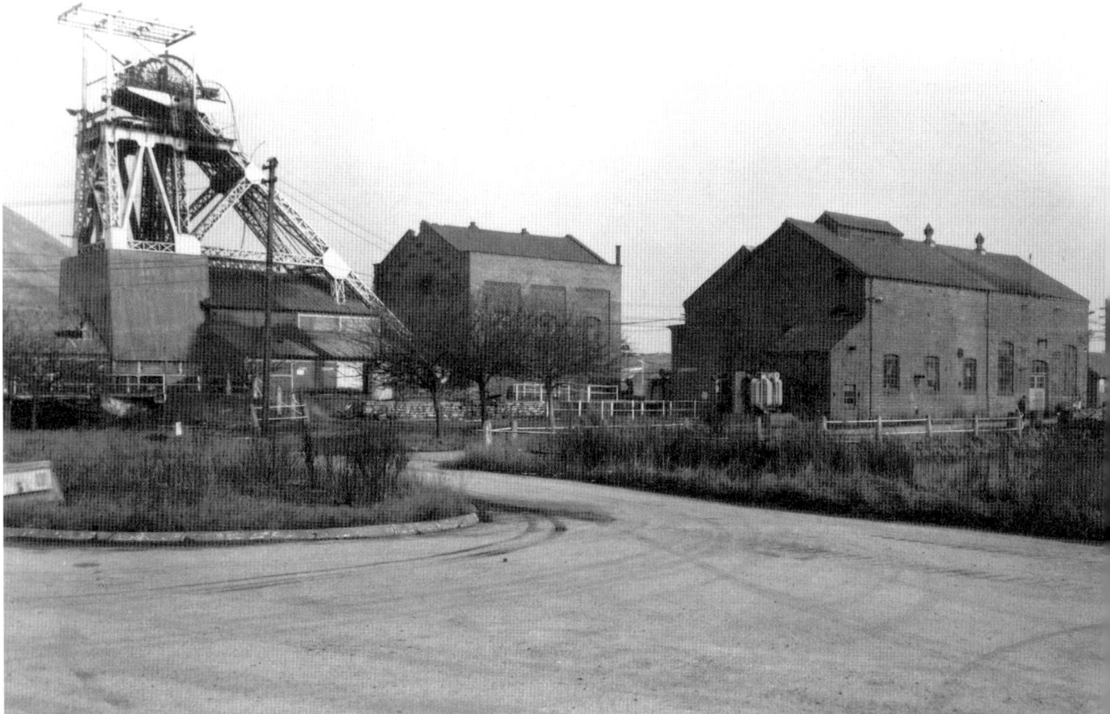

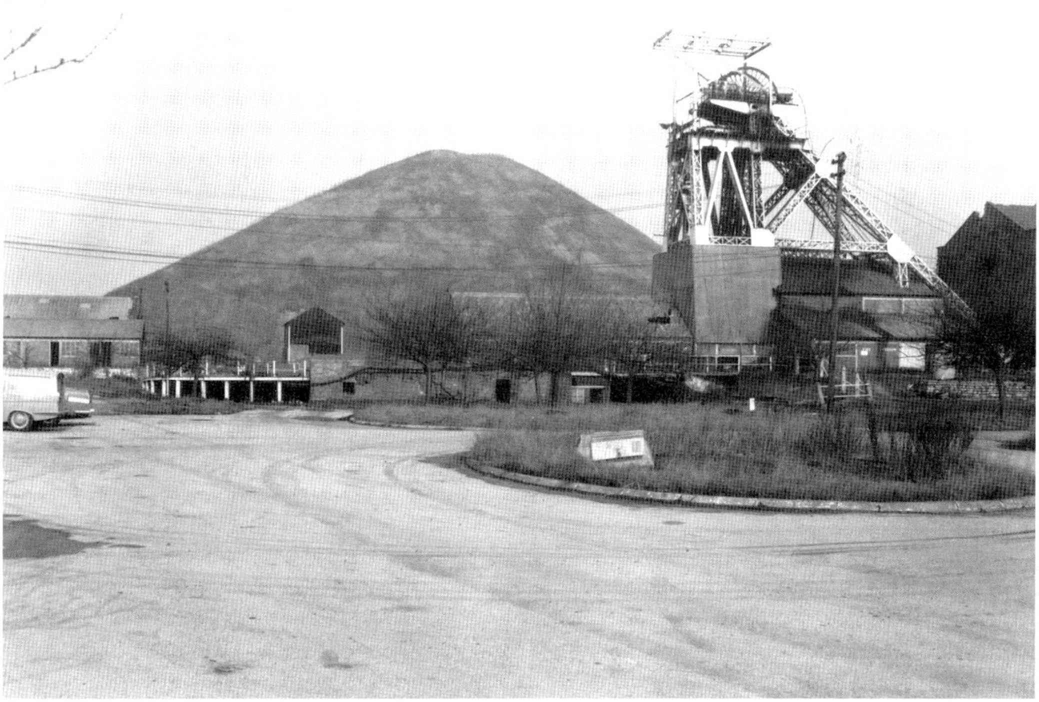

Opposite top, bottom & above: No. 3 shaft. It reached a depth of 477.5m. The scene opposite top was taken in 1951. The other two were taken in the 1960s. Note the removal of the chimney after the earlier photograph had been taken.

The site is now an industrial and commercial estate plus a brickworks.

Parklands No 5 Colliery

Parklands Colliery Co., Ltd.

Situated 500m north-east of Alsagers Bank

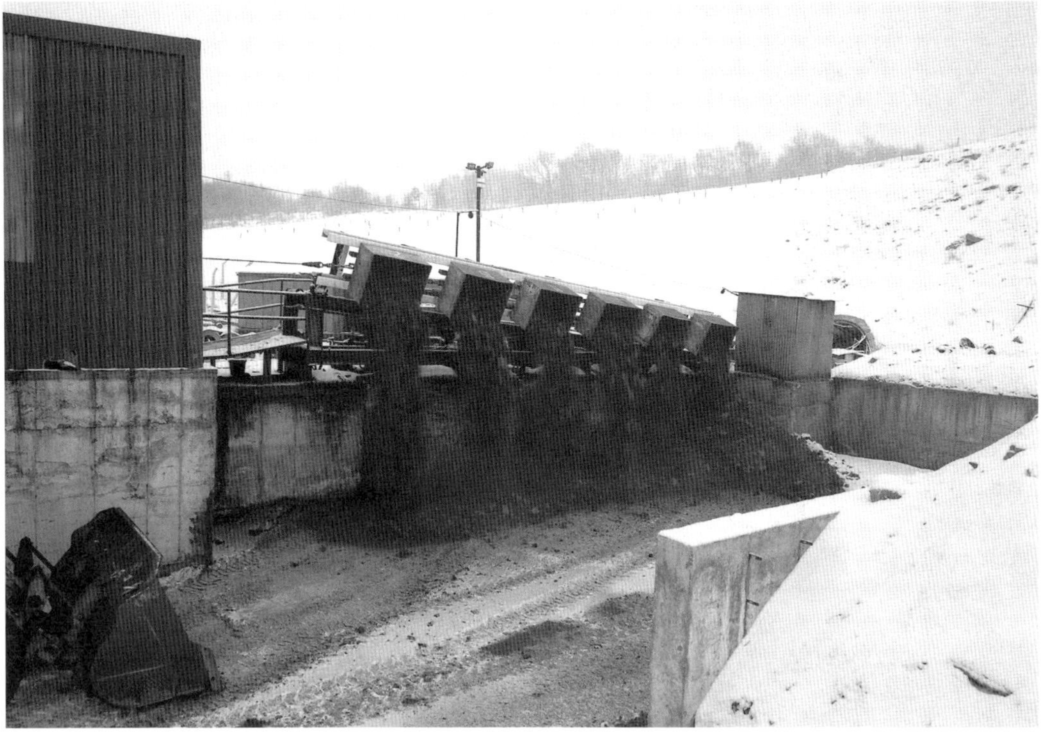

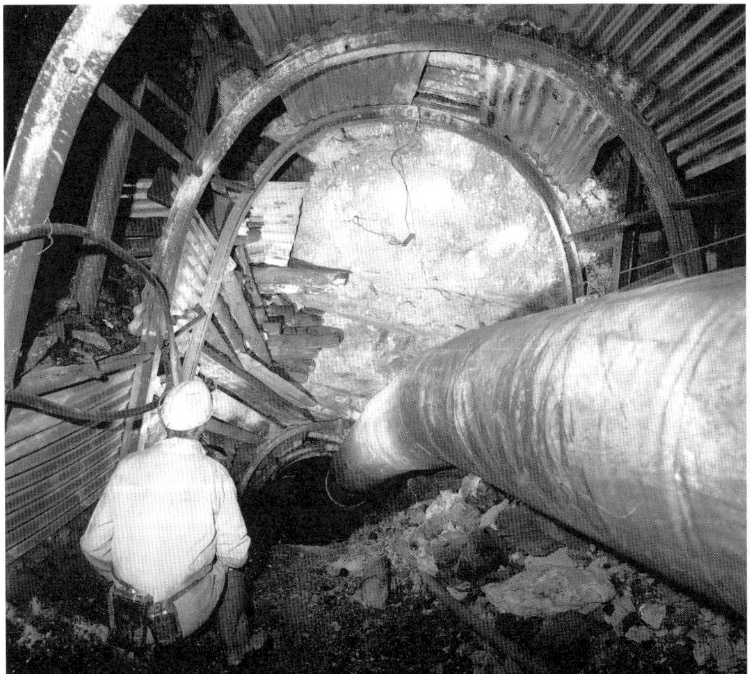

Above: The surface of Parklands No. 5. Emptying a train of six tubs using a hydraulic tipper, February 1991. PD

Left: A step in the roadway to negotiate a geological fault. Arthur Cartwright is to the left, September 1991. PD

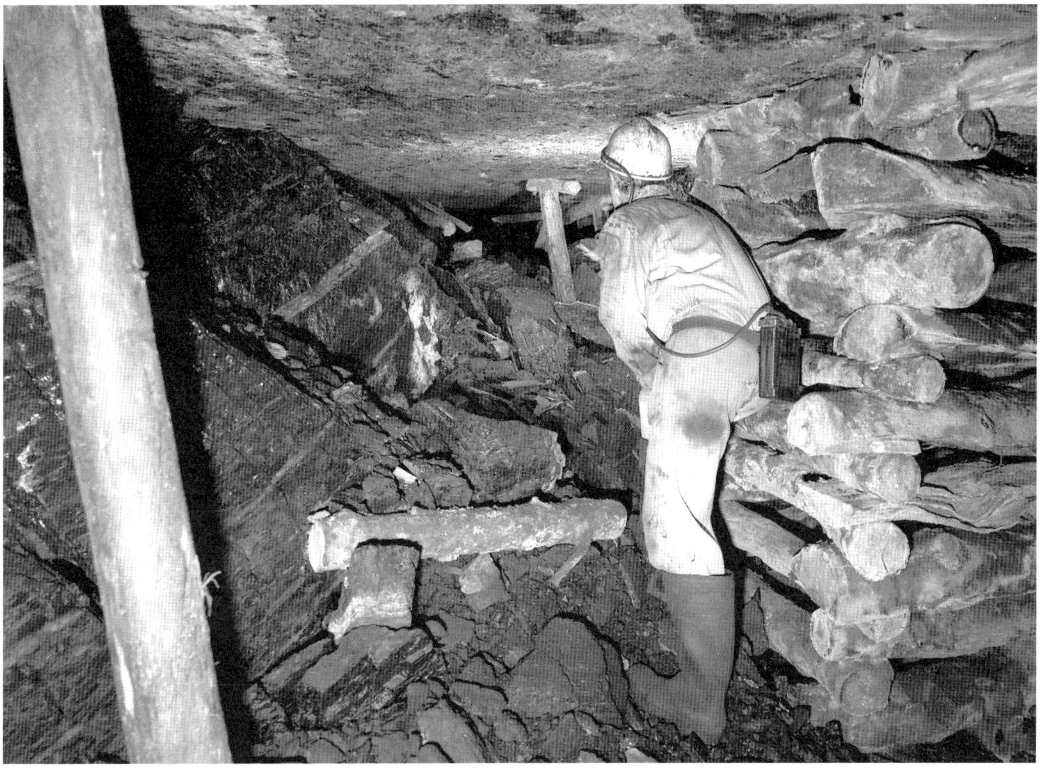

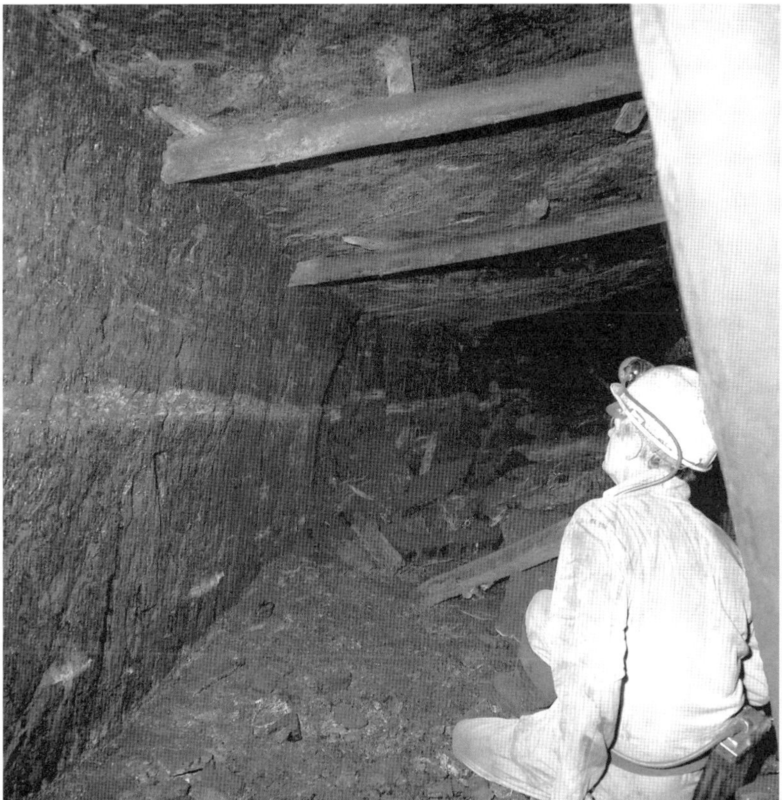

Above: Charlie Kirkham, the mine manager, inspecting an old roadway in the Miry Wood Colliery, which was abandoned in 1940 and had been broken into by recent workings. In the lower photograph, note the holes knocked into the coal to put in roof timbers and the wedges used to tighten the roof timbers against the roof, February 1991. PD

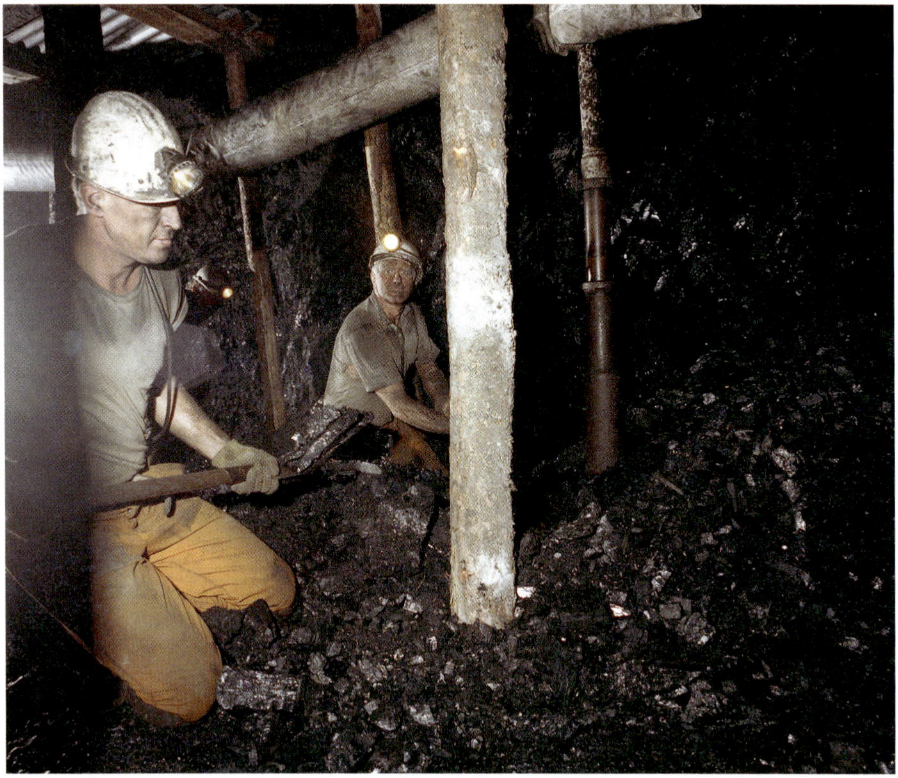

Colliers loading coal onto an AFC (armoured flight conveyor), February 1991. PD

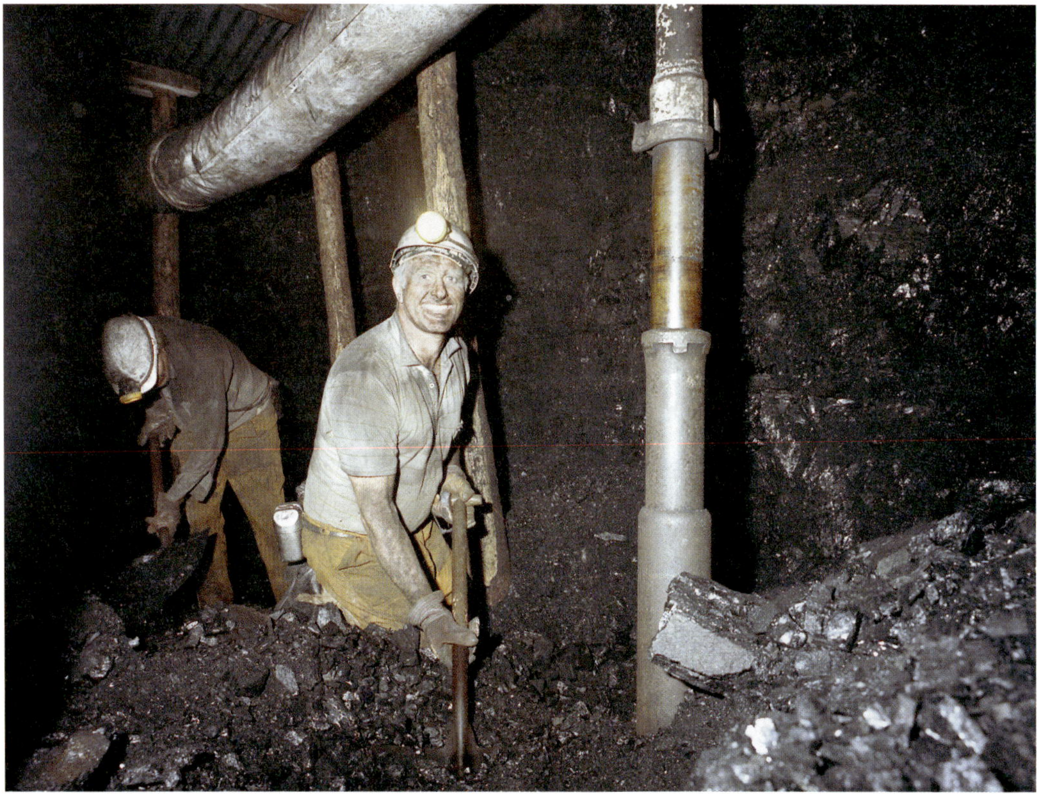

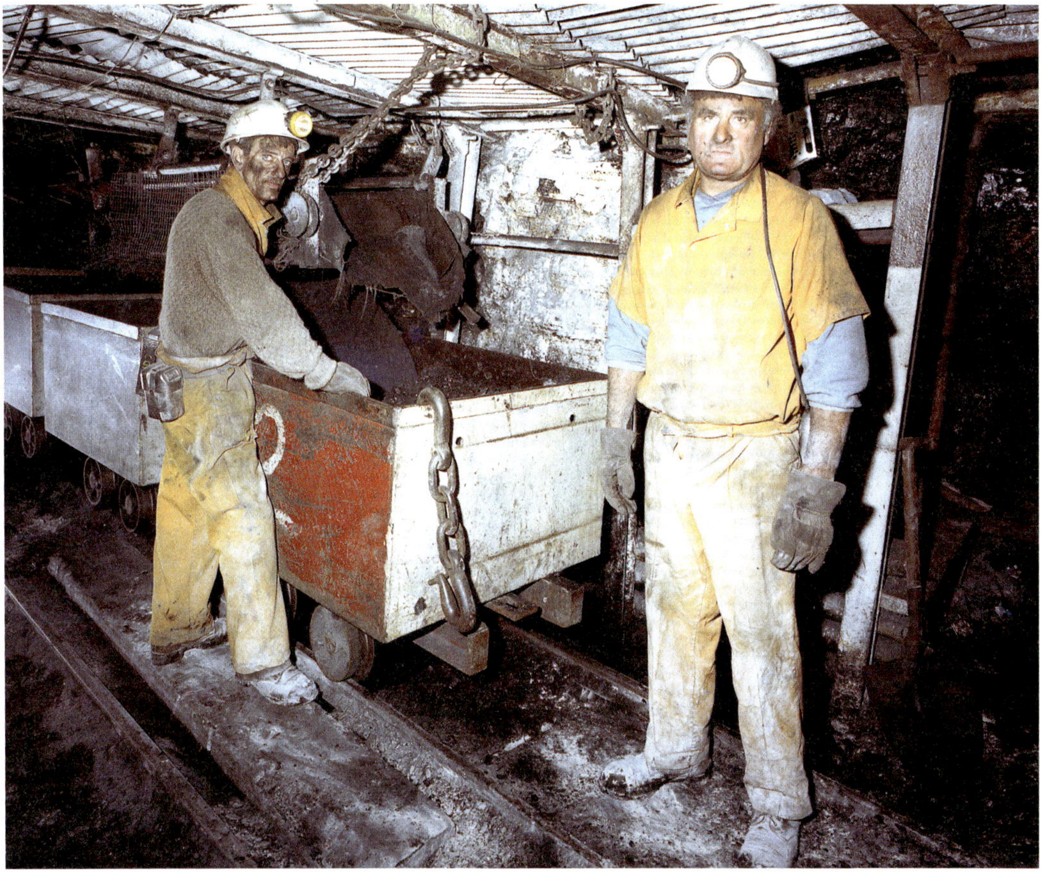

Filling tubs from a conveyor ready for hauling up the steep incline to the surface. In the lower photograph, the empties can be seen on the right, September 1991. PD

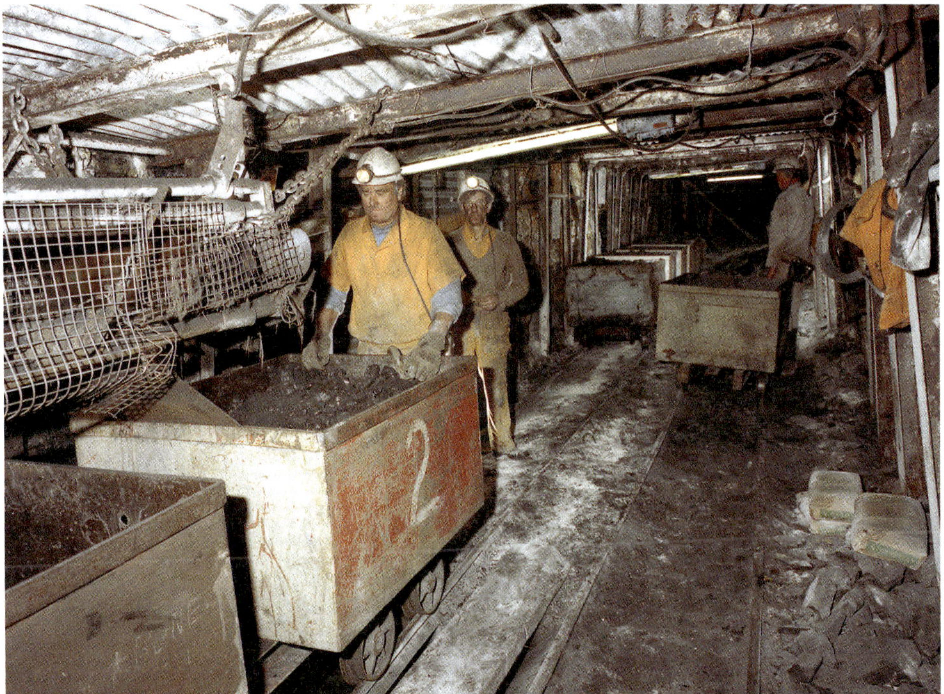

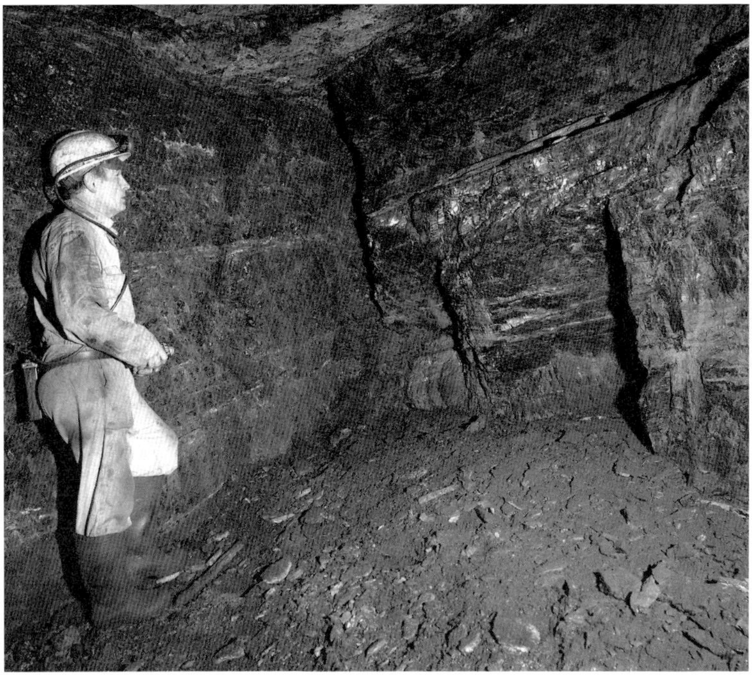

Left: Parklands No. 5 Colliery with Charlie Kirkham looking at the face of an old roadway encountered in the Bullhurst Seam from the Miry Wood Colliery. Note the hand picked slot on the left-hand side of the roadway, instead of the more usual undercut, which was used to provide a free face for blasting, February 1991. PD

Below: Men undertaking roadway repairs. September 1991 (PD)

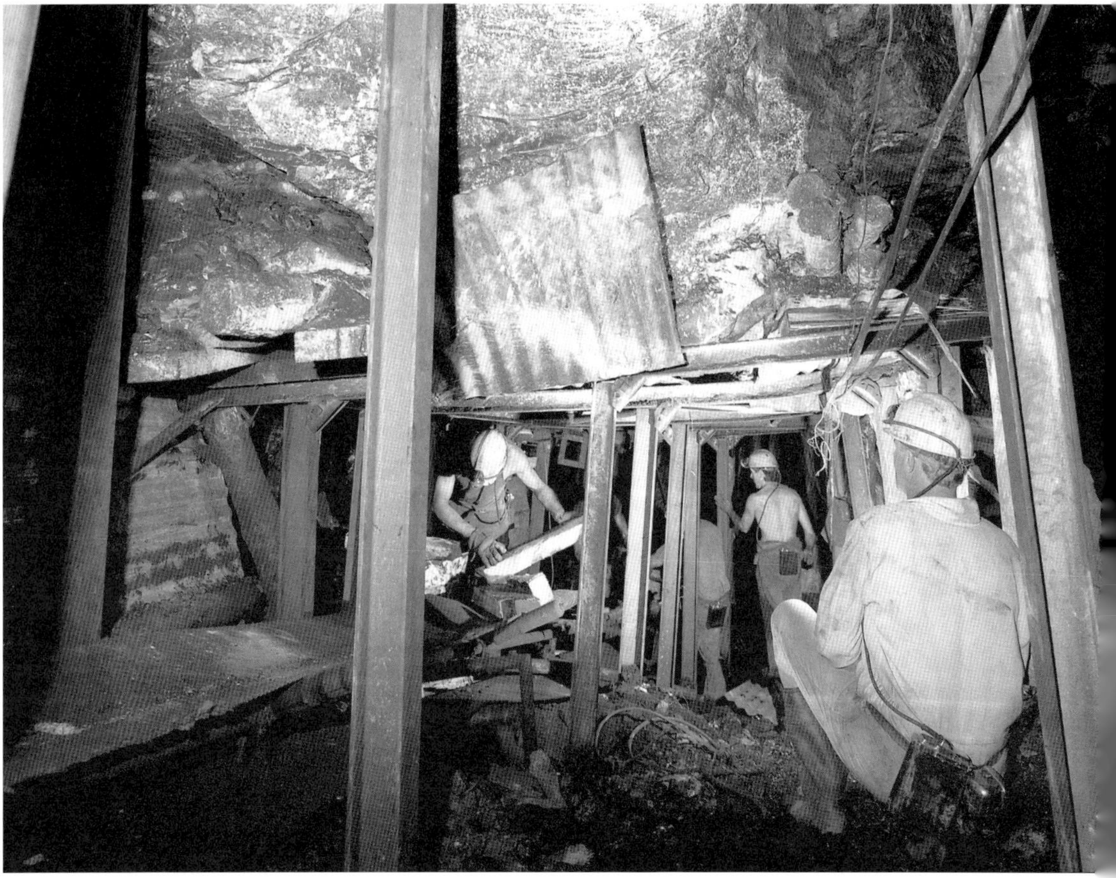

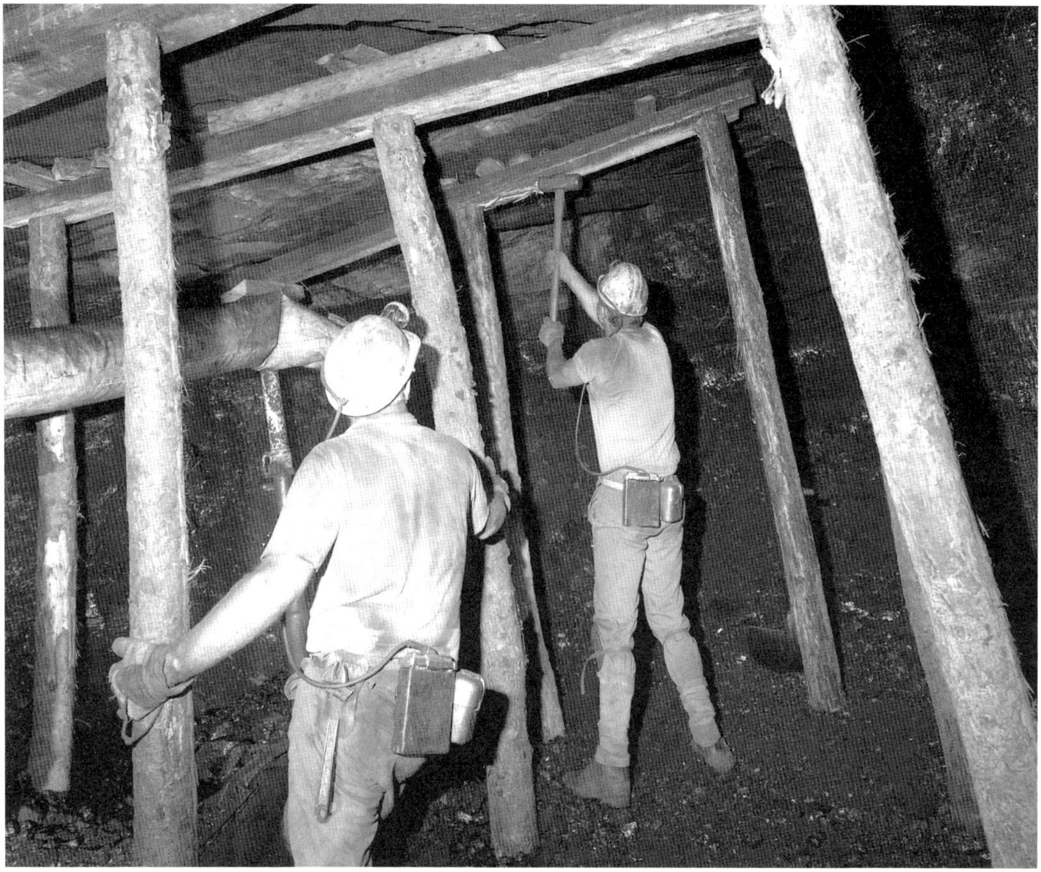

Above: Setting supports in a heading, February 1991. PD

Below: A train of empty tubs being lowered into the pit bottom. September 1991. PD

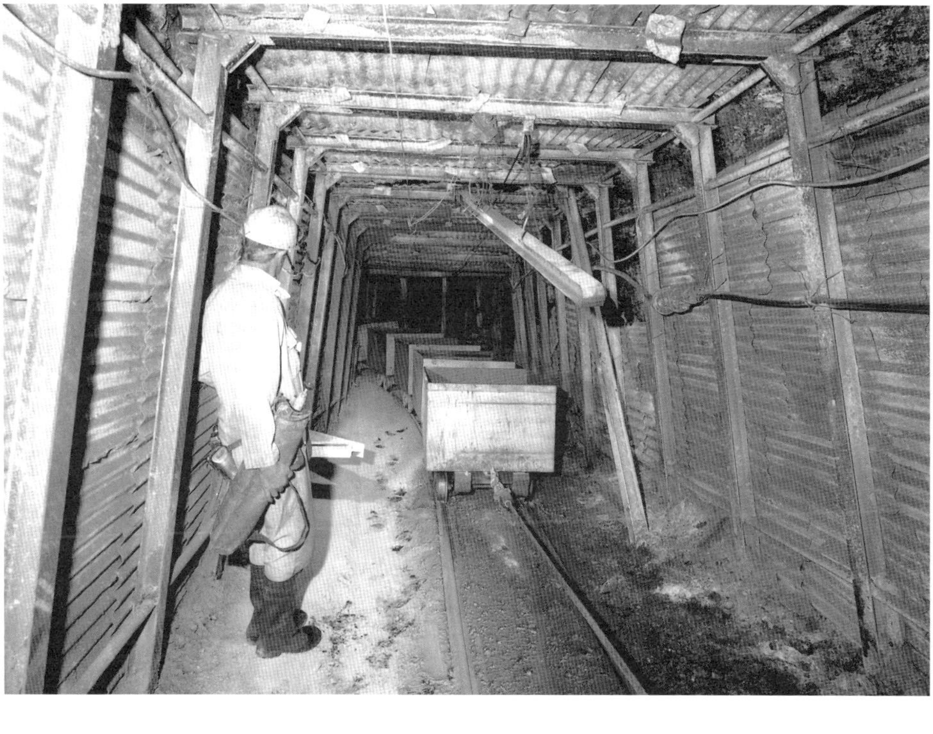

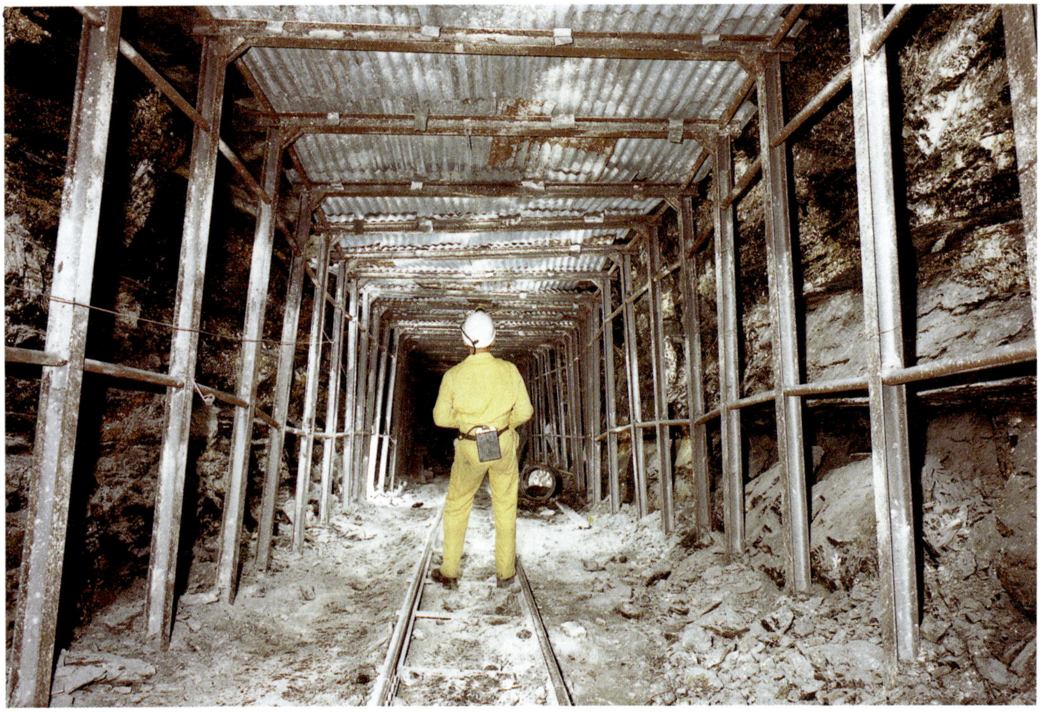

Above: Author Paul Deakin in the return air road. Note the thickness of the seam (c. 3m thick). September 1991

Below: Hand filling onto a small AFC (armoured flight conveyor), February 1991. PD

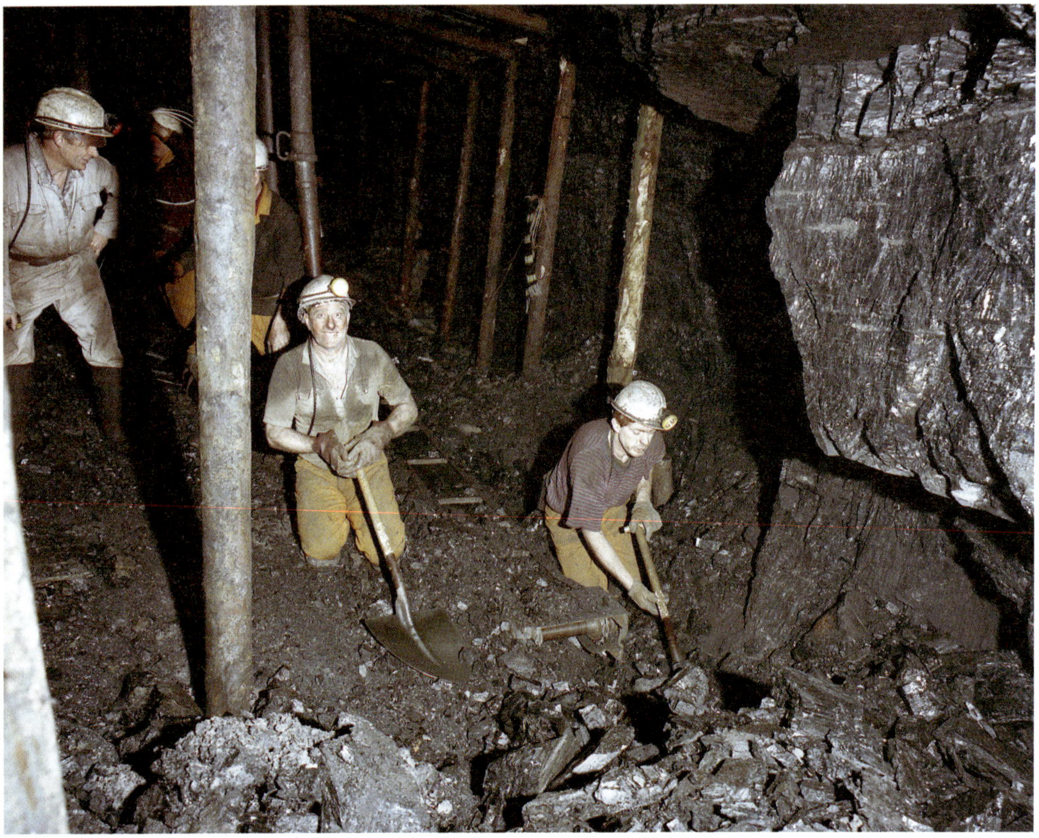

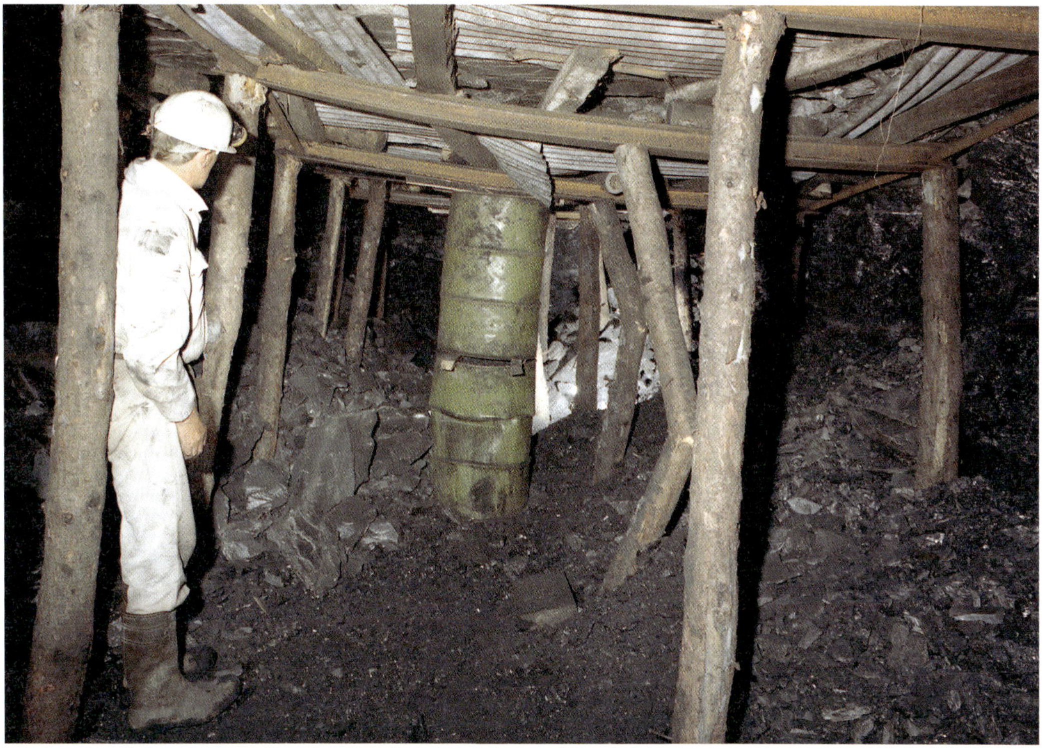

Above: Arthur Cartwright, the manager, inspecting a roadway suffering from convergence during pillar extraction. Note the oil drums filled with stone which make excellent chocks, September 1991. PD

Below: Failing supports during pillar removal, September 1991. PD

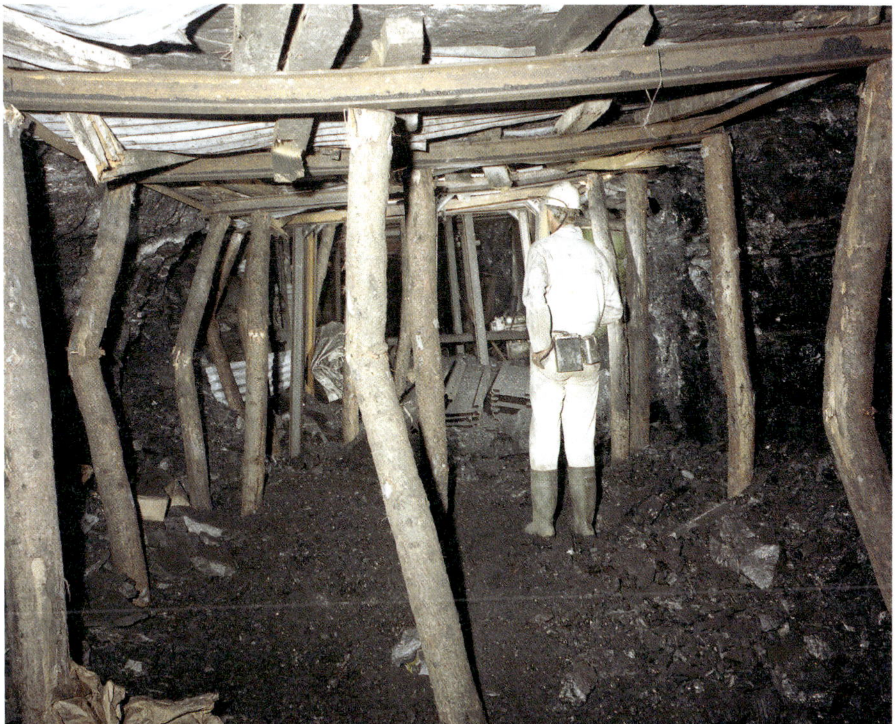

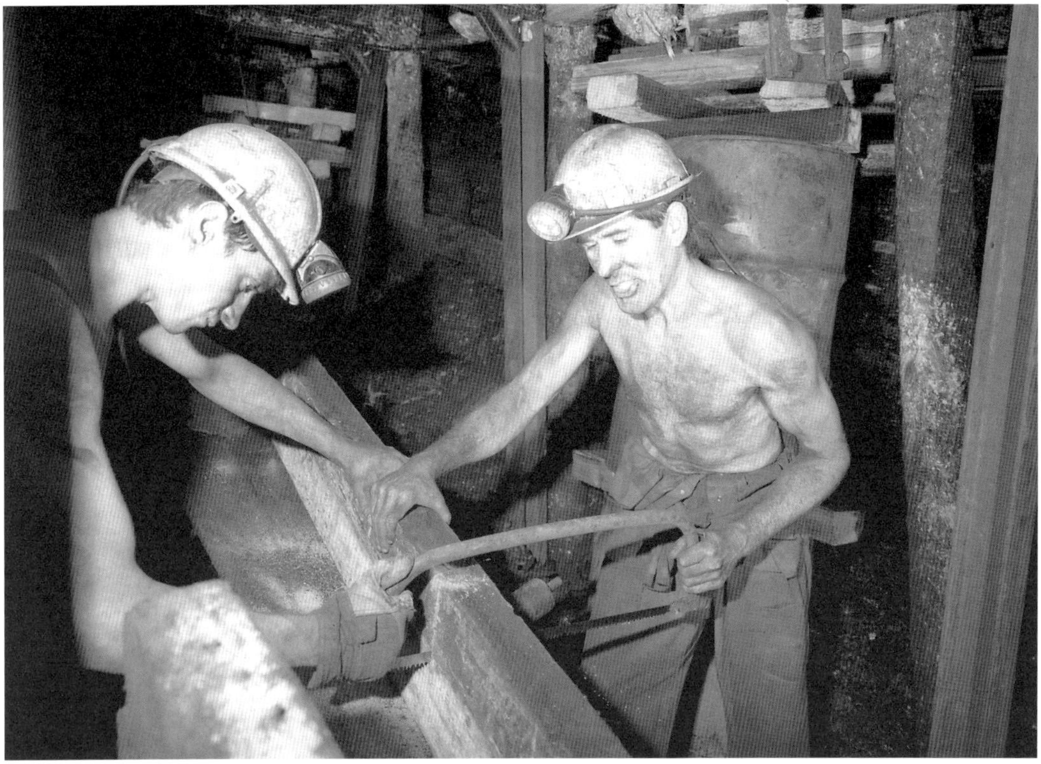

Above: Colliers sawing timber for supports, September 1991. PD
Below: Setting supports during pillar removal, September 1991. PD

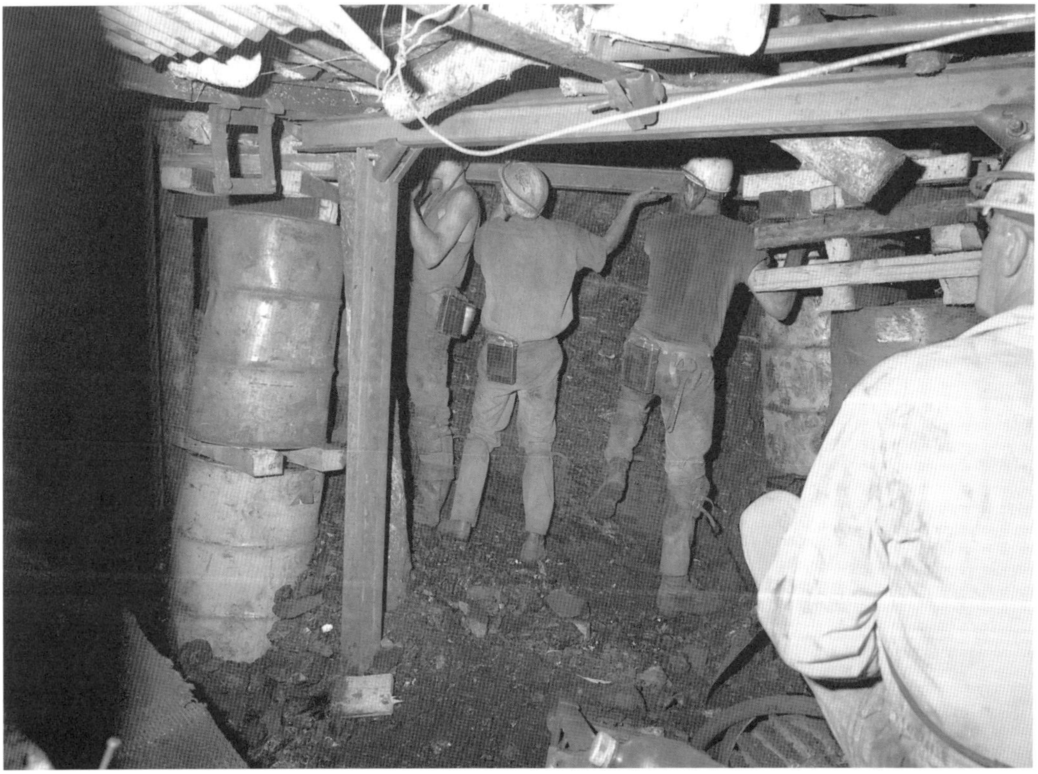

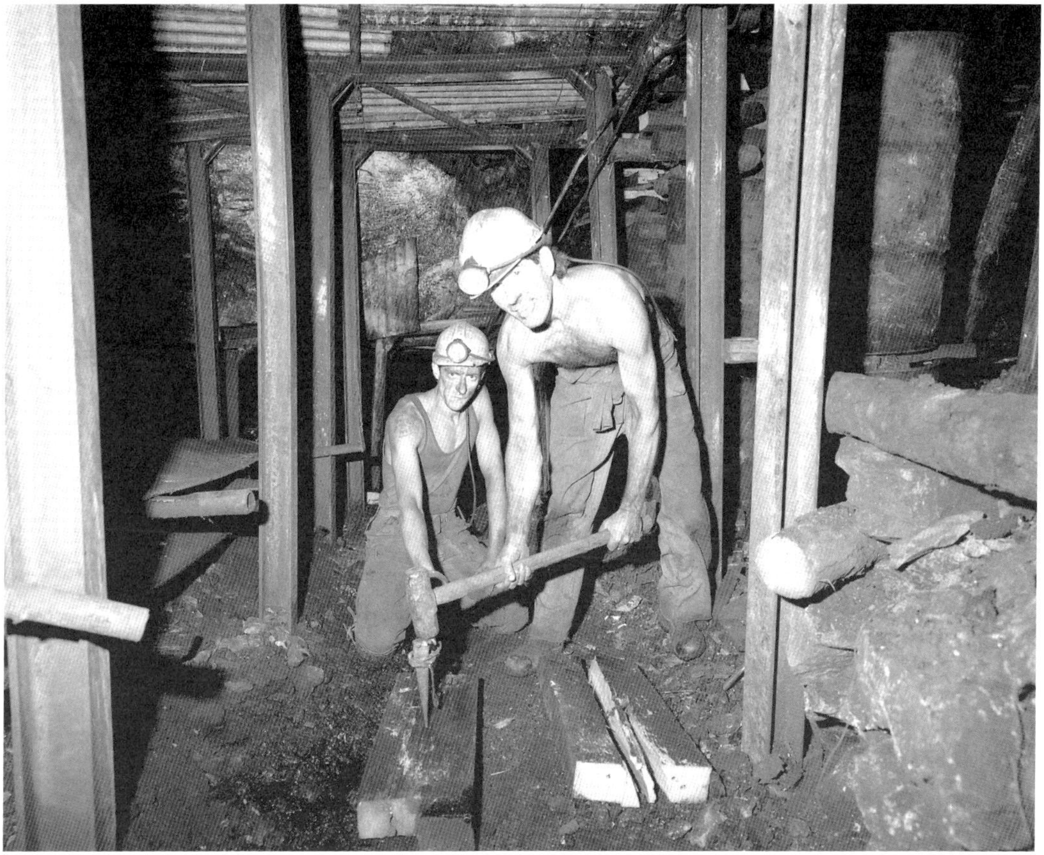

Above: Colliers splitting timber using a cold set and hammer, September 1991. PD
Below: Arthur Cartwright in a roadway converging during pillar extraction. Again note the oil drums filled with stone. These were used extensively in North Staffordshire's small mines to provide support. The site has now been restored to a paddock.

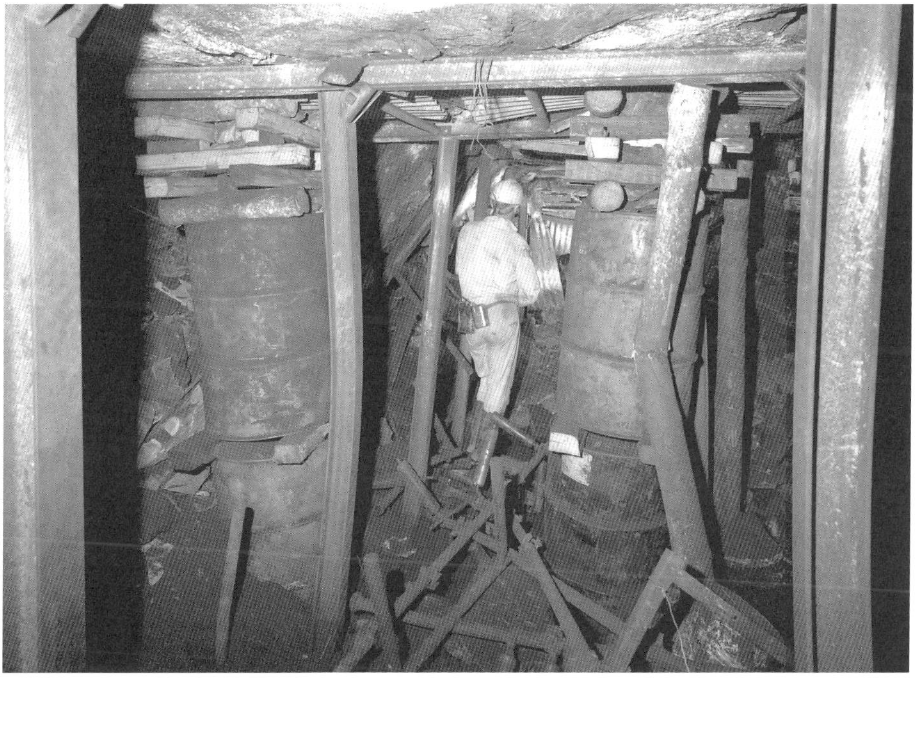

Silverdale Colliery

Shelton Iron, Steel and Coal Co. Ltd.

Situated 300m west of Silverdale Church. The mine was initially sunk in the 1830s, but coal and iron ore mining in the area are recorded in the 13th century. The recent colliery site was often referred to as Kents Lane.

After privatisation operated by Coal Investments PLC and later Midland Mining Ltd. Last coal worked 24 December 1998. The last underground coal worked in North Staffordshire.

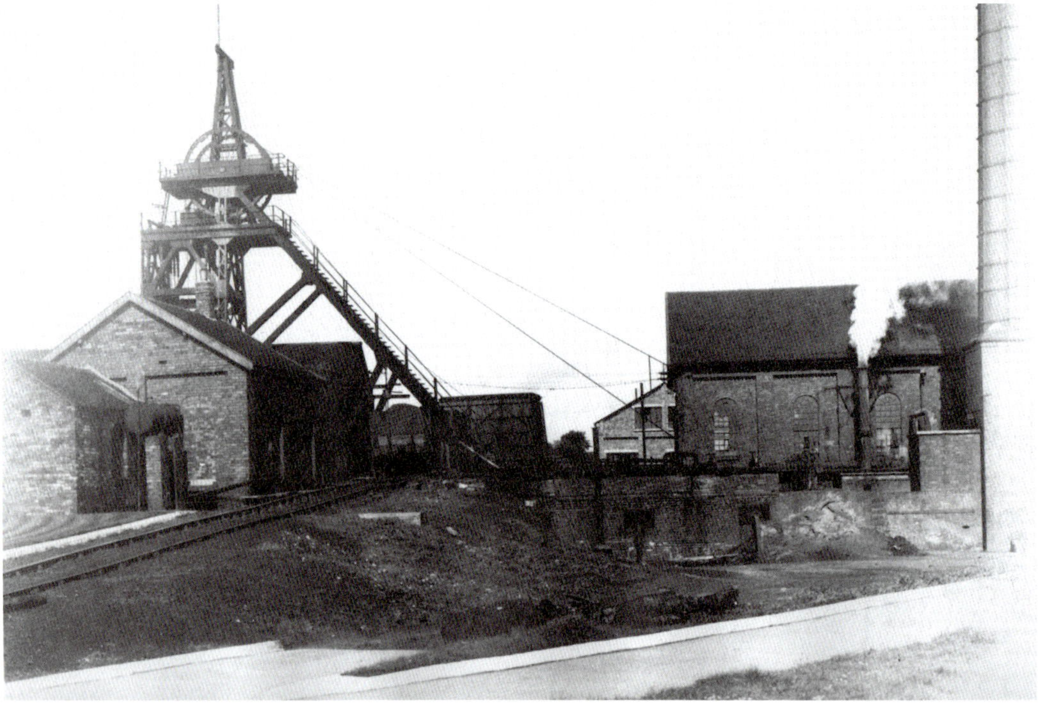

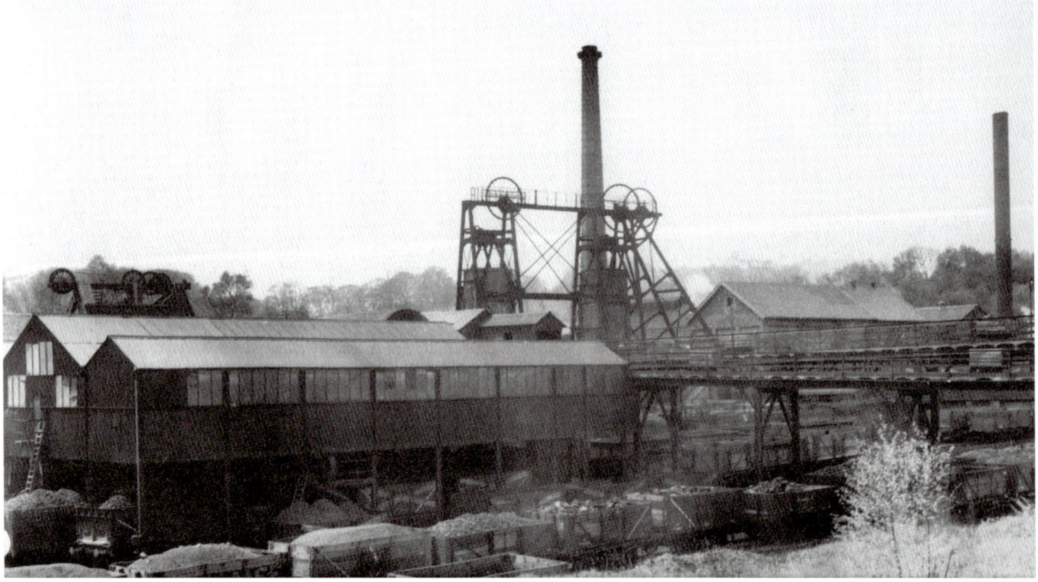

Opposite page; Top: A view of the 310.7m deep No. 17 shaft in 1951, M.C. The Kents Lane site had four shafts and four surface drifts. The original surface drift was driven in the 1960s to access the Spencroft Seam. A second, (confusingly referred to as No. 1) was driven in a south-west direction to provide access to the Winghay workings. In the late 1970s, No. 1 Surface Drift was extended and Nos. 2 and 3 Surface Drifts were driven parallel to No. 1. They accessed an area referred to as the Silverdale New Mine, lying to the south of the Keele Service Station on the M6 motorway. The drifts were about 3.8km in length, at a gradient of 1:4.

Below: A view of Nos. 14 and 15 shafts, sunk to 262.5m and 262.4m in depth respectively. There was a cage in each shaft with a single winding engine working both cages. The headgear to No. 16 shaft can be seen just above the buildings on the left-hand side. Taken in 1951. MC

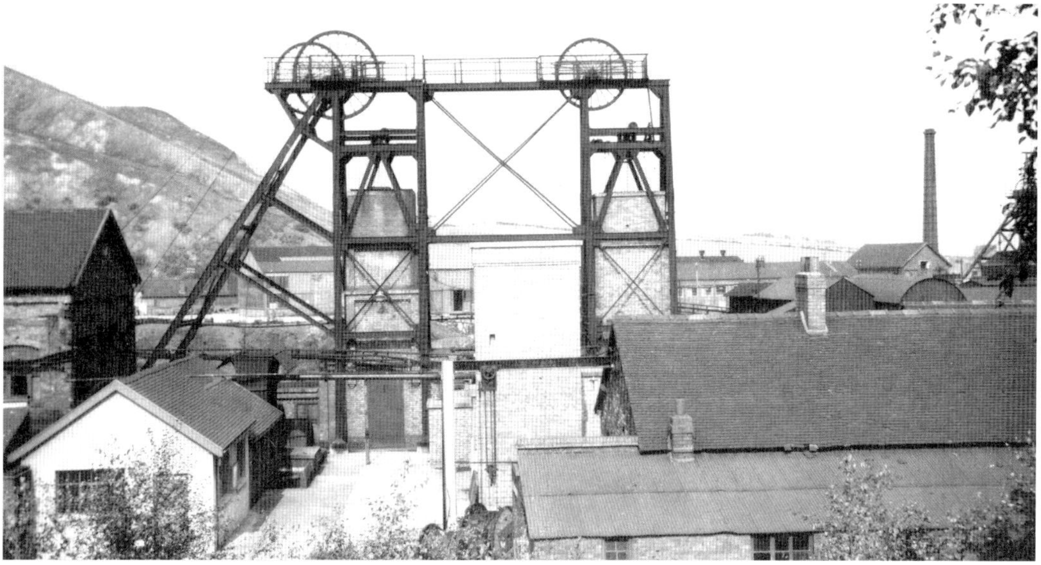

Above: Another view of Nos. 14 (on left) and 15 (on right) Shafts. Note the large tip on the left-hand side, now removed, 1951. MC.

Below: No. 17 Shaft, taken in 1980. PD

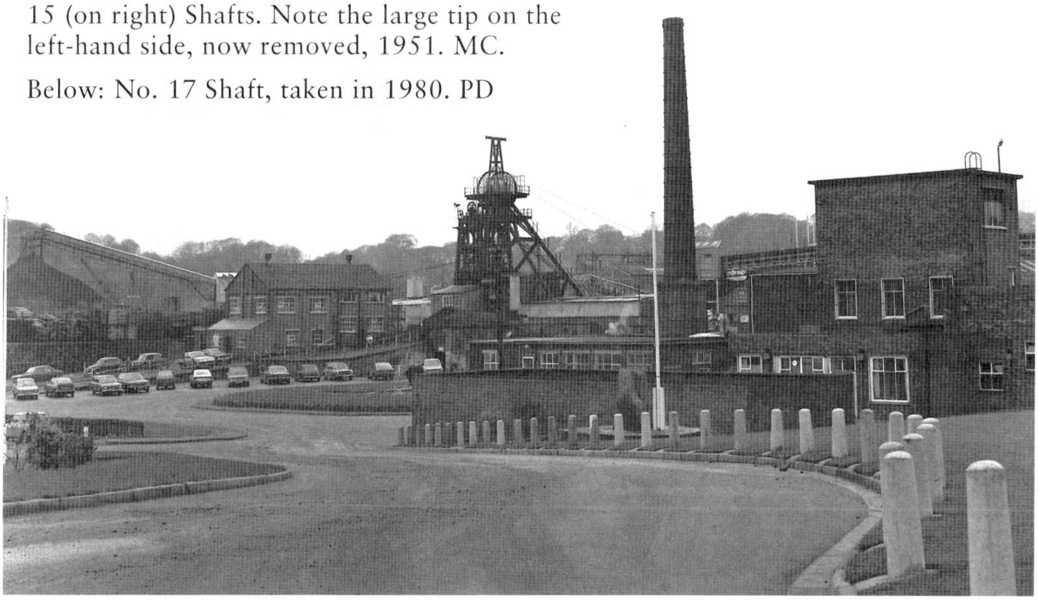

The brick headgear of the 258m deep No. 16 shaft. Taken in 1980. PD

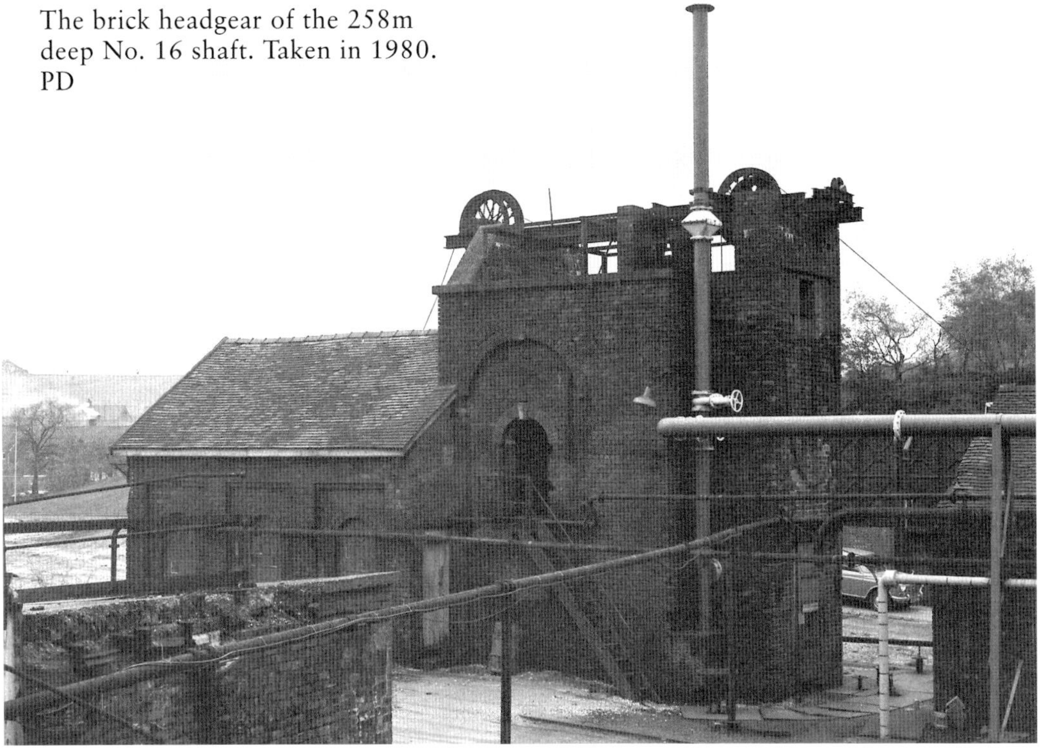

Below: The top of No. 2 drift into Silverdale new mine. Although sunk in the early 1800s, a new mine was developed to work the area south of Keele Service Station during the 1970s-80s. This photograph shows the man-riding cars at the top of the drift, 23 December 1997. PD

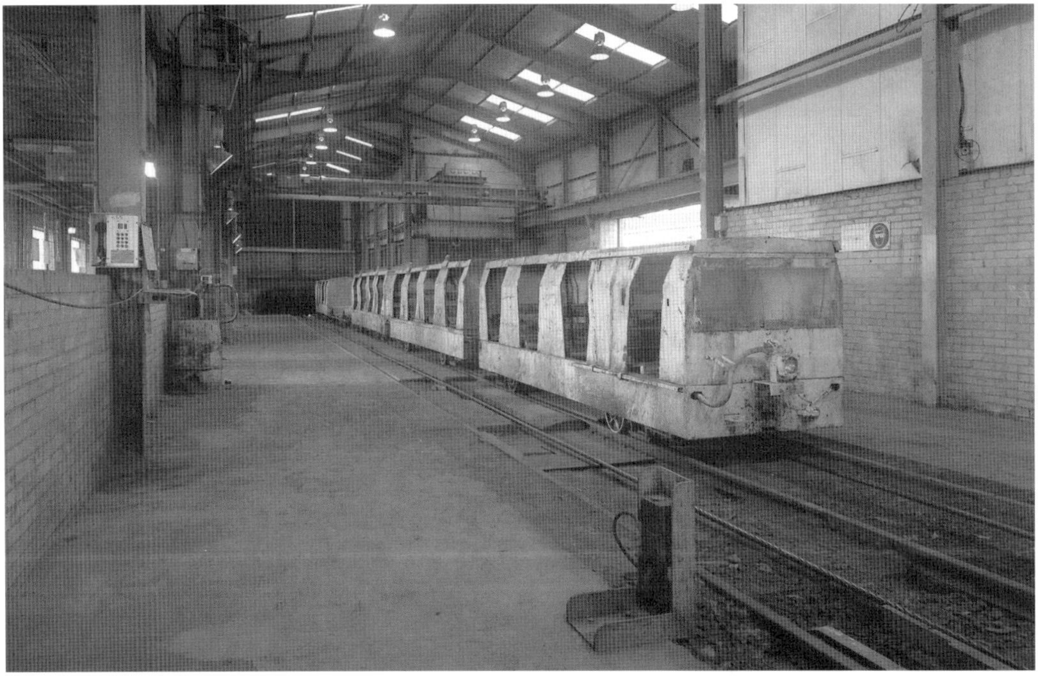

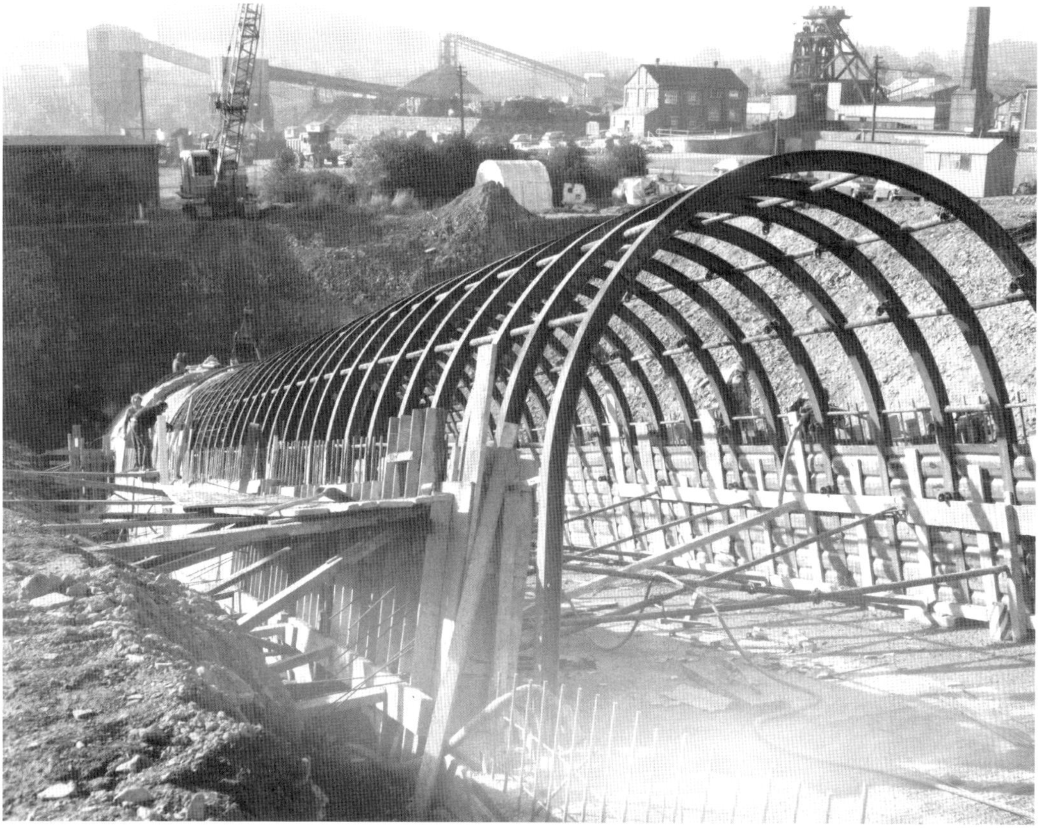

Construction of the initial part of the No 1 Drift by cut and cover, c. 1971. (No. 17 shaft is visible in the top right hand corner of the above photograph.) MC

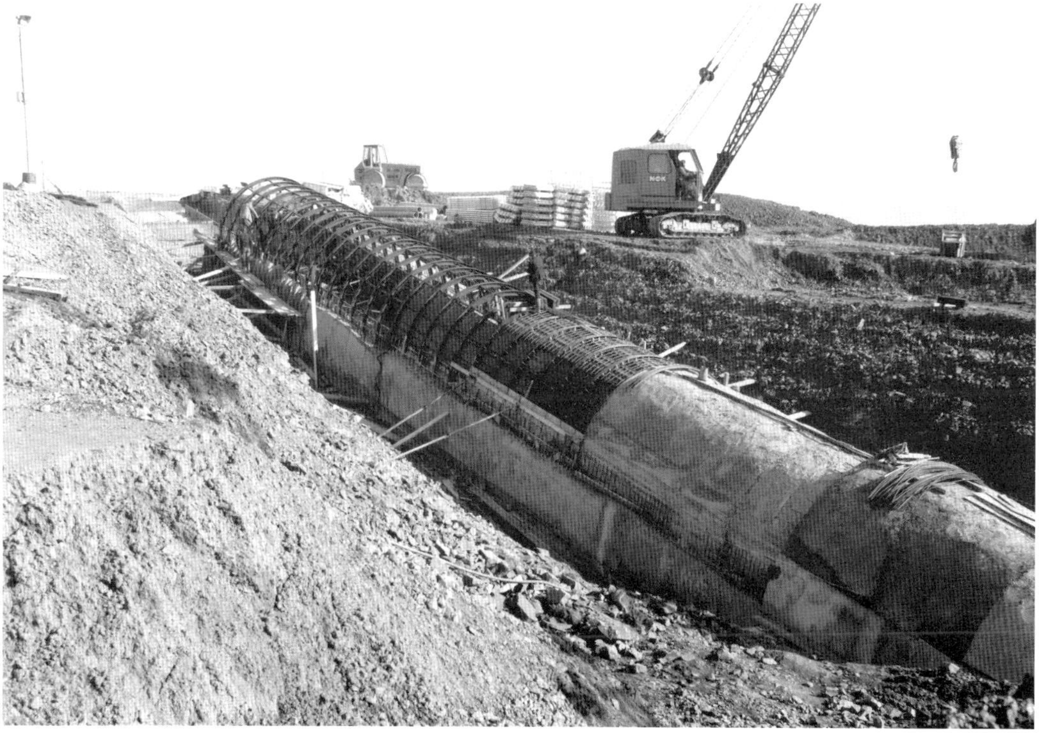

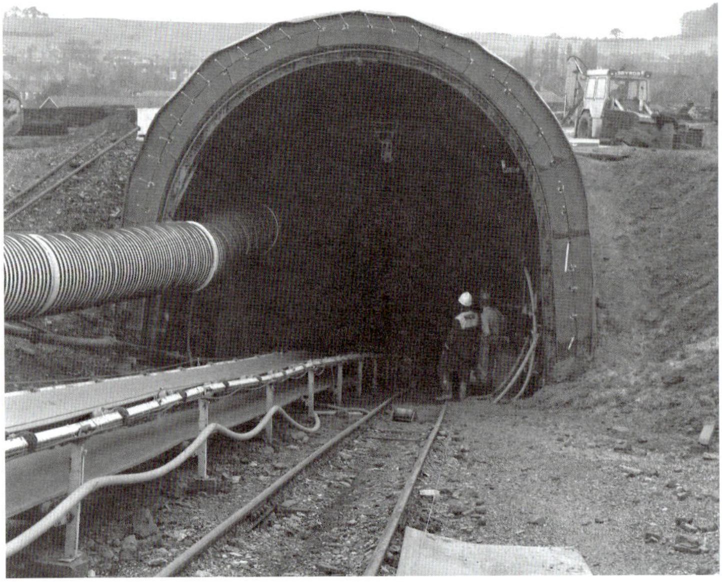

Left: Portal of No 2 Drift during construction, 1980. PD

Below: Rapid loading bunker and a train being loaded in 1996. PD

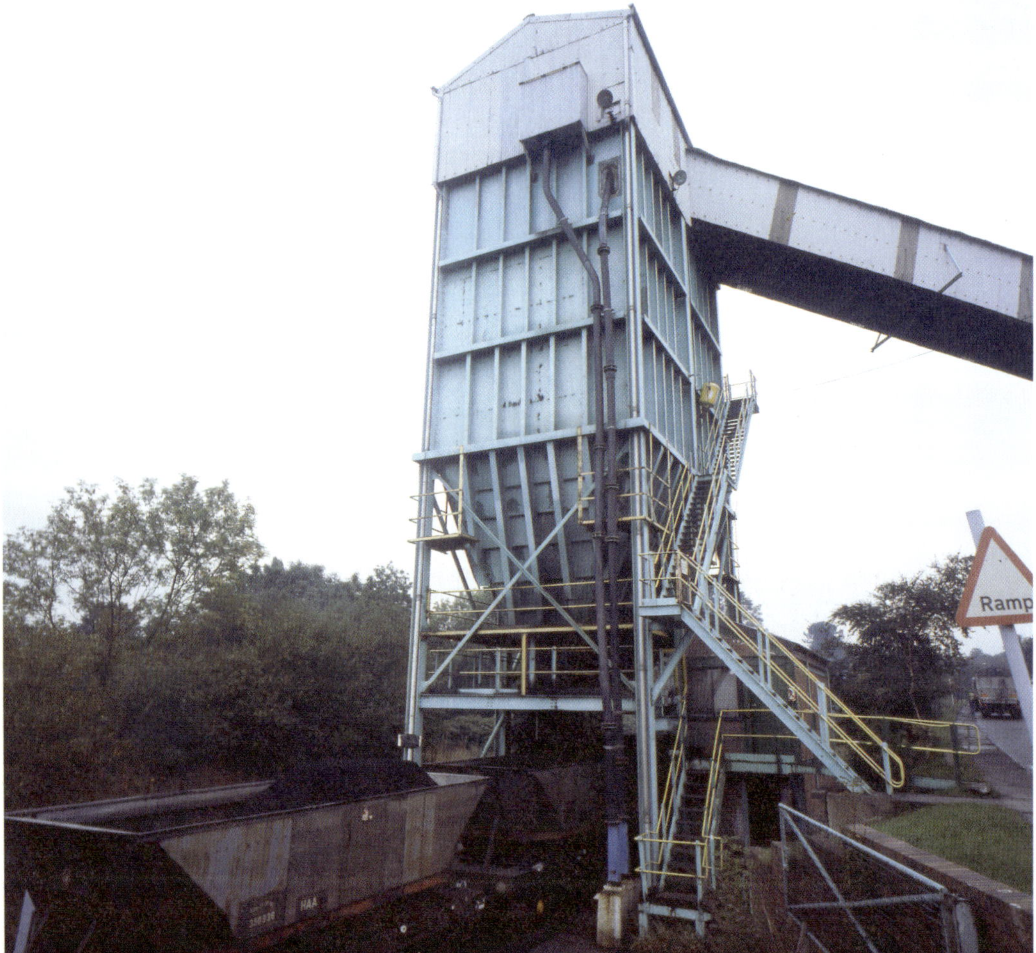

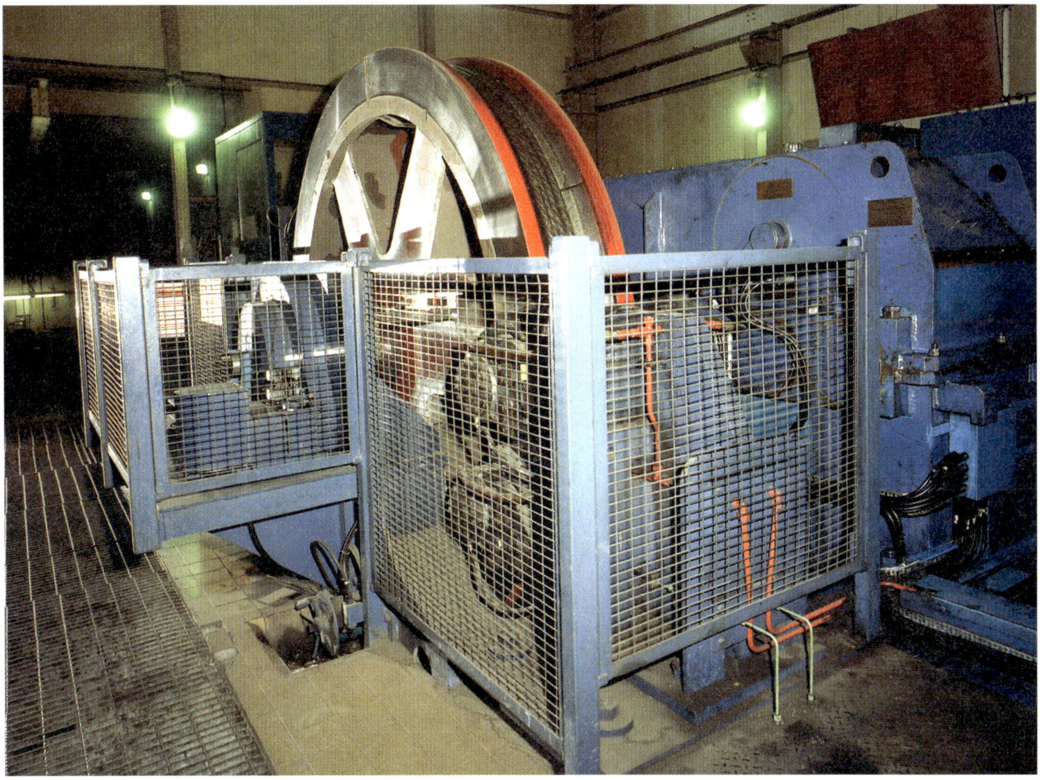

Above: The endless haulage engine on No. 2 drift, 23/12/97. PD
Below: Andy Mifflin (undermanager) at a junction between No 1 Drift and a connecting roadway to No 3, 1996. PD

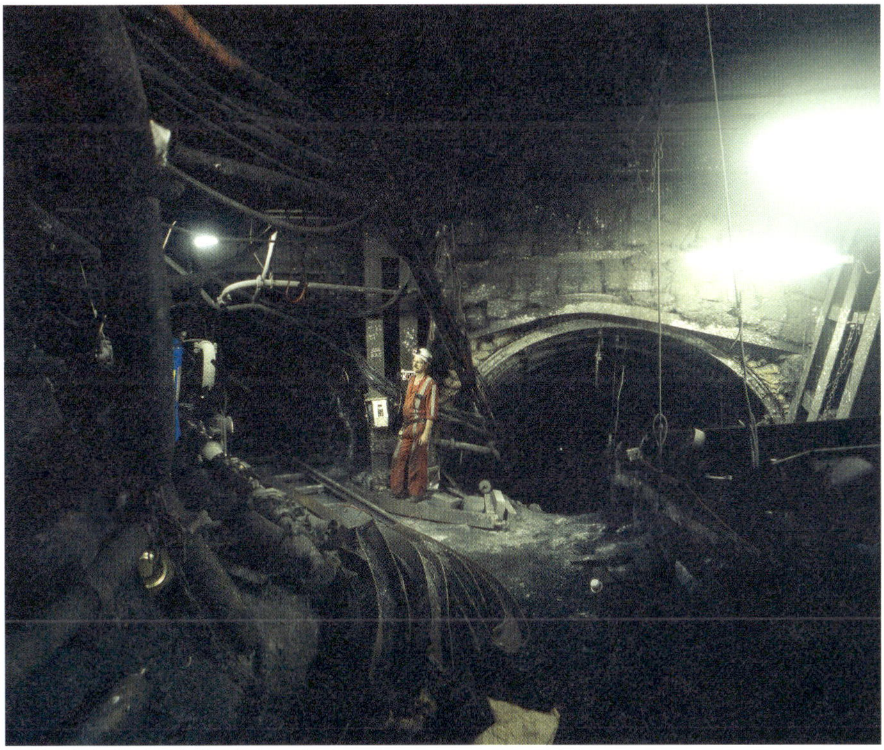

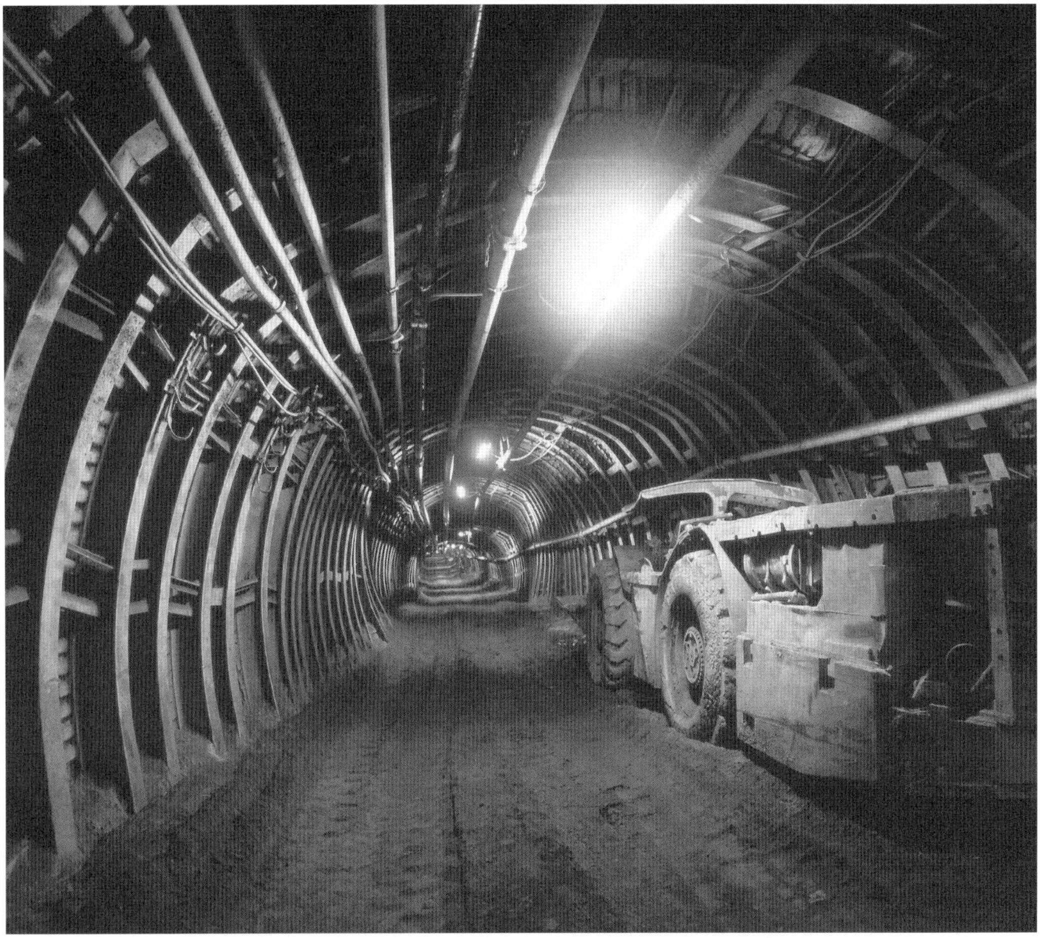

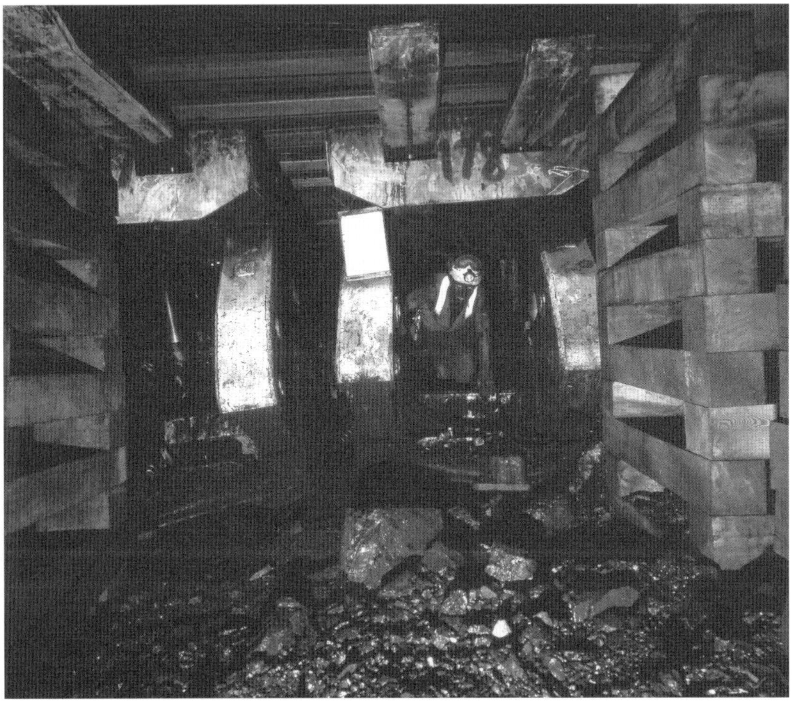

Above: Roadway immediately below the pit bottom with a FSV (Free Steered Vehicle) parked on the right, 1996. PD

Left: Danny Godwin coming off Ragman 4s coalface, 1996. PD

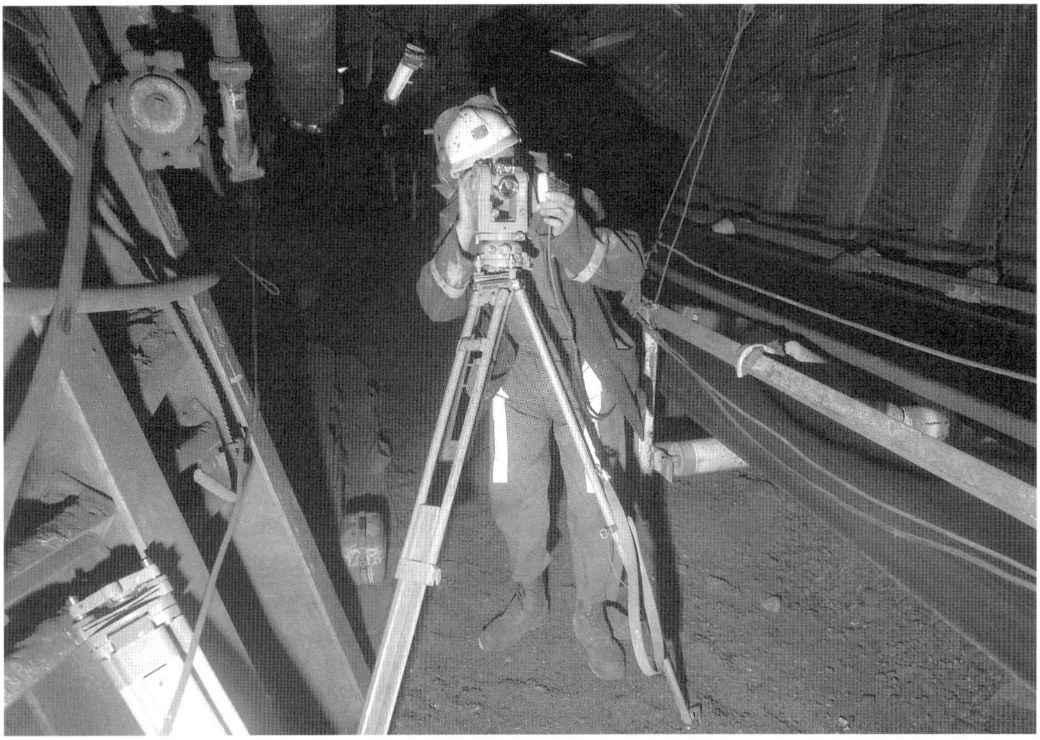

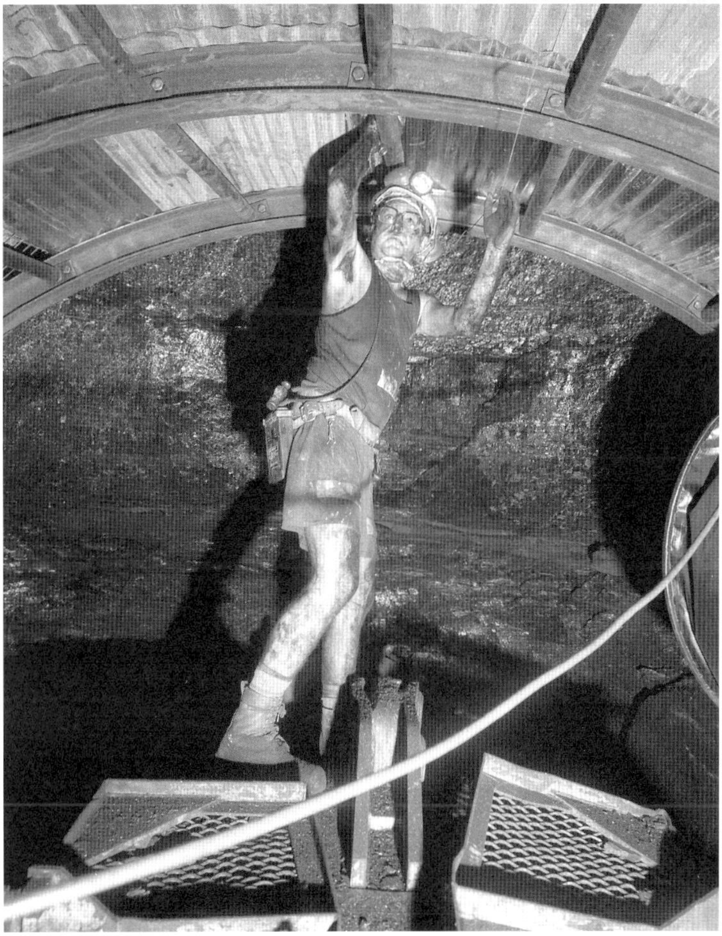

Above: Ray Hammersley surveying, using a Survey Wild T2 theodolite, 23 December 1997. PD

Right: Bob Barnett, a linesman with the survey office, extending the centre line in the Ragman 3s development, 23 December 1997. PD

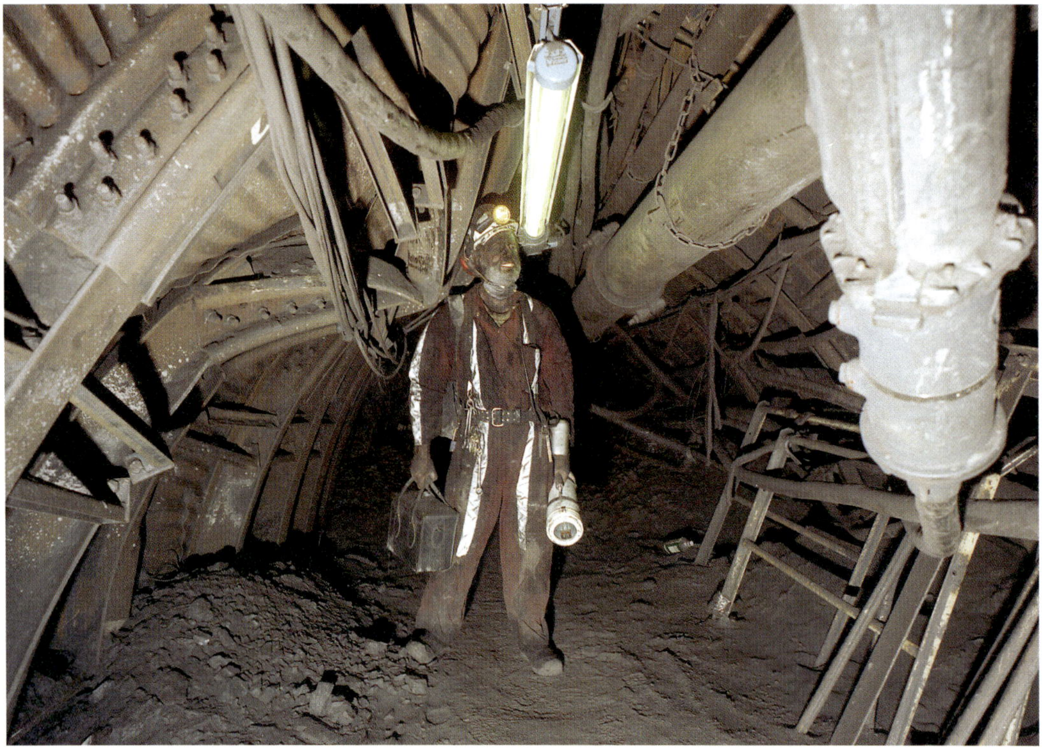

Above: Keith Trannah, fitter and keen amateur photographer, carrying his camera and flameproof flashgun. Note the twisted steelwork, 22/12/98. PD

Below: Rudi Leteller driving the shearer into the main gate on Ragman 3s, 22 December 1998. PD

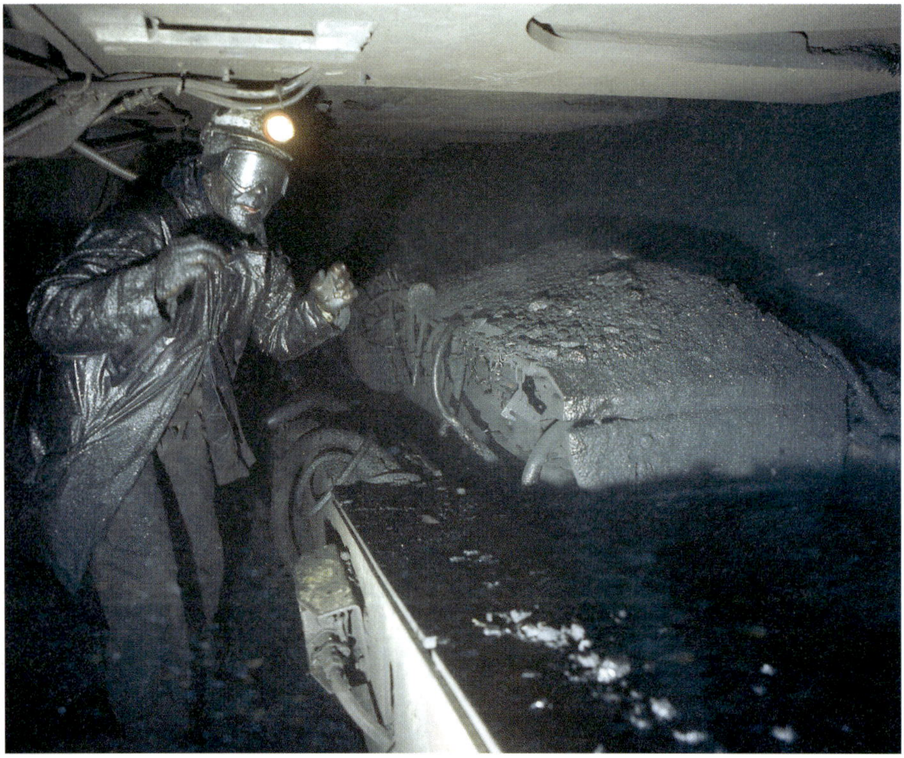

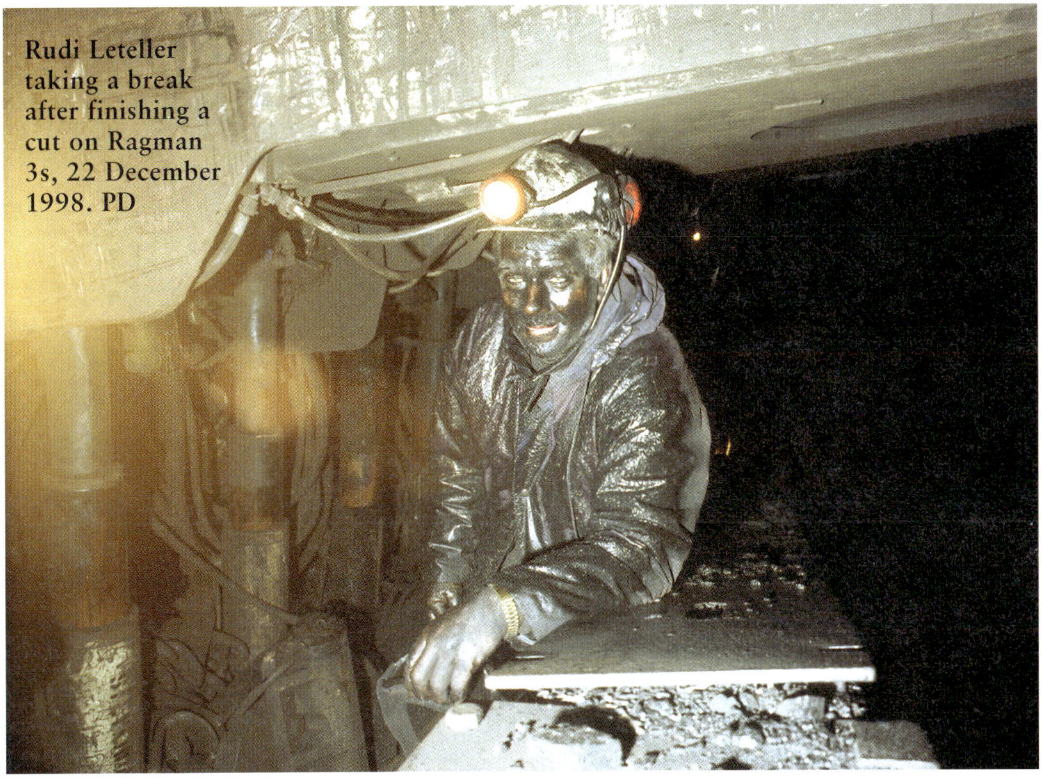

Rudi Leteller taking a break after finishing a cut on Ragman 3s, 22 December 1998. PD

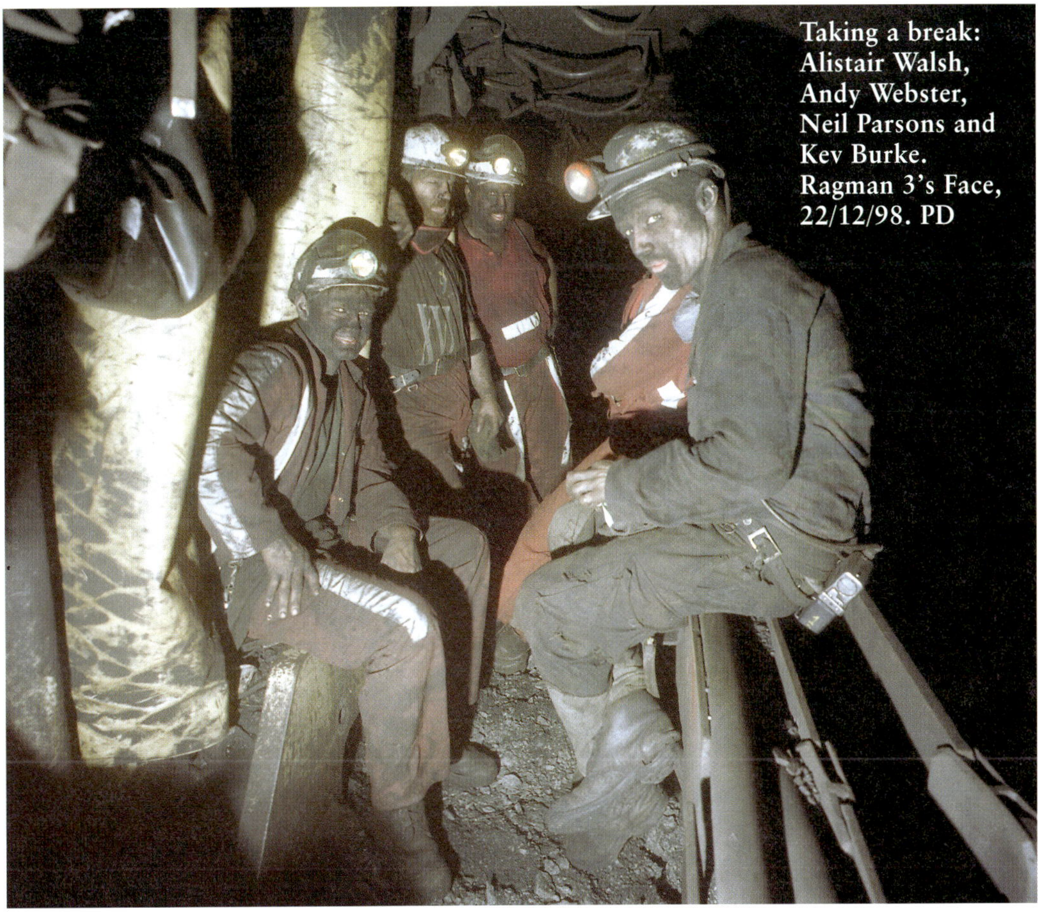

Taking a break: Alistair Walsh, Andy Webster, Neil Parsons and Kev Burke. Ragman 3's Face, 22/12/98. PD

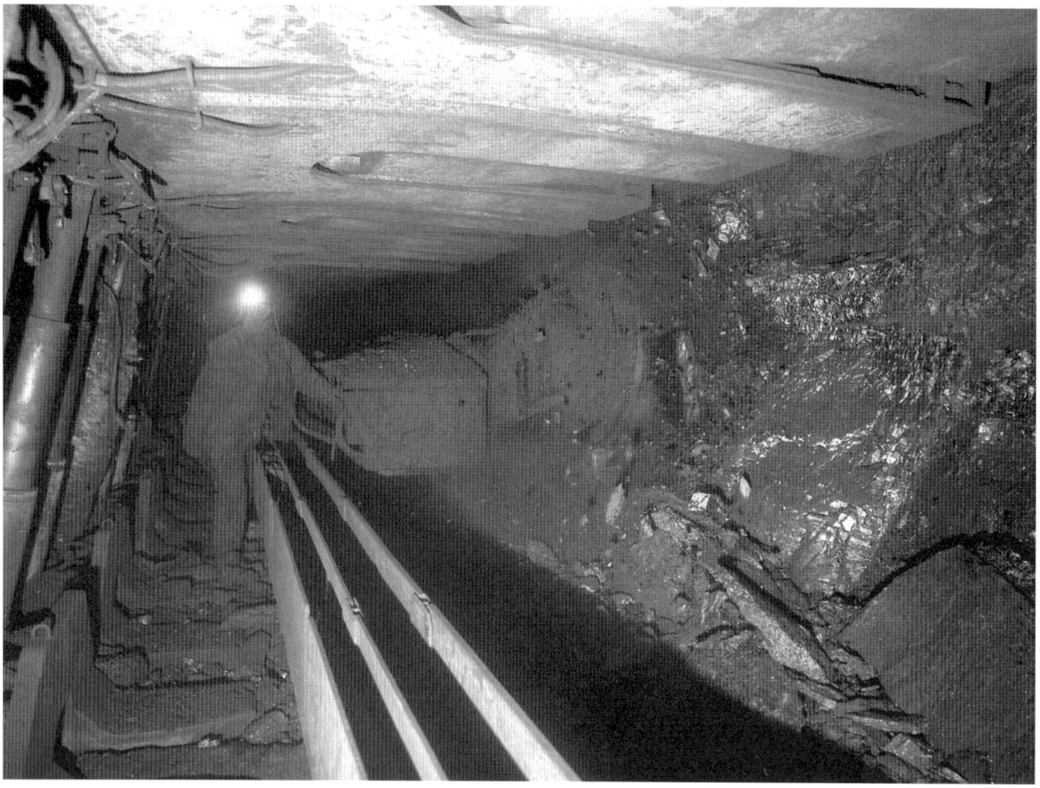

Above: Rudi Leteller driving the shearer, Ragman 3s face, 22/12/98. PD

Below: Kev Pritchard driving the shearer on the Ragman 3s face. Note the roof canopies on the self-advancing supports at roof height, 22/12/98. PD

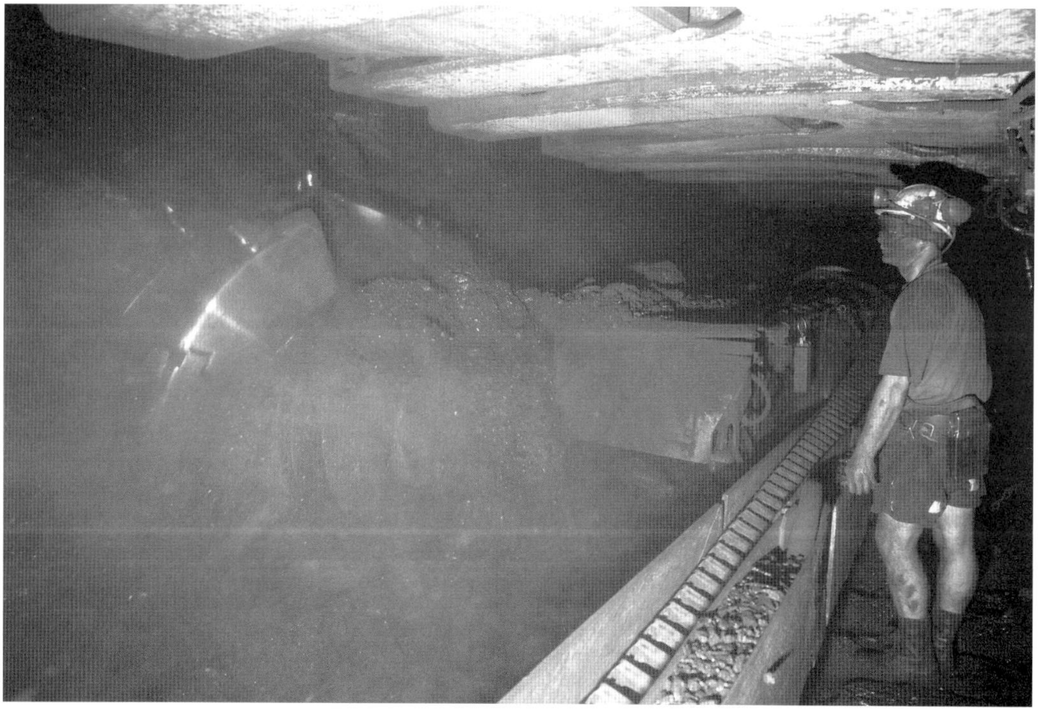

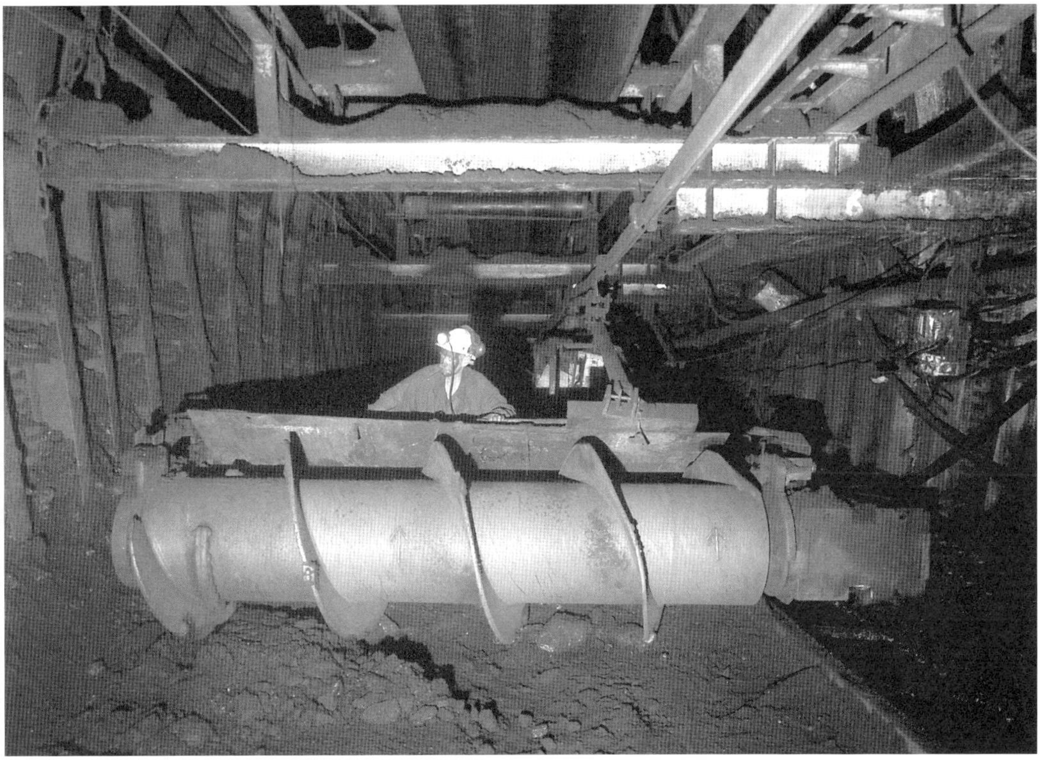

Above: Malcolm Walker standing behind the scrowl used for lifting coal from the floor of a strata bunker. Bunkers are commonly used in coal mines to temporarily store coal and to even out surges in production, 22/12/98. PD

Below: Keith Trannah silhouetted against a stopping (a seal put in a roadway to block off old workings and restrict the egress of gas), 22/12/98. PD

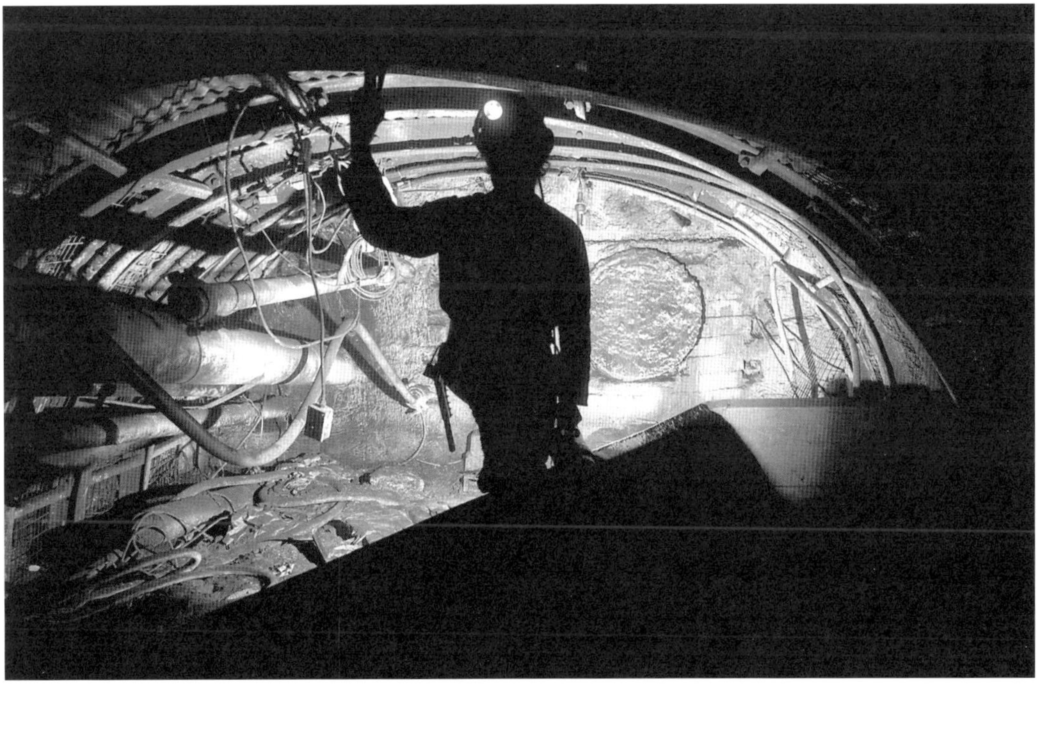

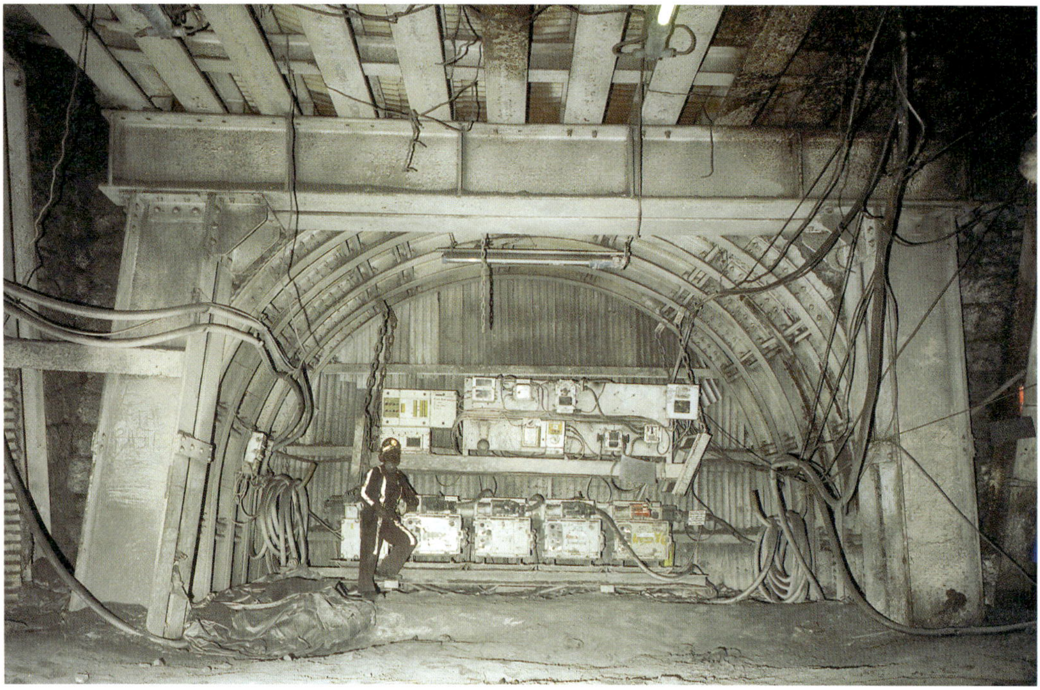

Above: Keith Trannah standing by electrical switchgear. Note the substantial steel supports, 22 December 1998. PD

Below: Here he is driving a FSV, 22/12/98. PD

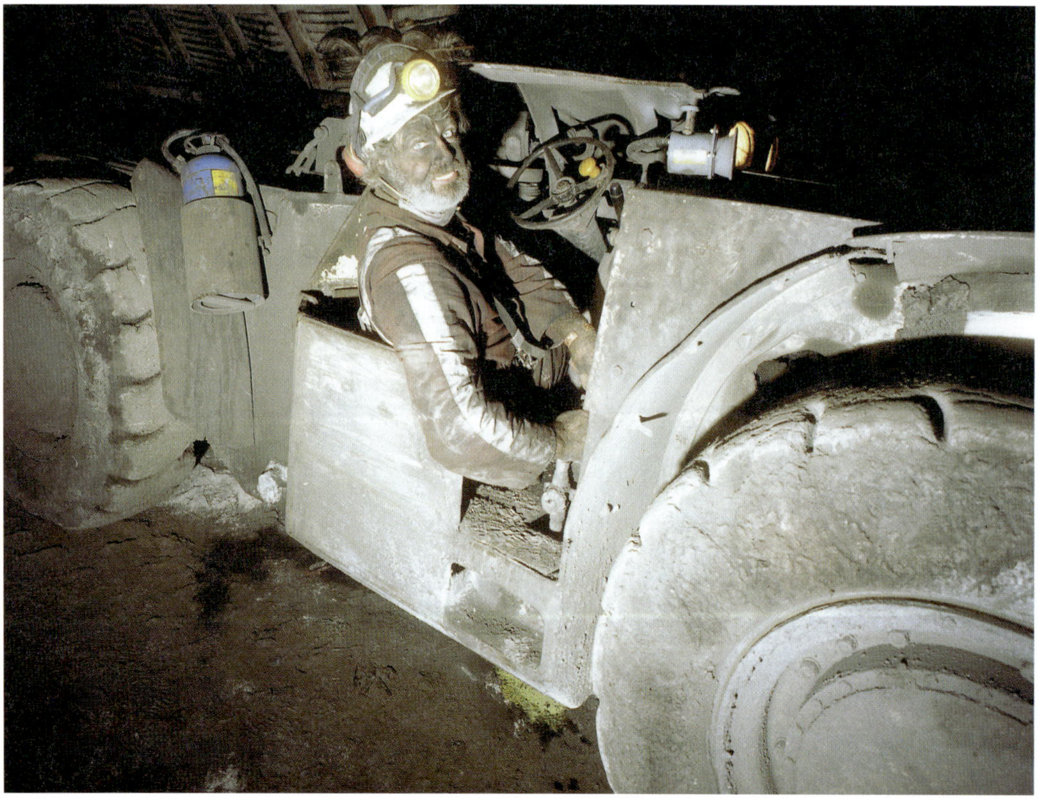

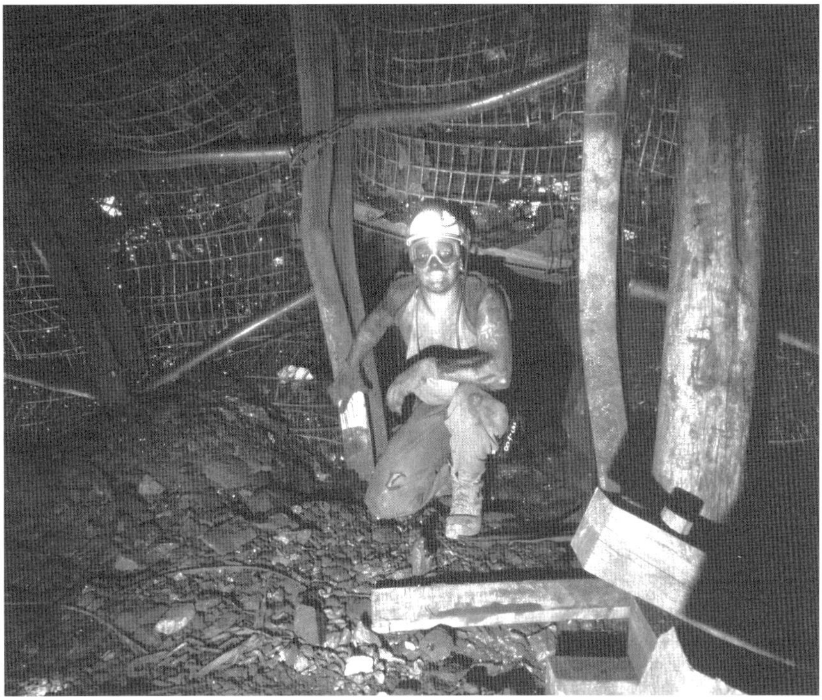

Right: Danny Goodwin coming out of the small connecting roadway between the dummy gate and the return. Ragman 4s face. 1996 (PD)

Below: Danny Goodwin checking methane gas levels in Ragman 4s Return. Seals were placed at regular intervals whilst the coalface retreated to restrict the flow of gas from the goaf (ie the area from where the coal has been extracted) into the working area.

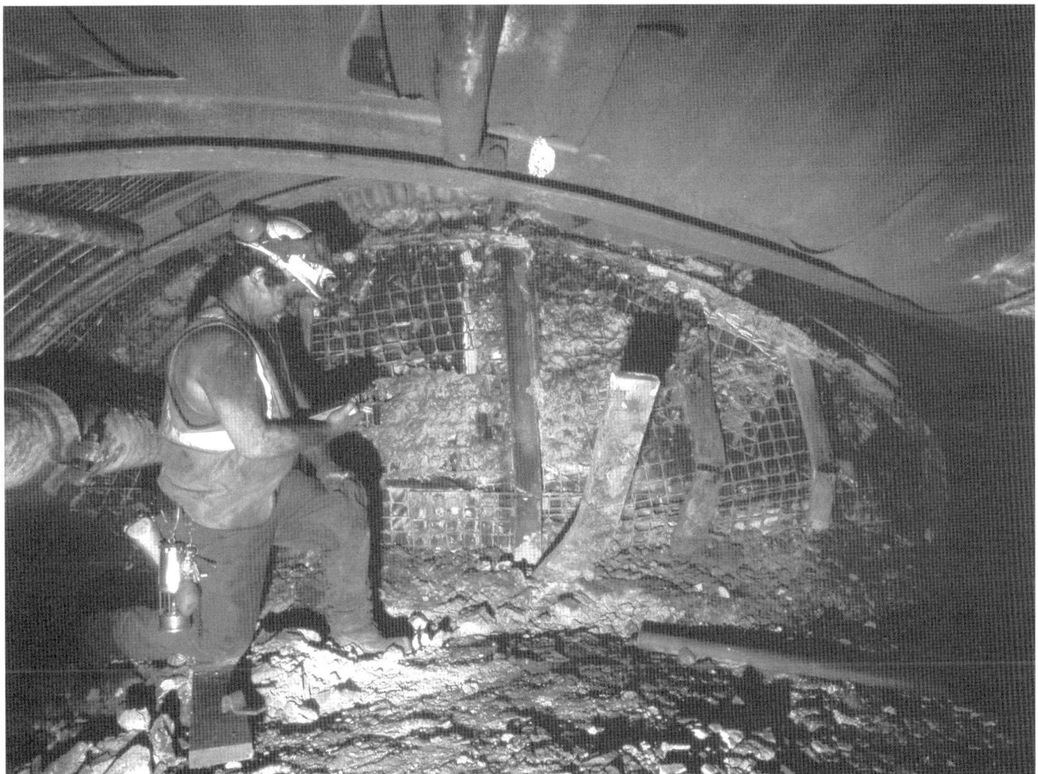

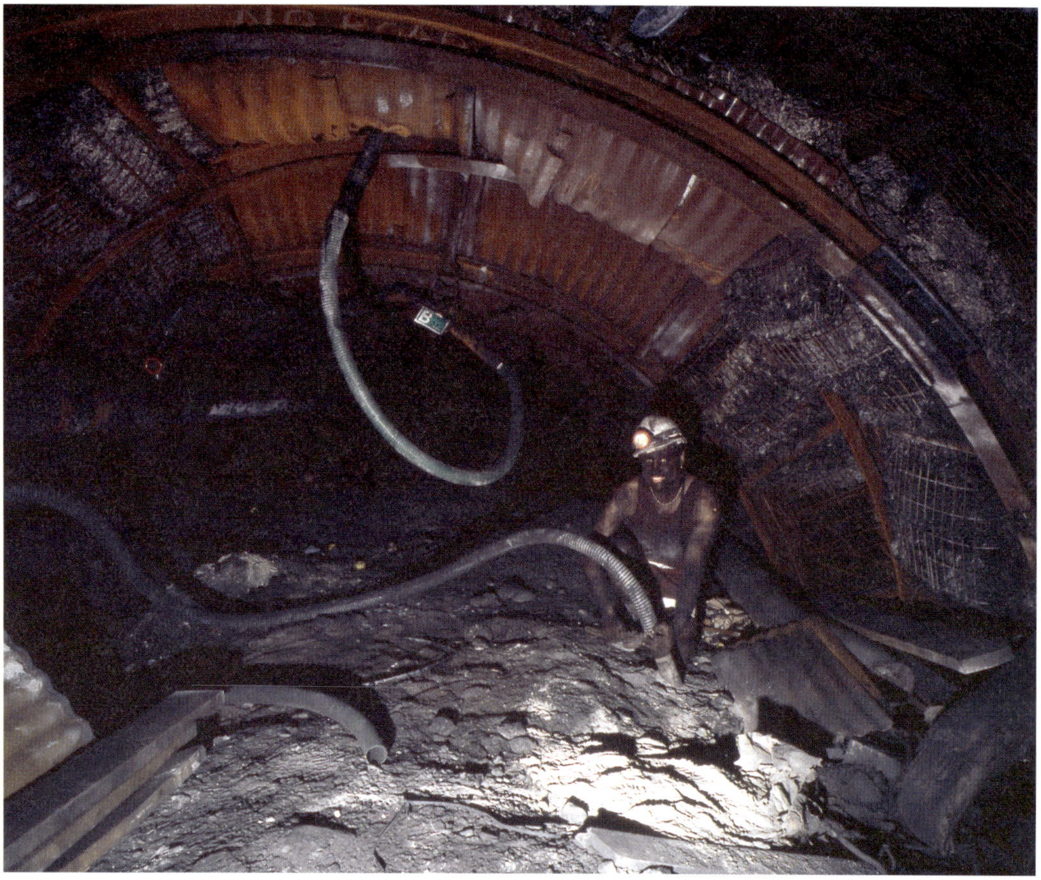

Above: : A collier by a methane drainage standpipe. Boreholes were drilled into the roof and floor of the seam as the coal face retreated and the methane gas was pumped out to the surface. Without methane drainage the mine would not have been able to work. The gas is still being drained from the colliery and used for electricity generation, 1996. PD

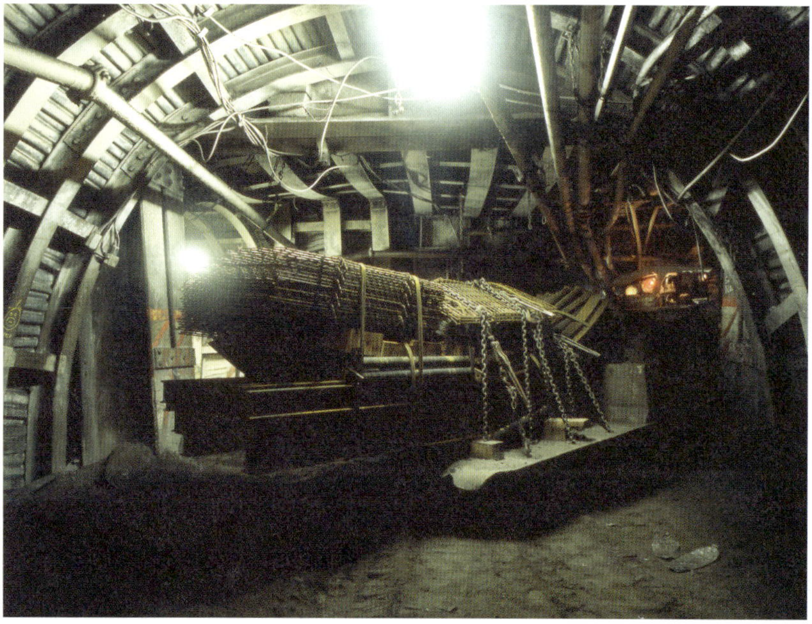

Left: A FSV carrying a palletised set of supports in 1996. PD

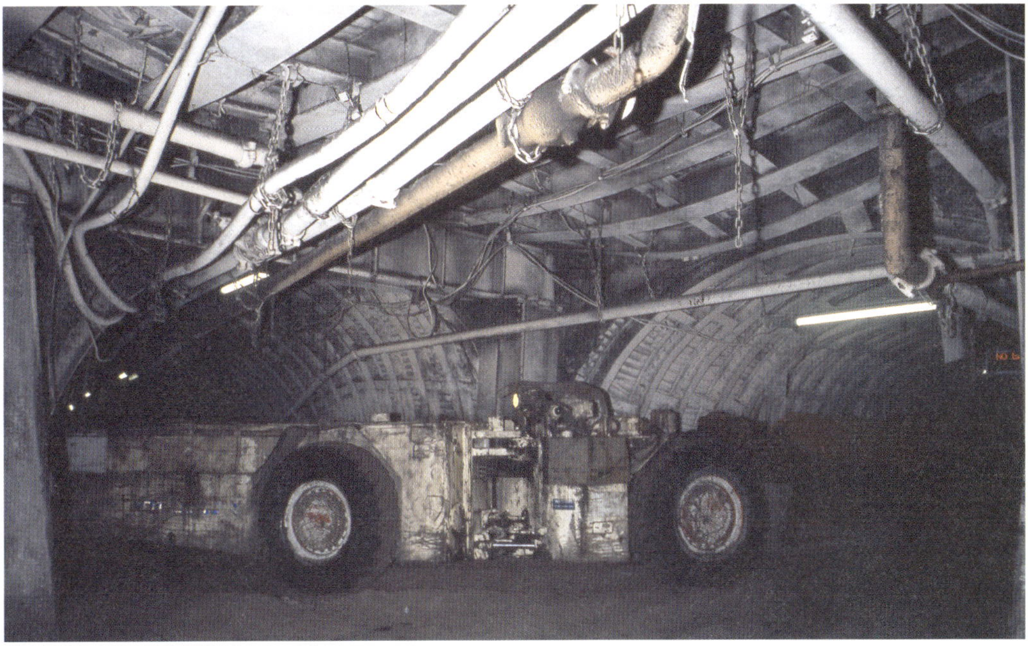

Above: A FSV negotiating a junction, 1996. PD
Below: Andy Mifflin standing by the dint in the main return No. 3 surface drift, 1996. PD

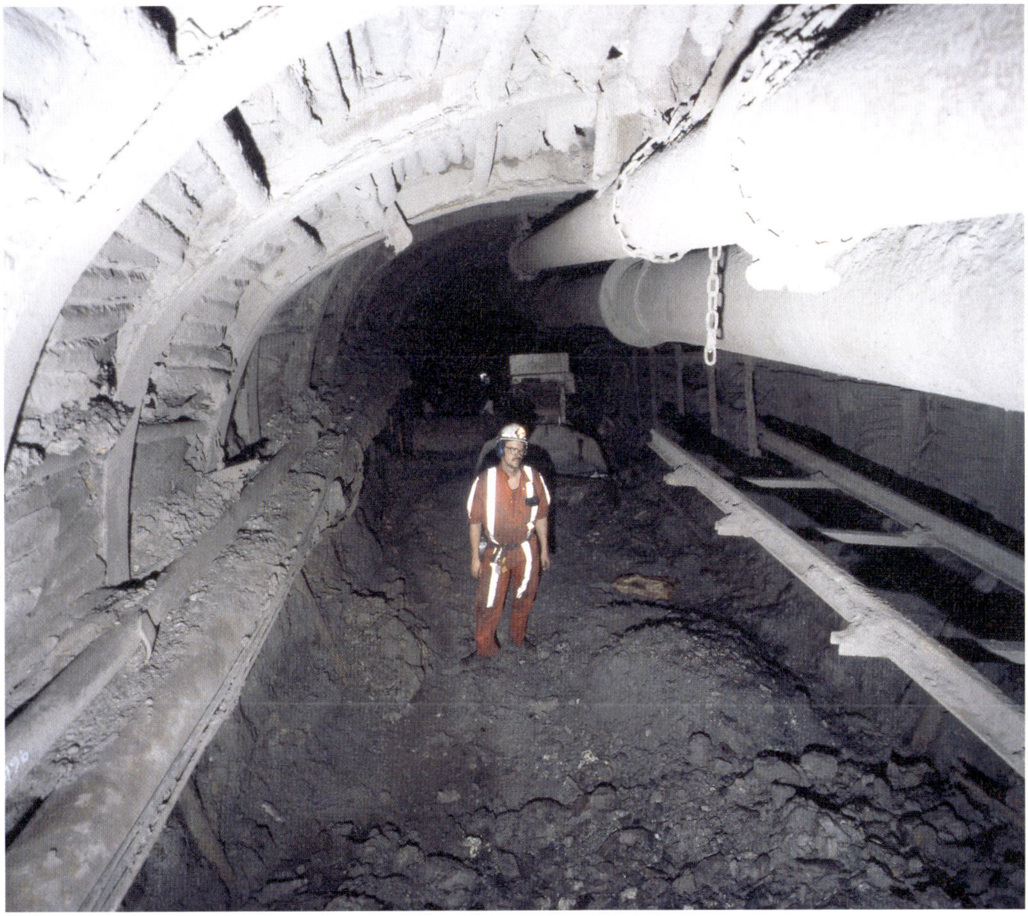

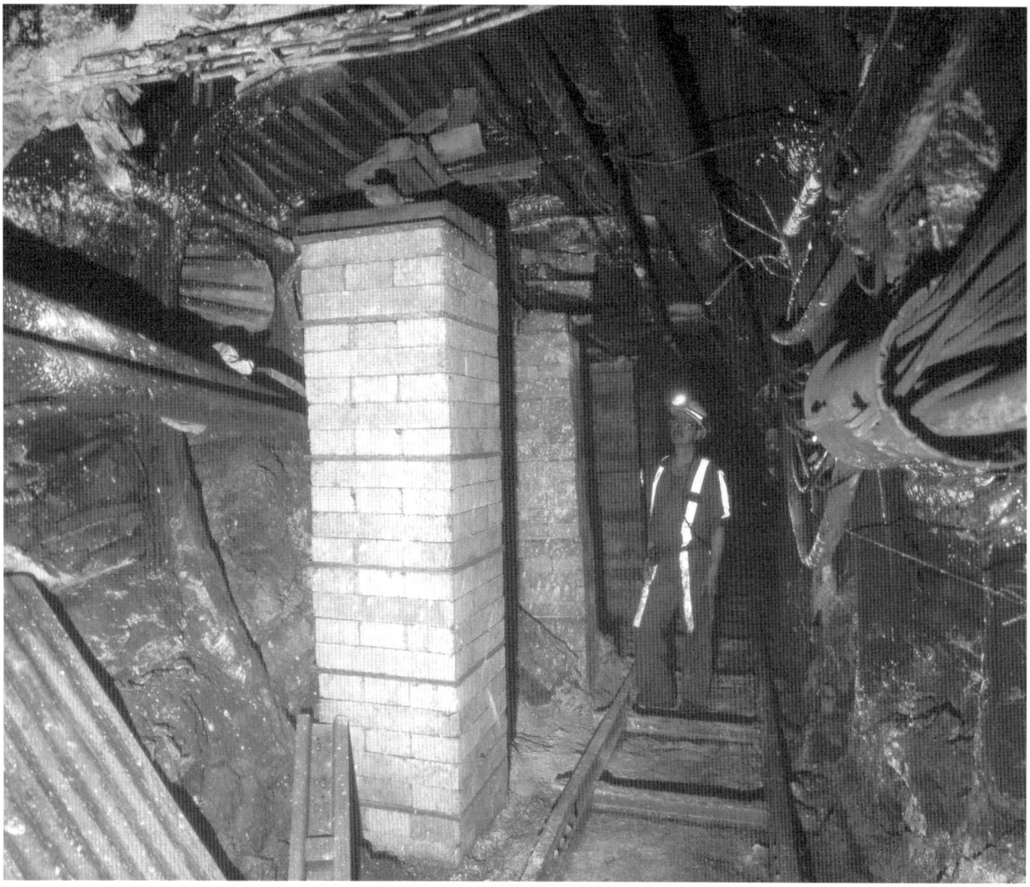

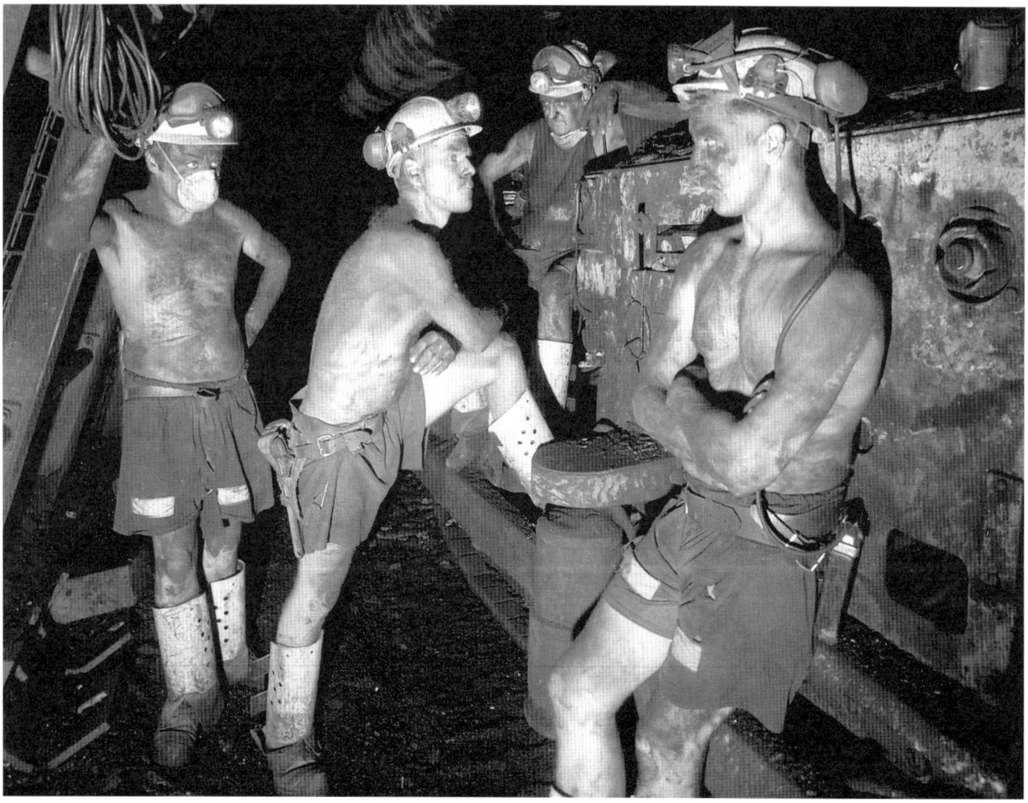

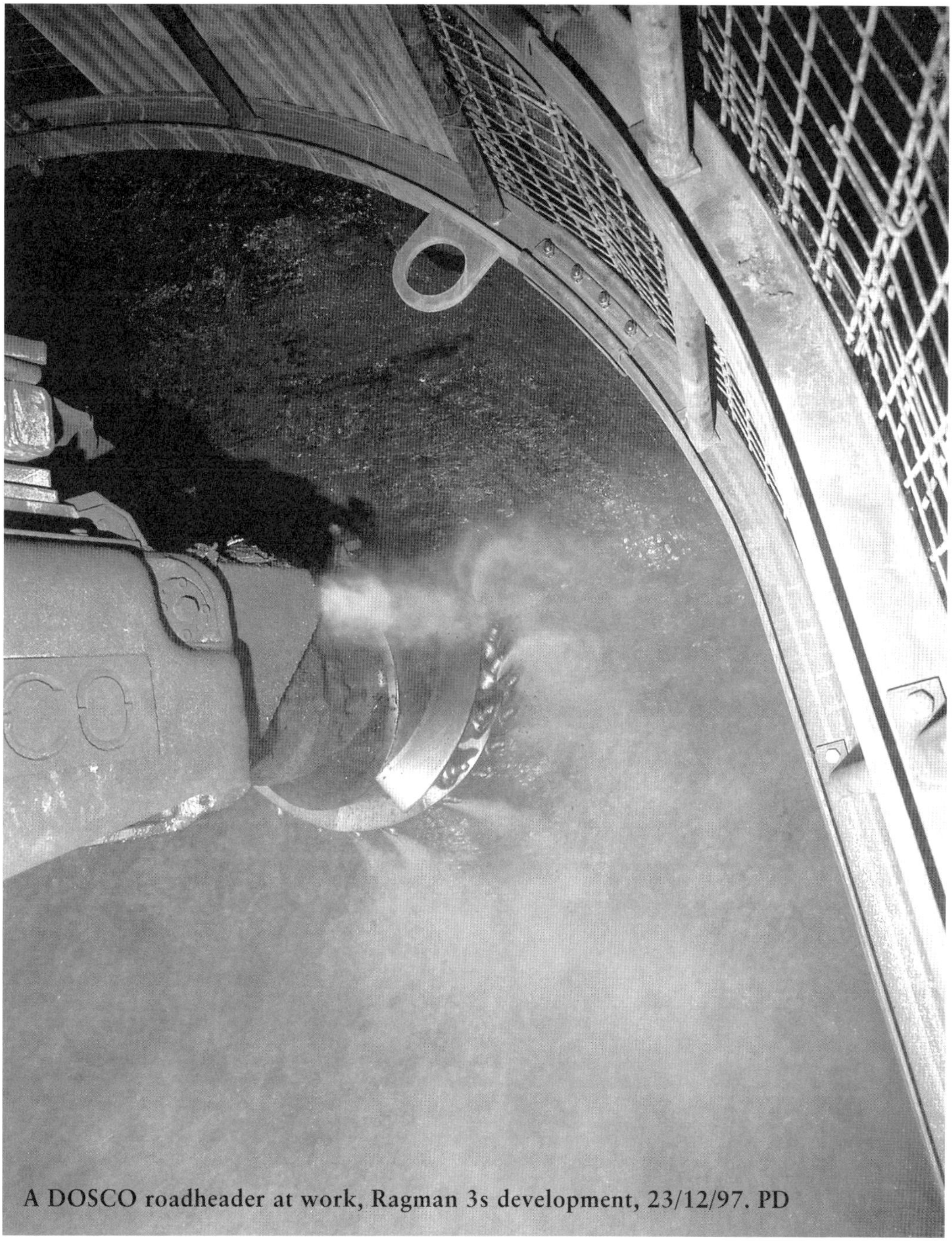

A DOSCO roadheader at work, Ragman 3s development, 23/12/97. PD

Opposite page above: Andy Mifflin by chocks in the connecting roadway between the No.1 and No.3 surface drift. Note the chocks are constructed on concrete blocks. These chocks help to resist the ever present pressure from the floor and roof, 1996. PD

Opposite page below: Group of colliers taking a break by the road-heading machine. Don Smith, Archie ?, John ? (Deputy), Pat Wooton, 23/12/97. PD

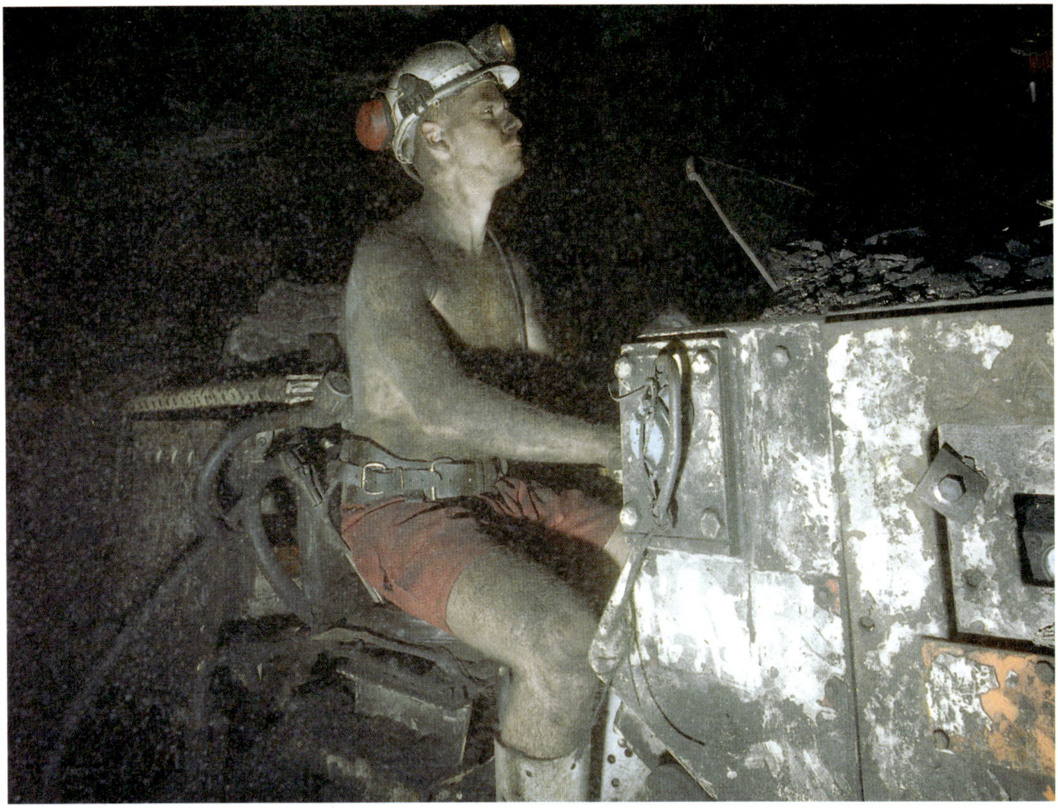

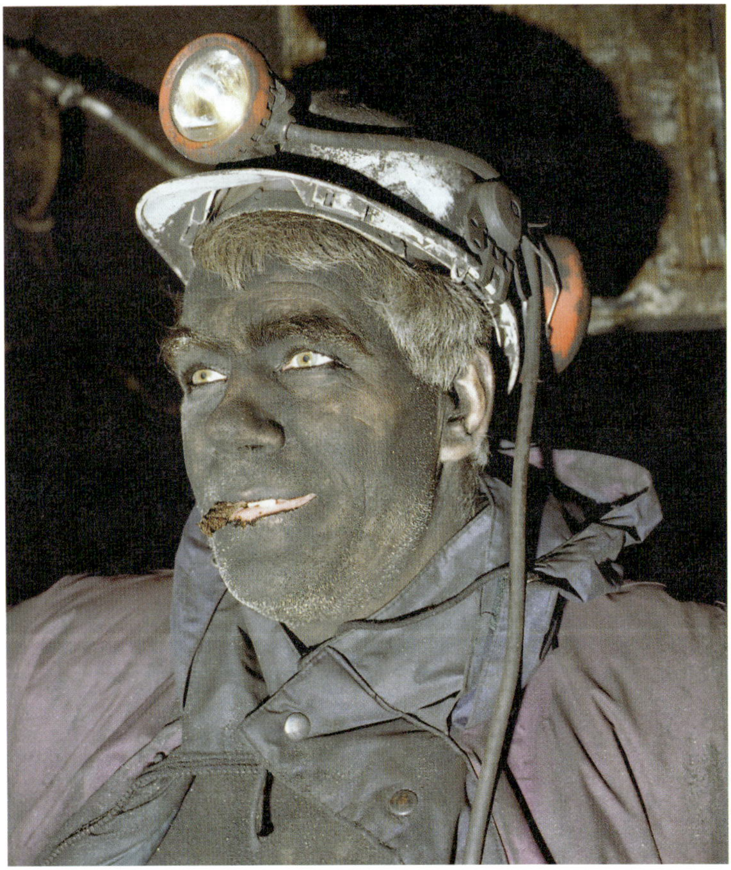

Above: Archie ? driving the DOSCO road header. Ragman 3s face development, 23 December 1997. PD

Left: Rudi Leteller chewing tobacco. He was the shearer driver on the Ragman 3s face, 22 December 1998. PD

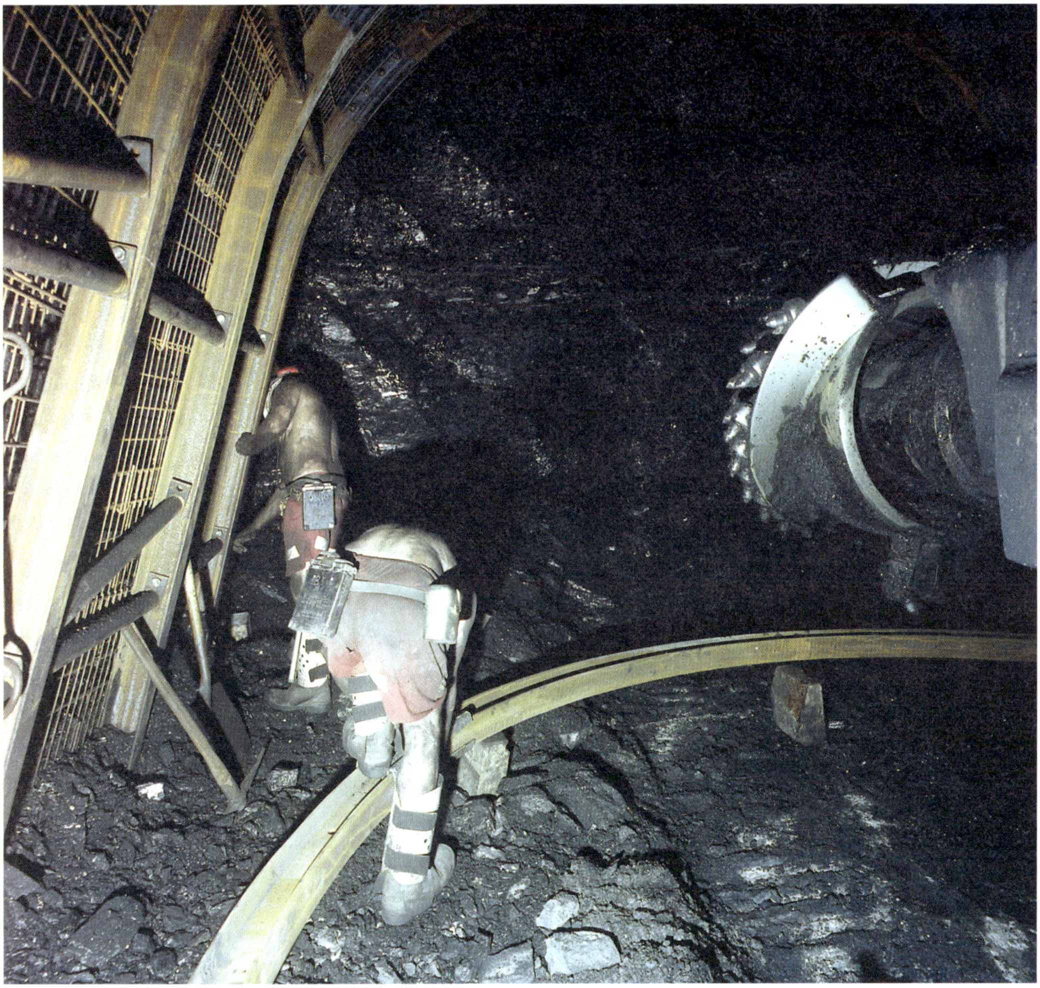

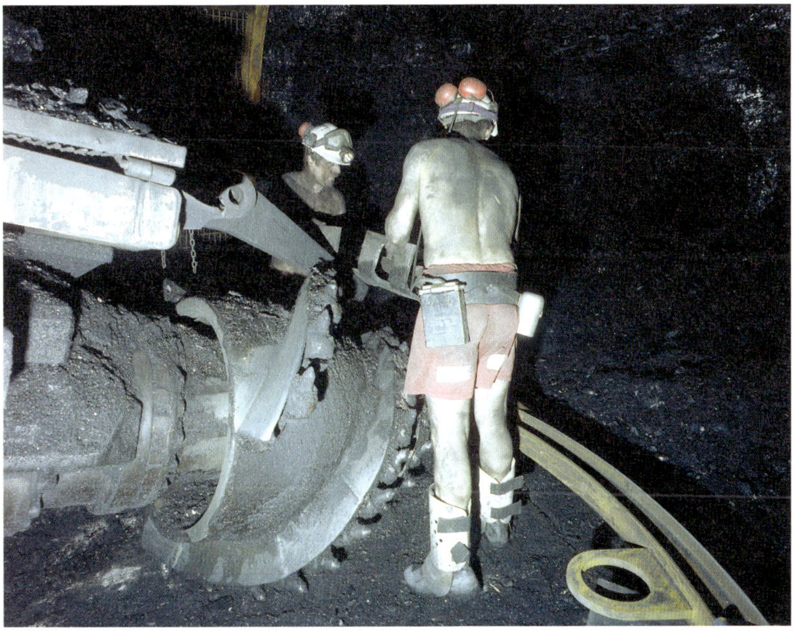

Above: Preparing an arch ready for lifting into position. Ragman 3s Development, 23 December 1997. PD
Left: Colliers attach the arch to the front of the cutting head, of the DOSCO roadheader, ready for lifting into position, 23 December 1997. PD

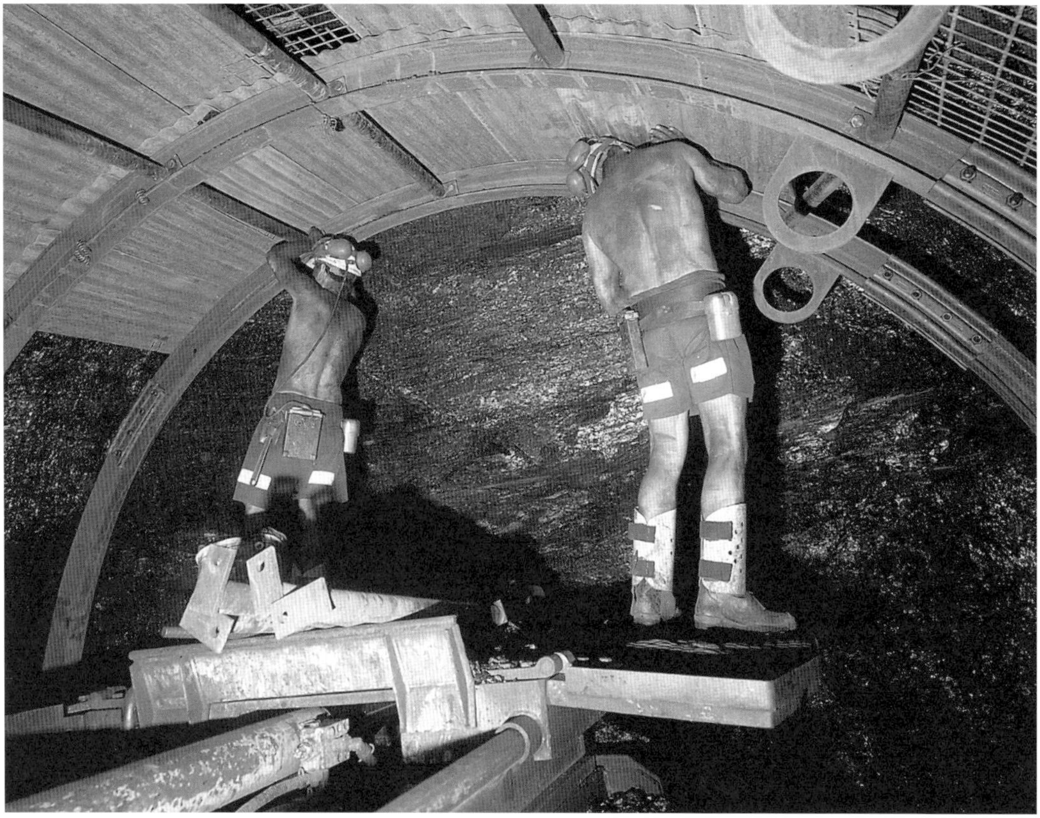

Above: Colliers standing on a platform on the DOSCO whilst setting supports. Ragman 3s Development, 23 December 1997. PD

Below: Dave Robinson and Nick Goodwin standing by the cutting head of the DOSCO roadheader. Ragman 3s Development, 23 December 1997. PD

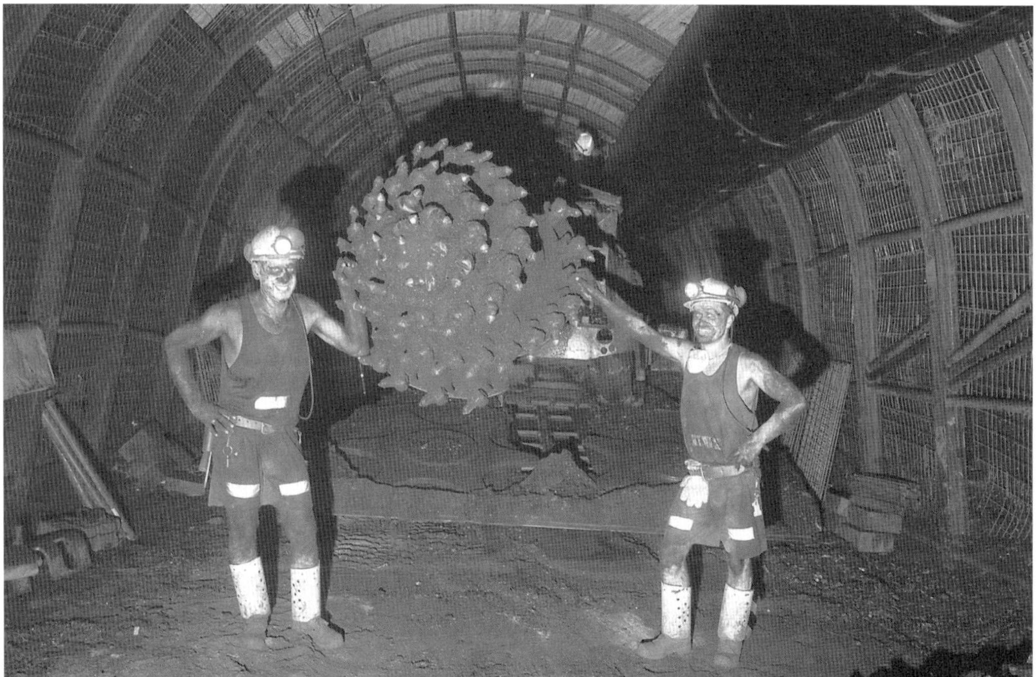

Right: Keith Tranner at work underground with his camera and flame proof flashgun.

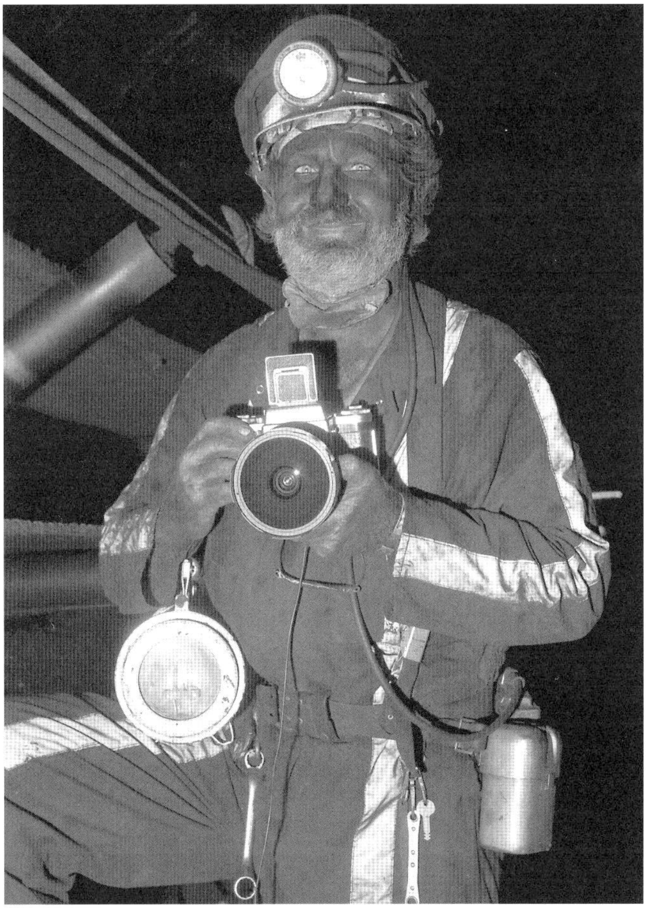

Below: The monument to Silverdale Colliery. On the plinth are all the names of the men and boys who lost their lives in the mine. 1998 PD

The colliery site is now being developed as an industrial estate.

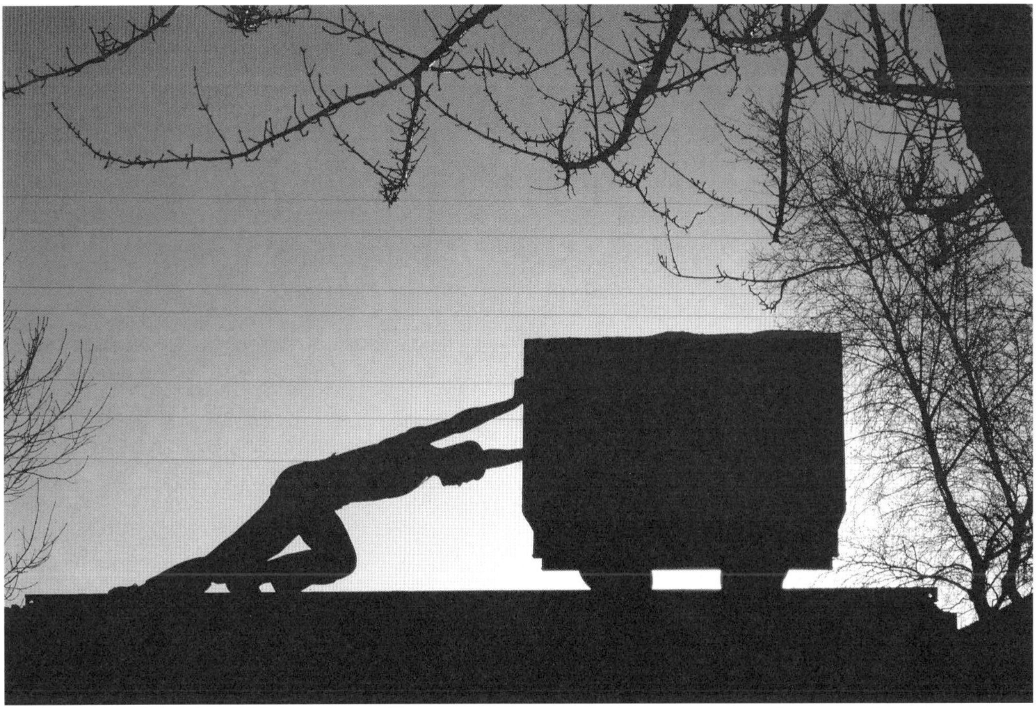

Sneyd Colliery

Sneyd Collieries Ltd.

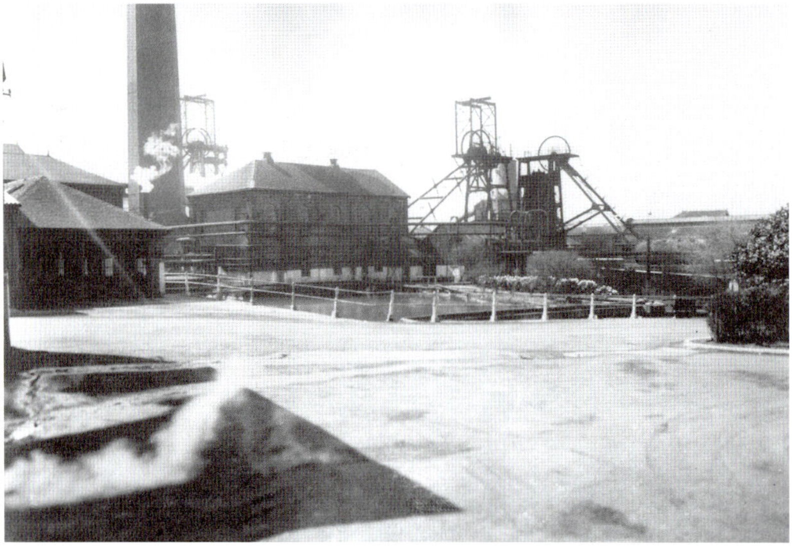

There were four shafts but only three in use at the time of closure: No. 1, 784.6m deep; No. 2, 556m; and No. 4, 807.6m. The mine was situated about 1km east of Burslem town centre. The shafts commenced sinking in 1887. It merged with Wolstanton in July 1962 and one of the shafts was retained for ventilation by Wolstanton until its closure in October 1985. On 1st January 1942, an explosion killed 57 men. It was the last time an accident claimed double-digit victims in this coalfield.

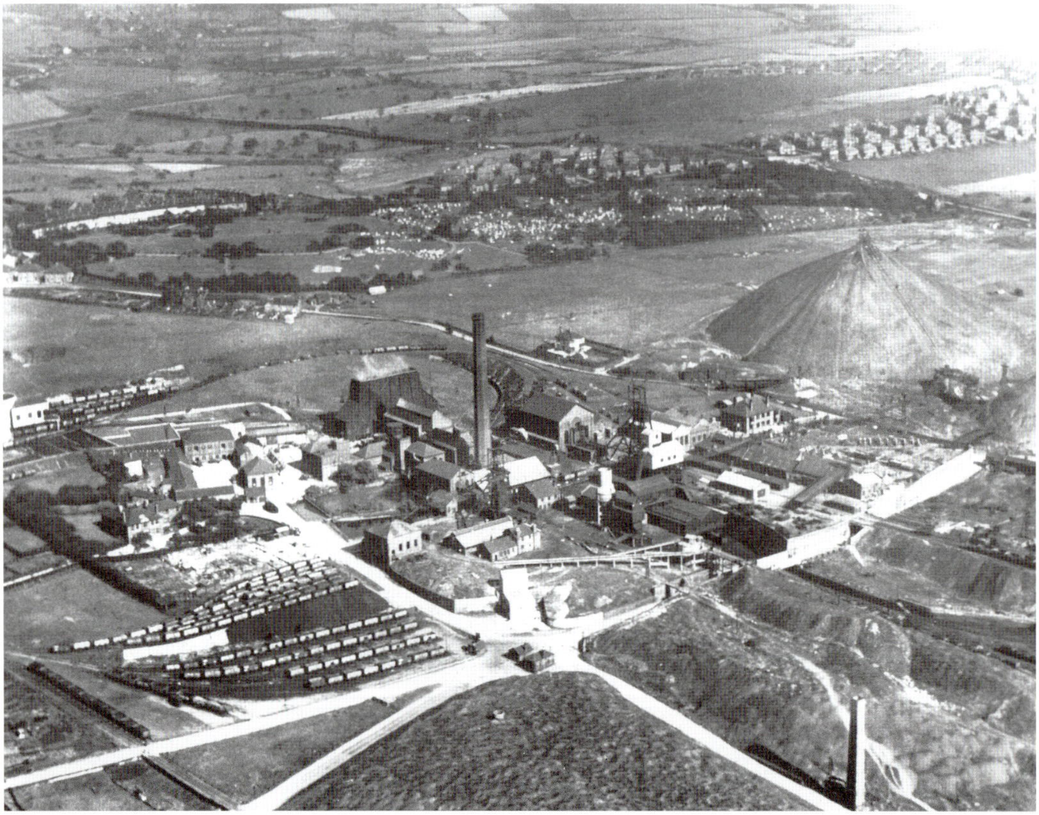

Another view of the colliery in the 1950s. MC

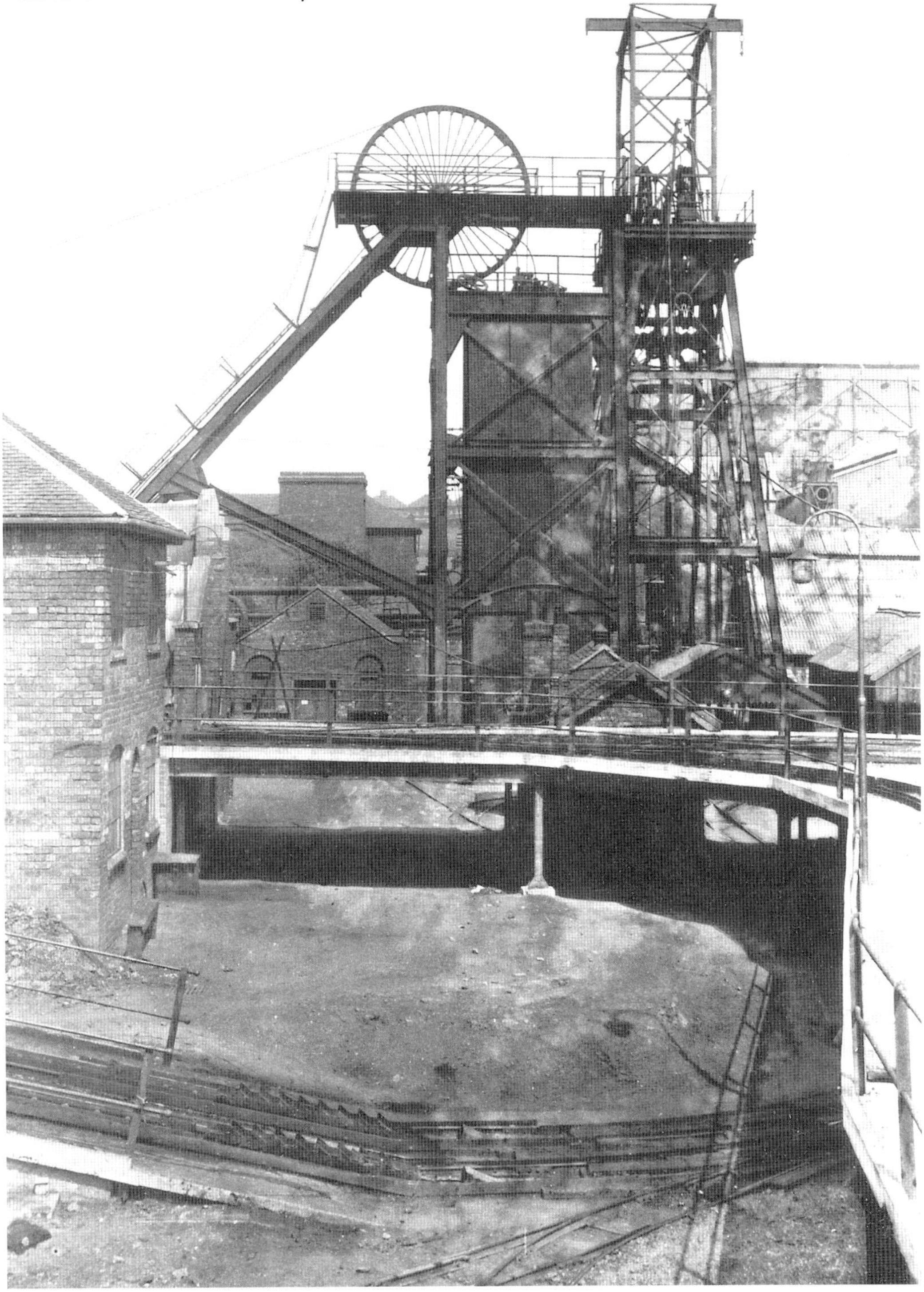

Opposite page; Top: A view of the colliery in 1951. MC

Opposite page; Bottom: An aerial photograph of the mine taken in the 1930s. Note the then almost rural setting of the mine. MC

Stafford Colliery

Stafford Coal & Iron Co. Ltd.

This colliery was situated on the Britannia Stadium site and was sunk in 1873. The mine was also known as Great Fenton or Stafford No. 2 (Stafford No. 1 being the Kemball Colliery which was situated at 500 metres to the north-east). There were two shafts: Homer being 853.1 metres deep and Sutherland 1,011.3m.

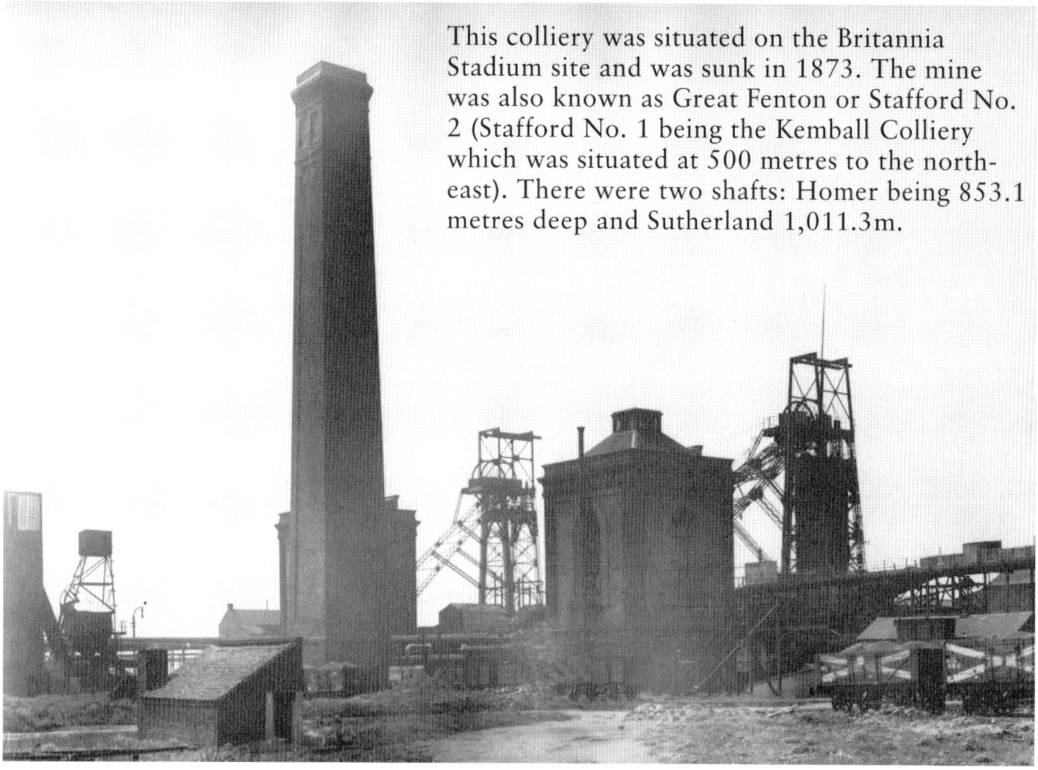

This page & opposite: General views of Stafford Colliery, taken in 1951. MC

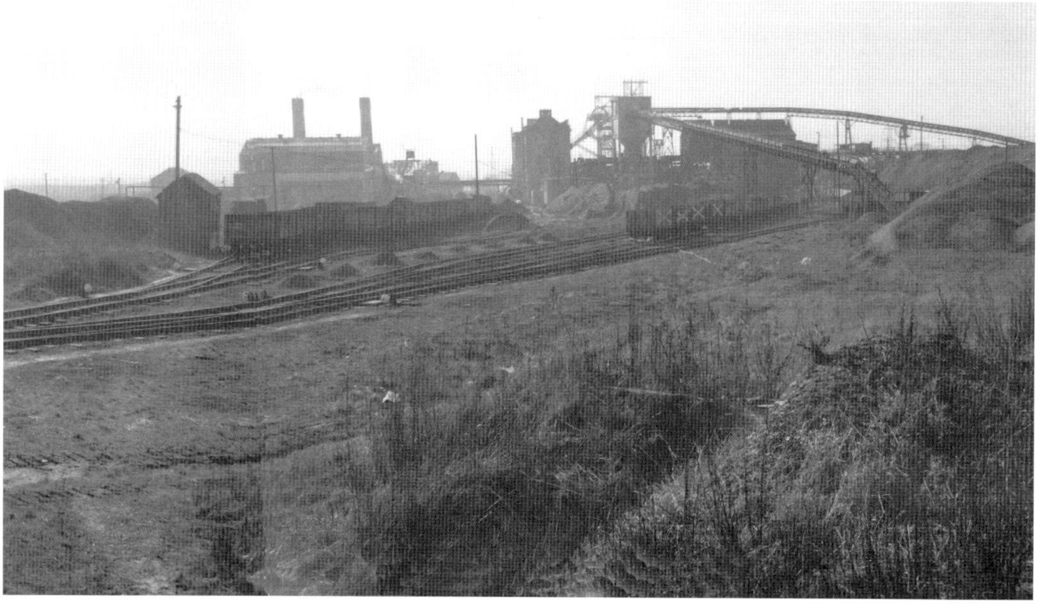

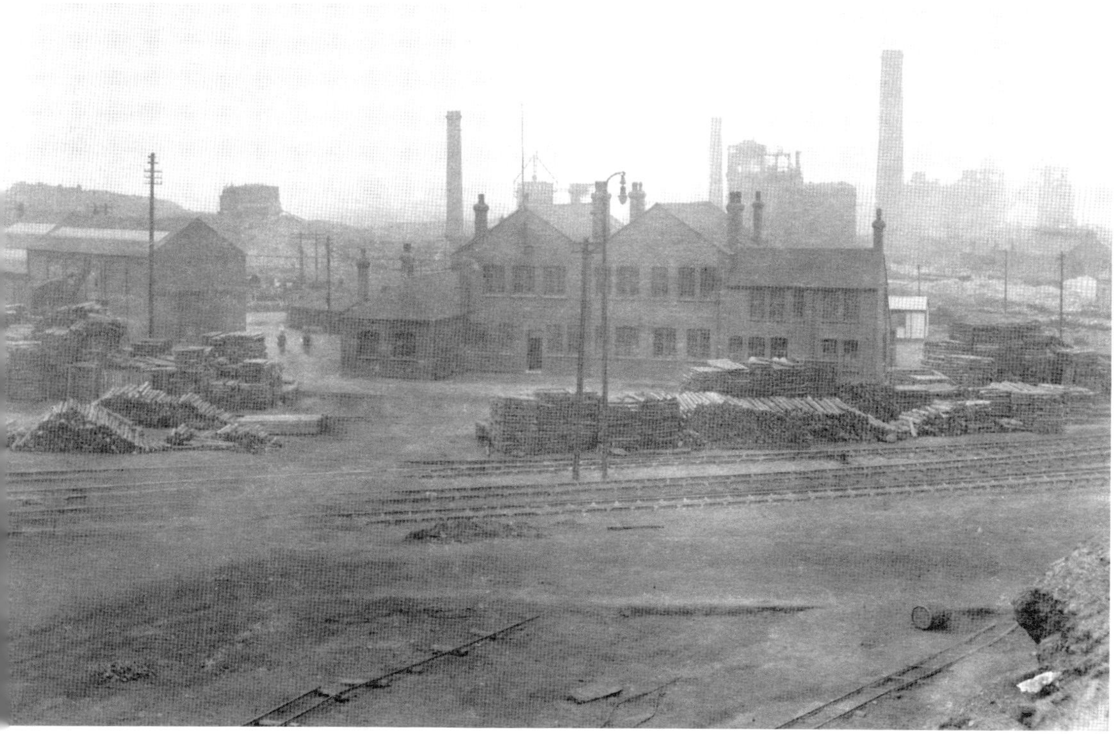

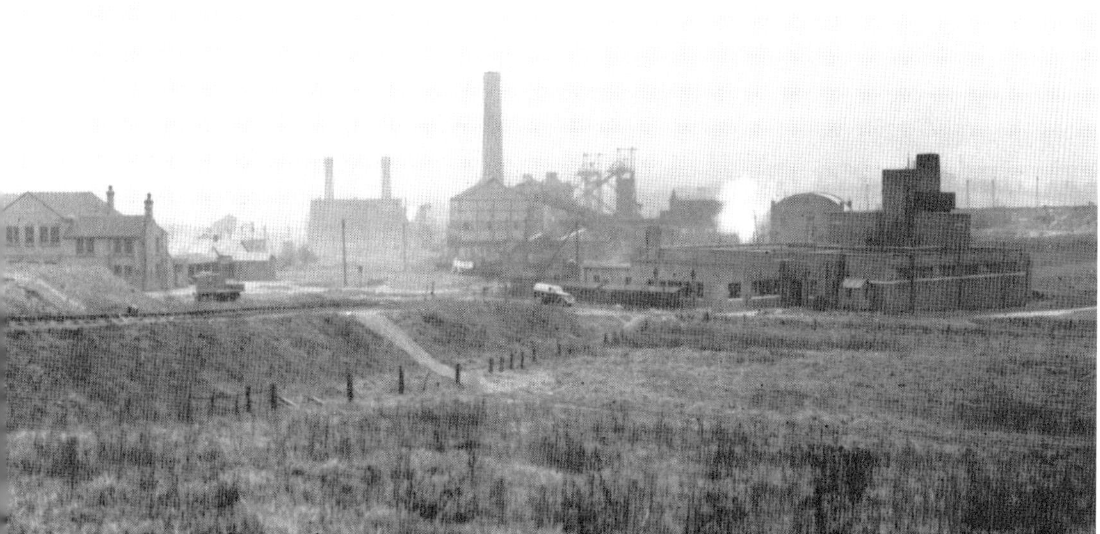

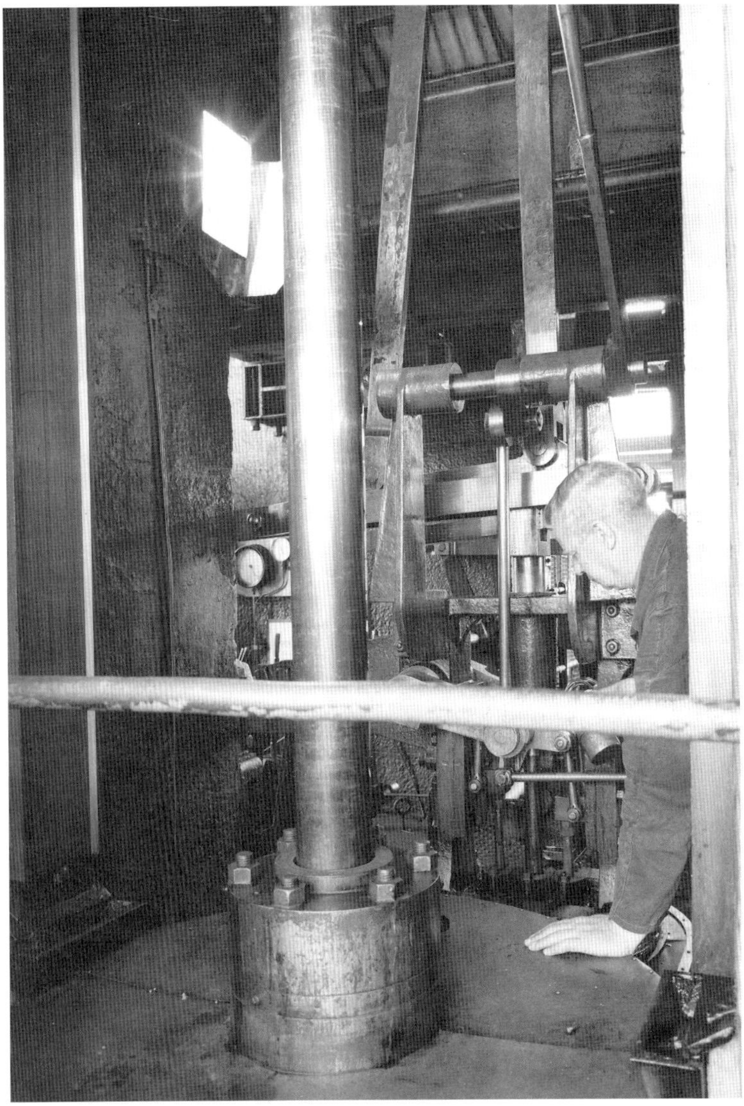

Left: One of the two piston rods on the Colliery's vertically aligned steam winder. MC

Below: The Colliery from Sideway, taken from near to the site of the incinerator. MC

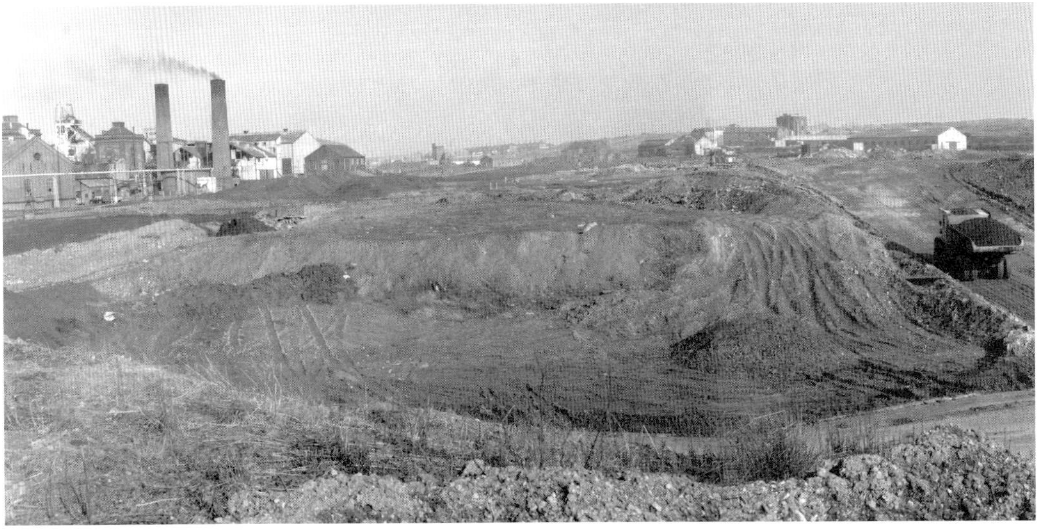

Victoria Colliery (Black Bull)

Norton & Biddulph Collieries Ltd.

Victoria Colliery was situated at Brindley Ford, about 2km south of Biddulph and immediately to the west of the main road. Sinking began in the 1850s-60s. There were five shafts: Victoria, 449.8m deep; Havelock, 451.9m; Salisbury 233.5m and two at the Brownlees site (which was situated about 350m west of the main site), 223.9m deep. The mine closed in July 1982.

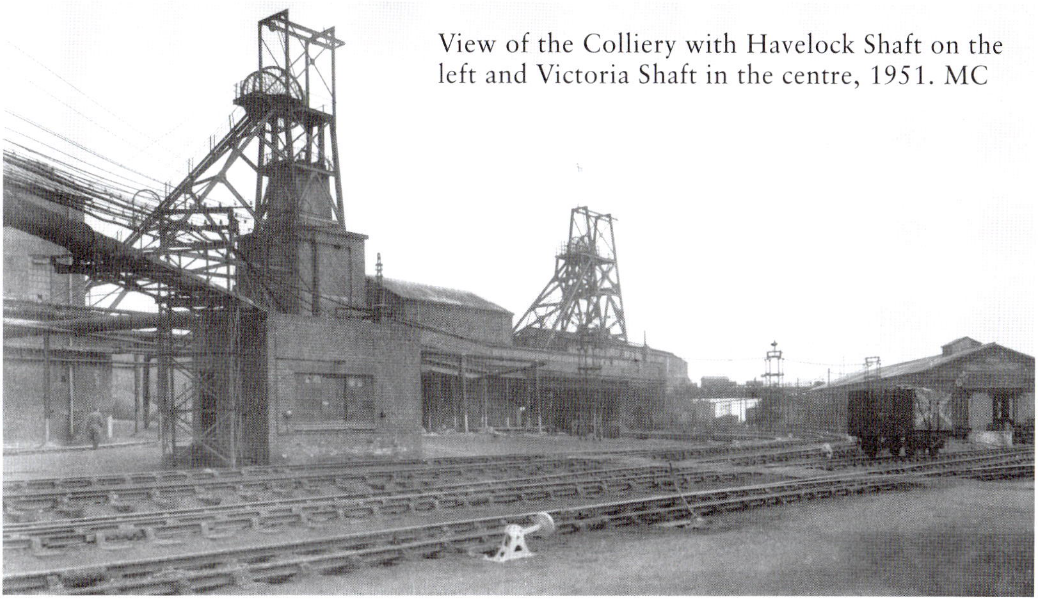

View of the Colliery with Havelock Shaft on the left and Victoria Shaft in the centre, 1951. MC

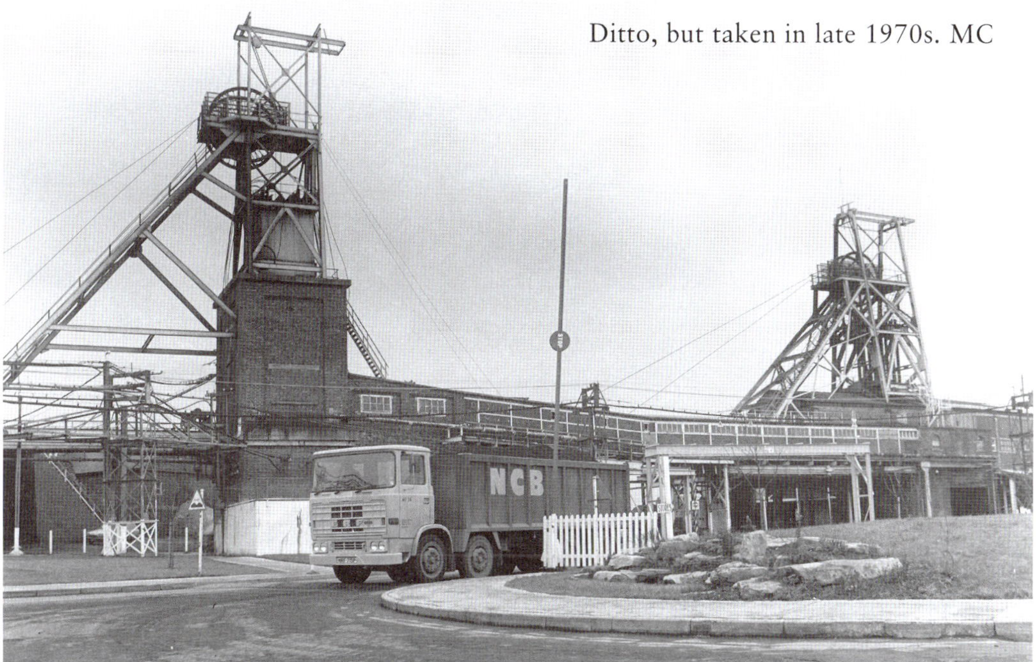

Ditto, but taken in late 1970s. MC

Salisbury Shaft headstocks. MC

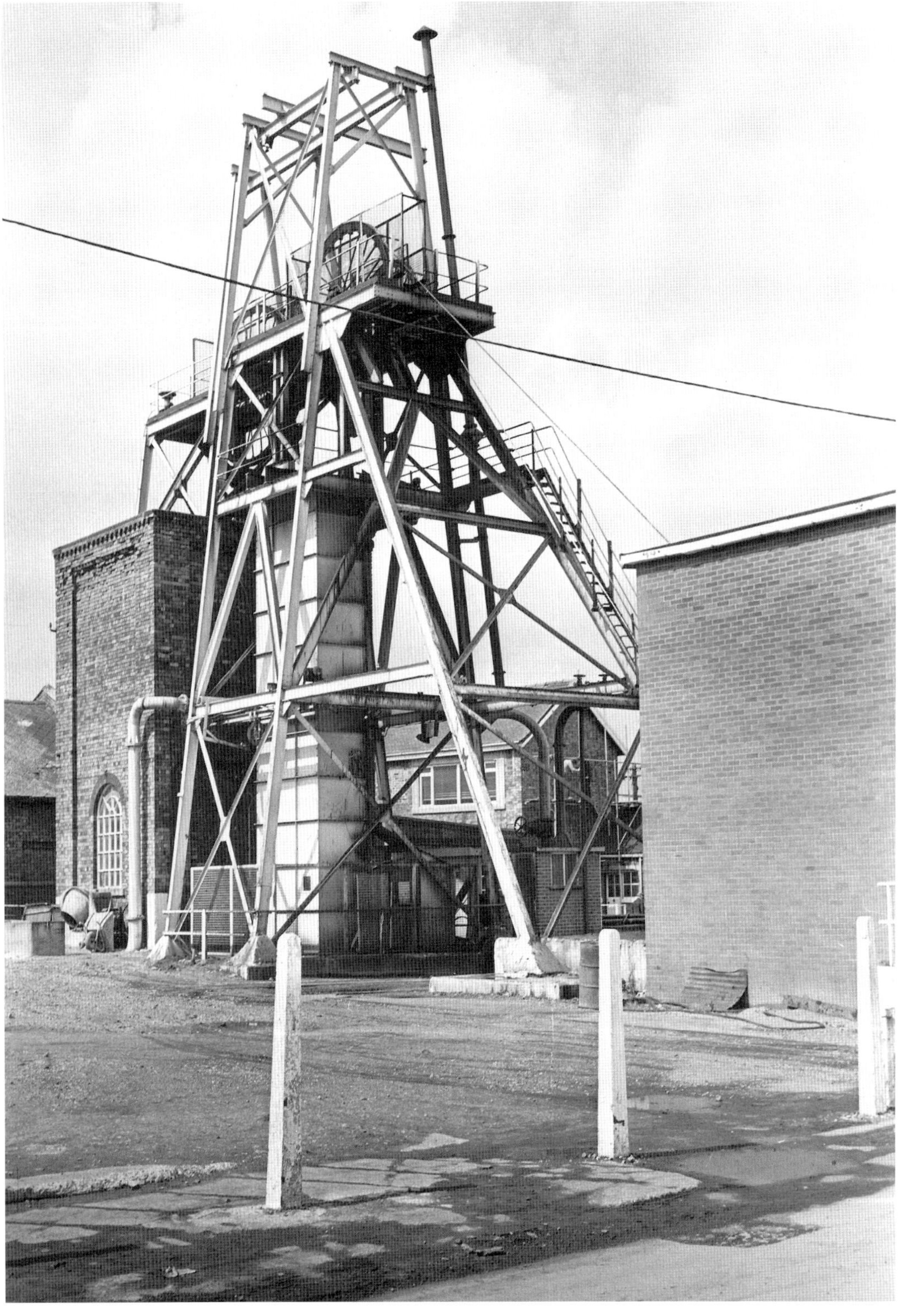

Havelock Shaft headstocks. MC

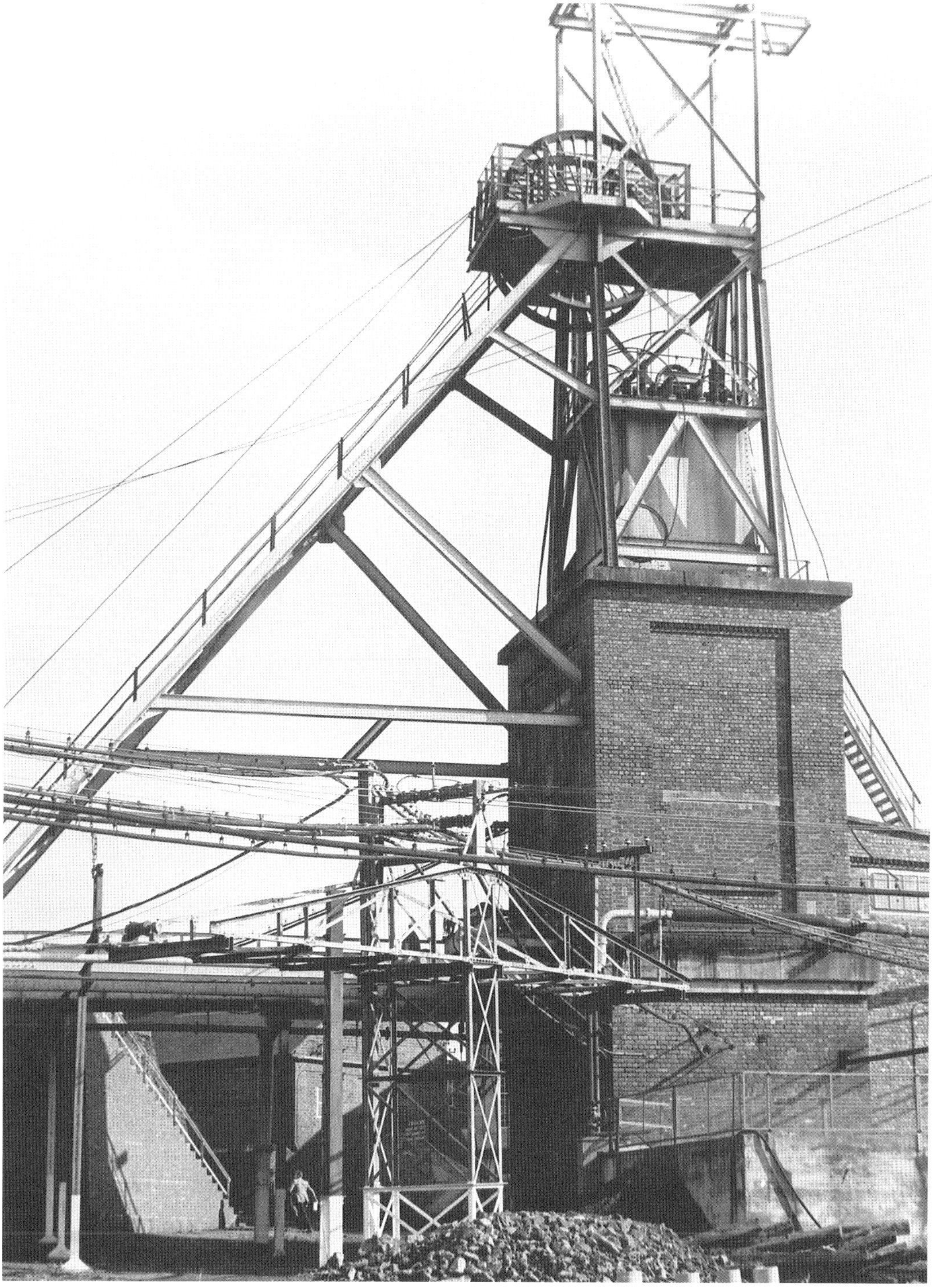

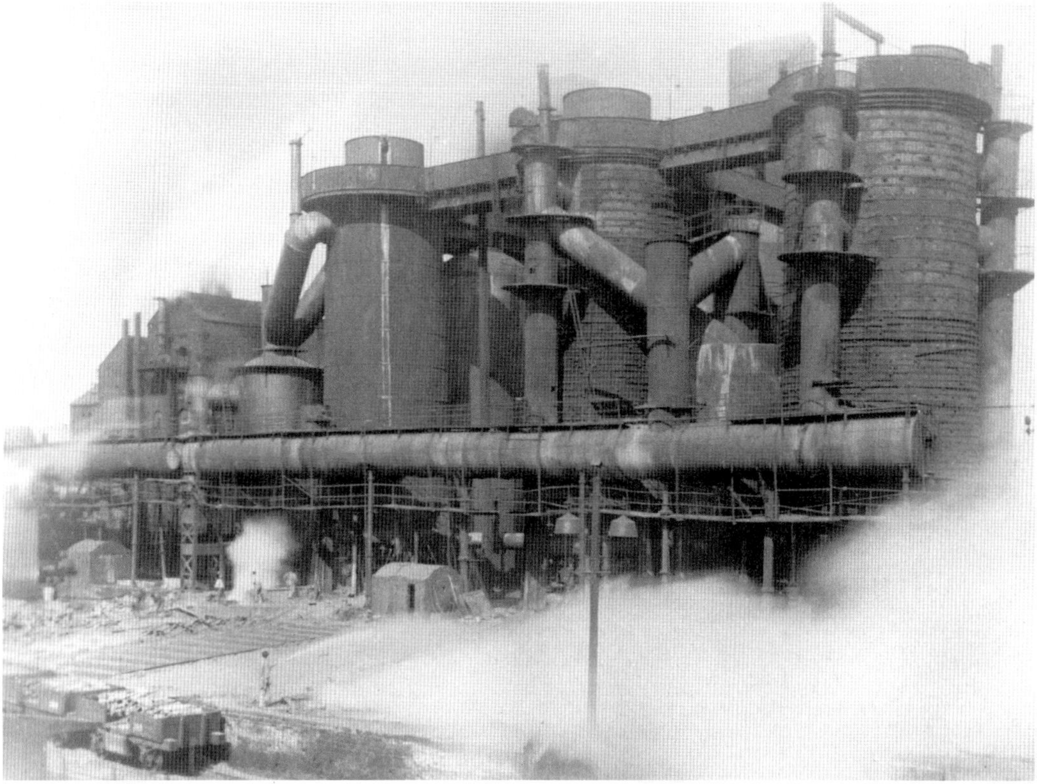

Above: Black Bull Ironworks, adjacent to the colliery, was also owned initially by Robert Heath of Biddulph Grange. The photograph shows blast furnaces and a man quenching the pig iron with water. MC

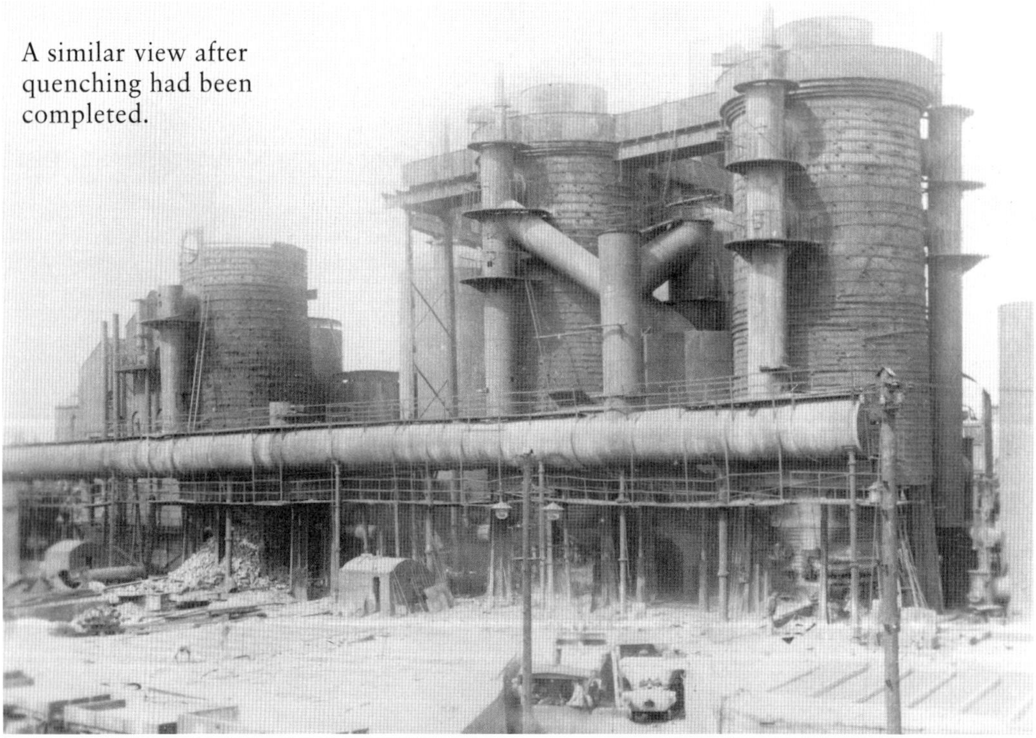

A similar view after quenching had been completed.

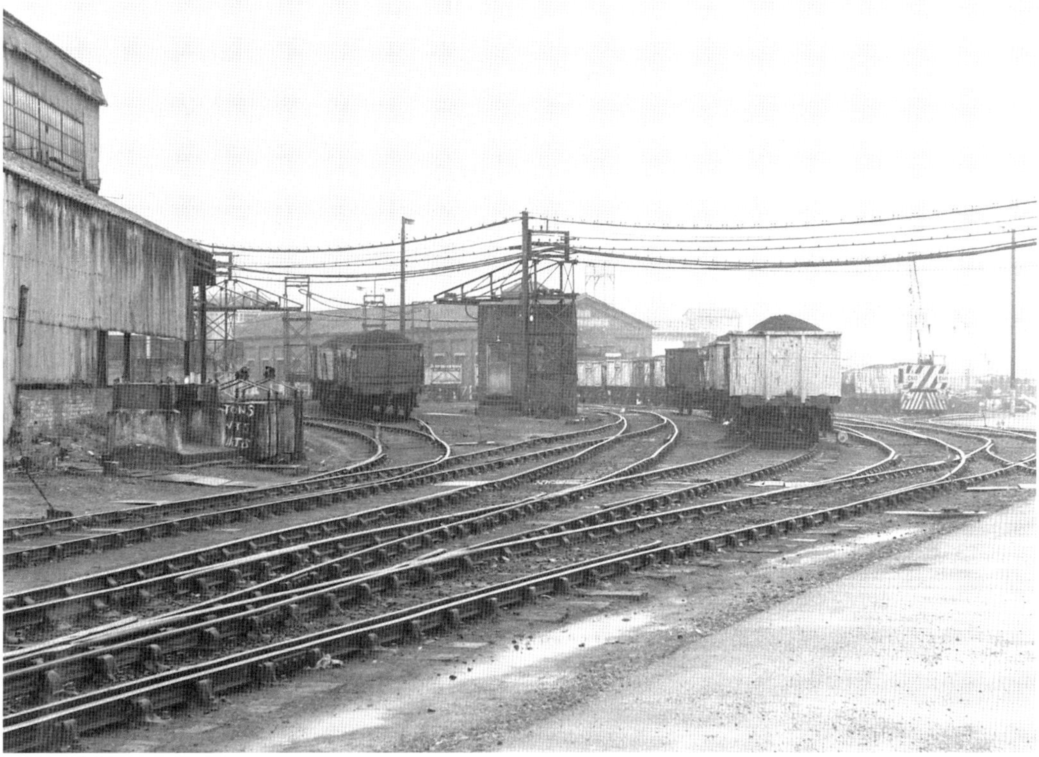

Above: View of the colliery sidings with railway waggons full of coal. MC

Below: Receipt for a sale of coal. Mr Knight bought 9 cwt (0.457 tonne) for one and six pence ($7^1/_2$p) on 9th November 1876. MC

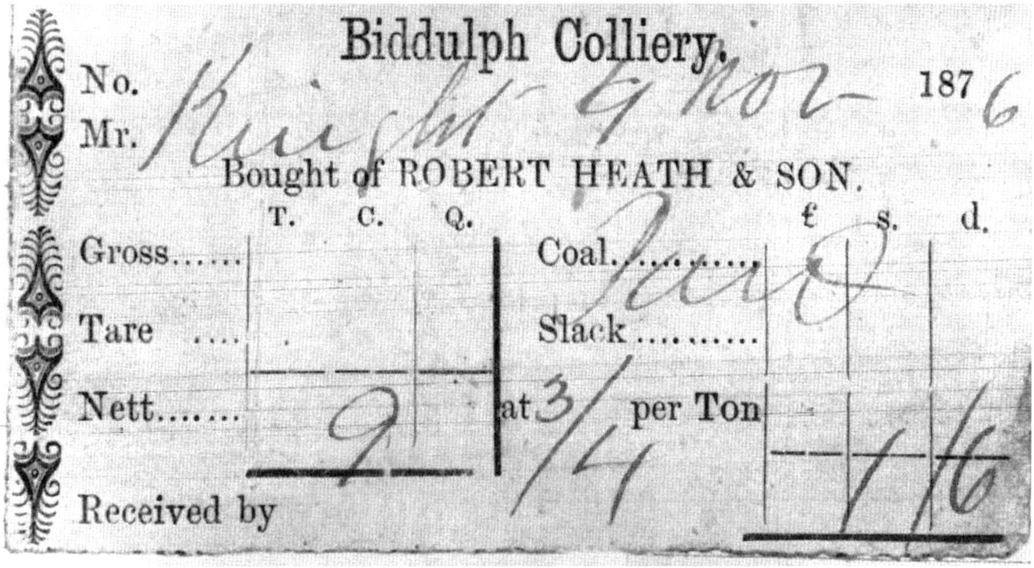

Biddulph Colliery.

No. 9 Nov 1876

Mr. Knight

Bought of ROBERT HEATH & SON.

	T.	C.	Q.		£	s.	d.
Gross......				Coal..........			
Tare				Slack..........			
Nett.......		9		at 3/4 per Ton		1	6

Received by

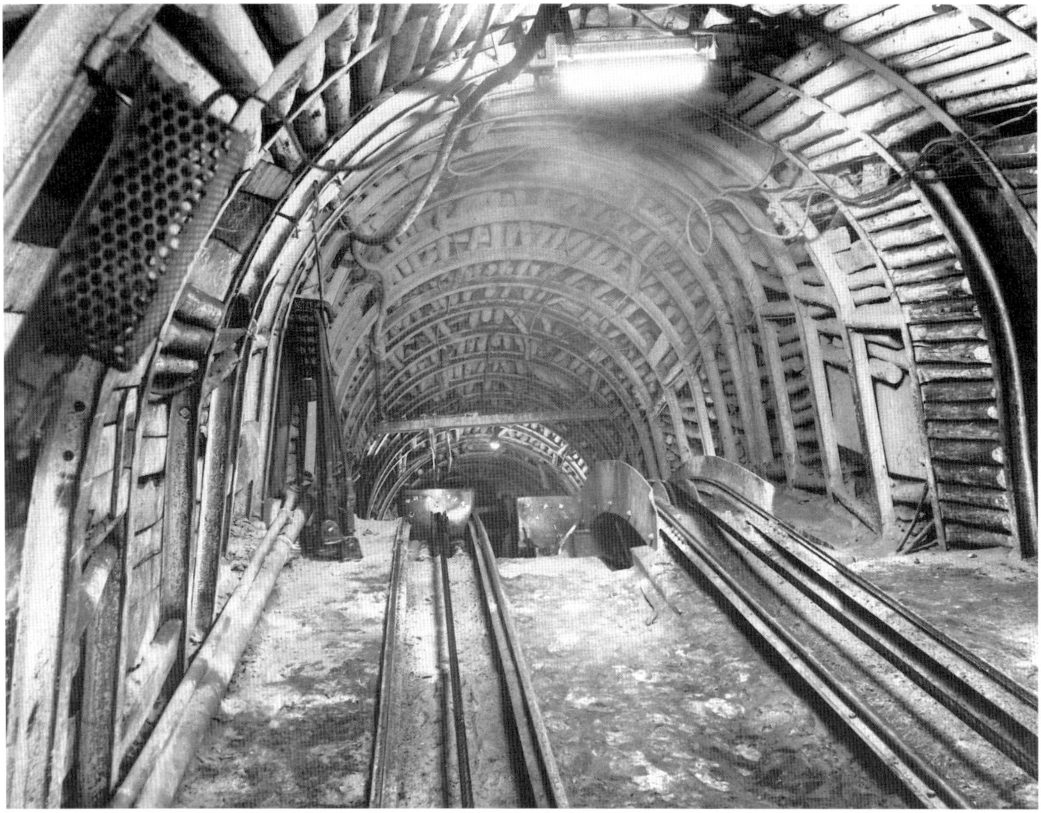

Above: Top of the Bullhurst Dip. Below: North Diamond, Loco Road. MC

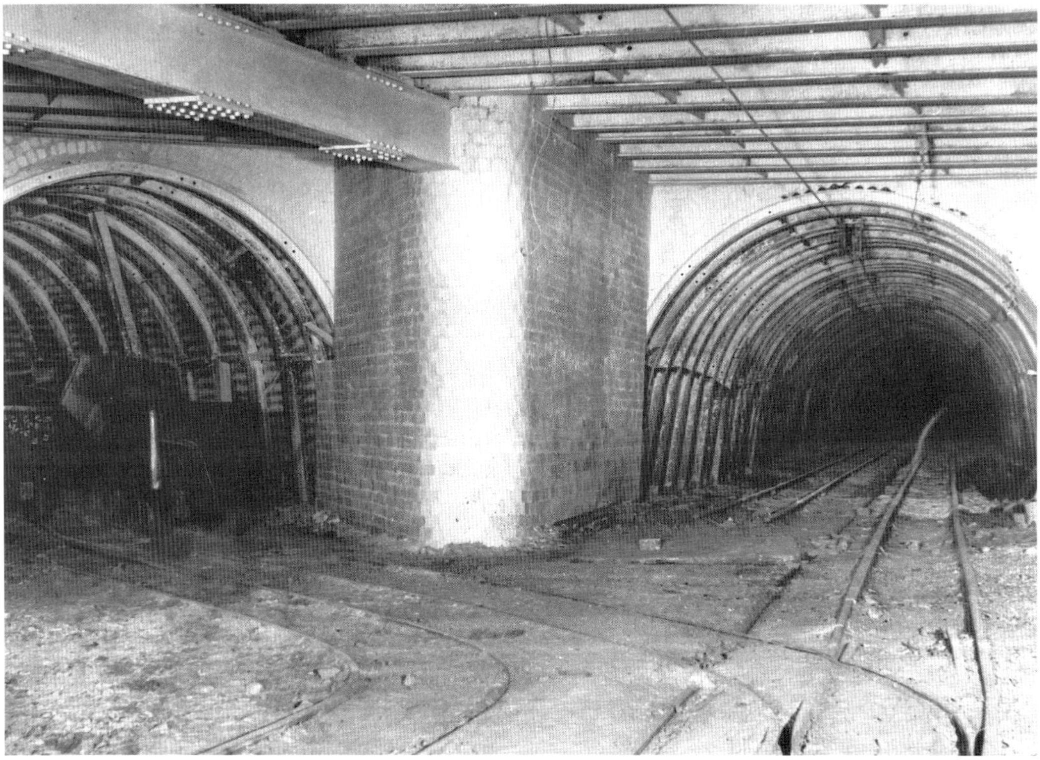

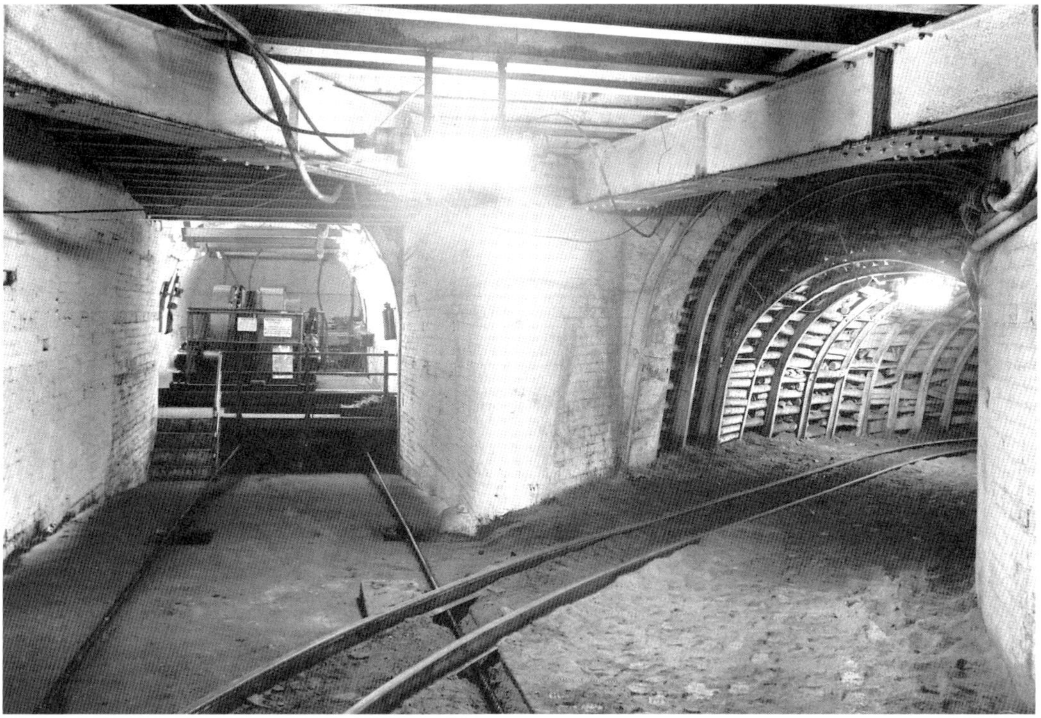

Above: Bullhurst Engine House. Note the two ropes on the floor on the left-hand side of the photo. The endless haulage engine can be seen in the left background. MC

Below: Colliers working at a heading face in the Ten Foot Seam. MC

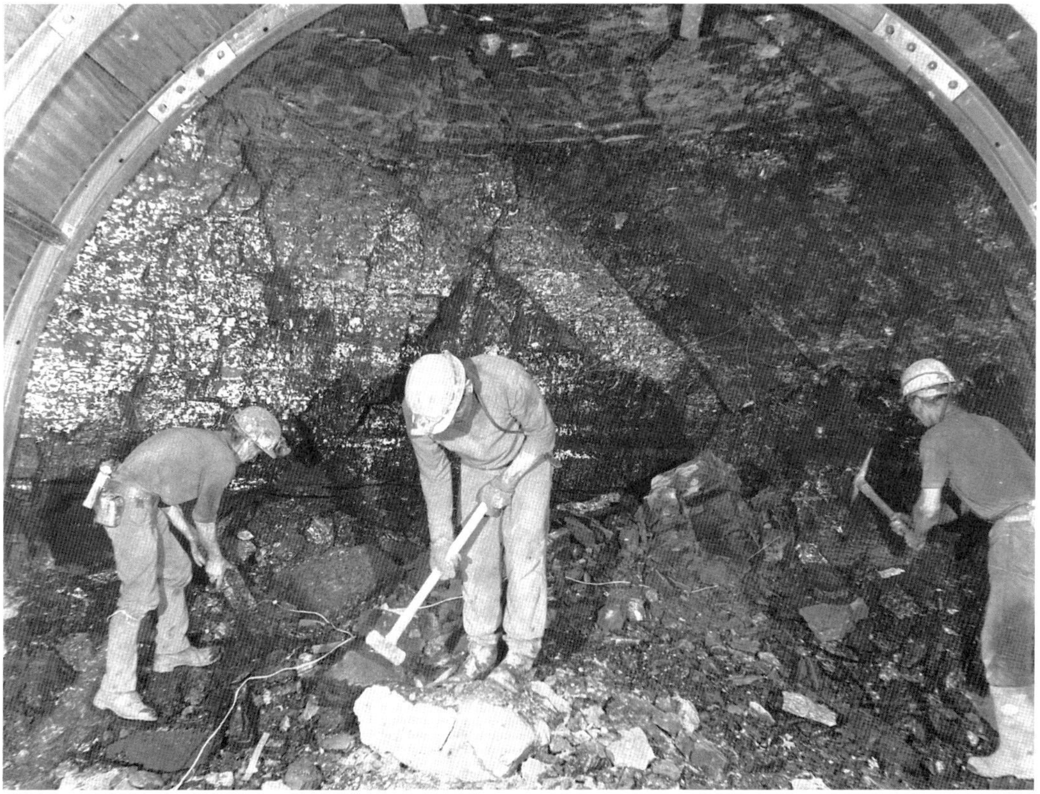

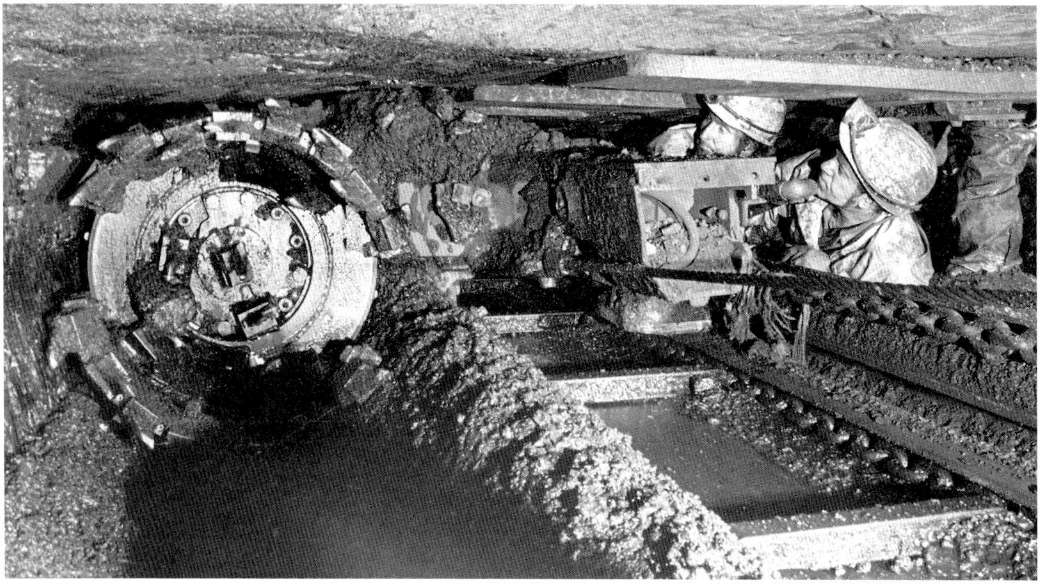

Above: Dick Archer and Johnie Jones driving a trepanner coal cutter on the South Diamond 1's coal face. The Diamond was about 0.85m thick, c. 1980. MC

Below: Ten Foot coal face. Note the self-advancing supports on the left and the coal on the AFC in the bottom right-hand corner, c. late 1970s. MC

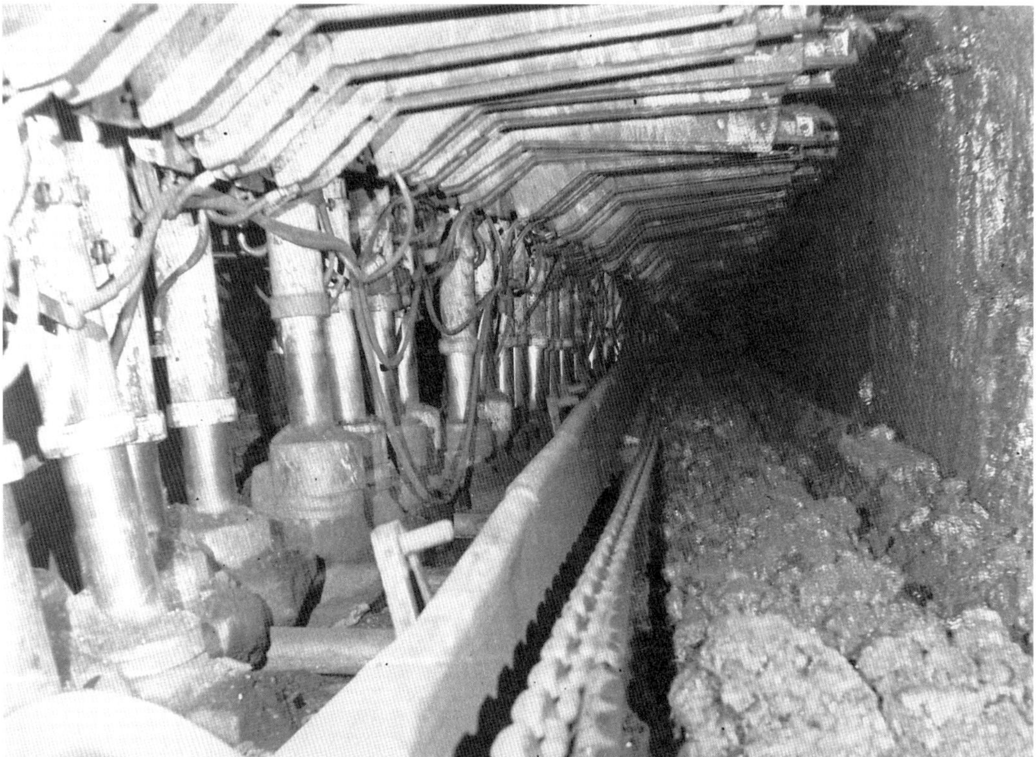

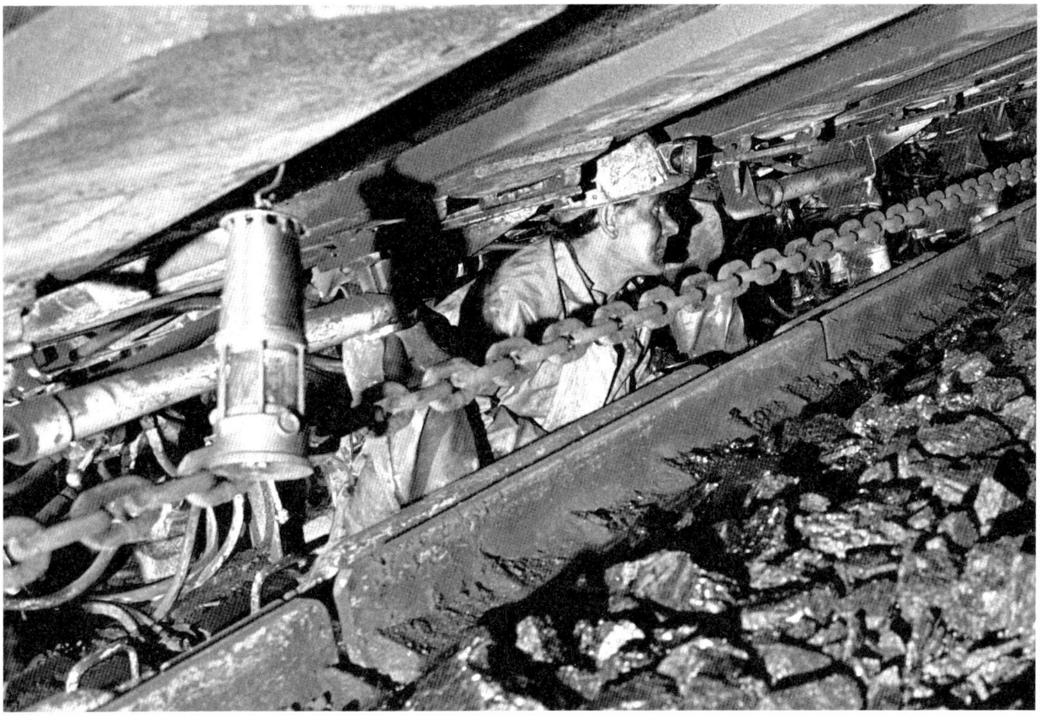

Above: Collier between the powered supports on South Diamond 1's face, c. 1980. MC

Below: Collier moving forward the self-advancing supports (chocks) on South Diamond 1's coal face, c.1980. MC

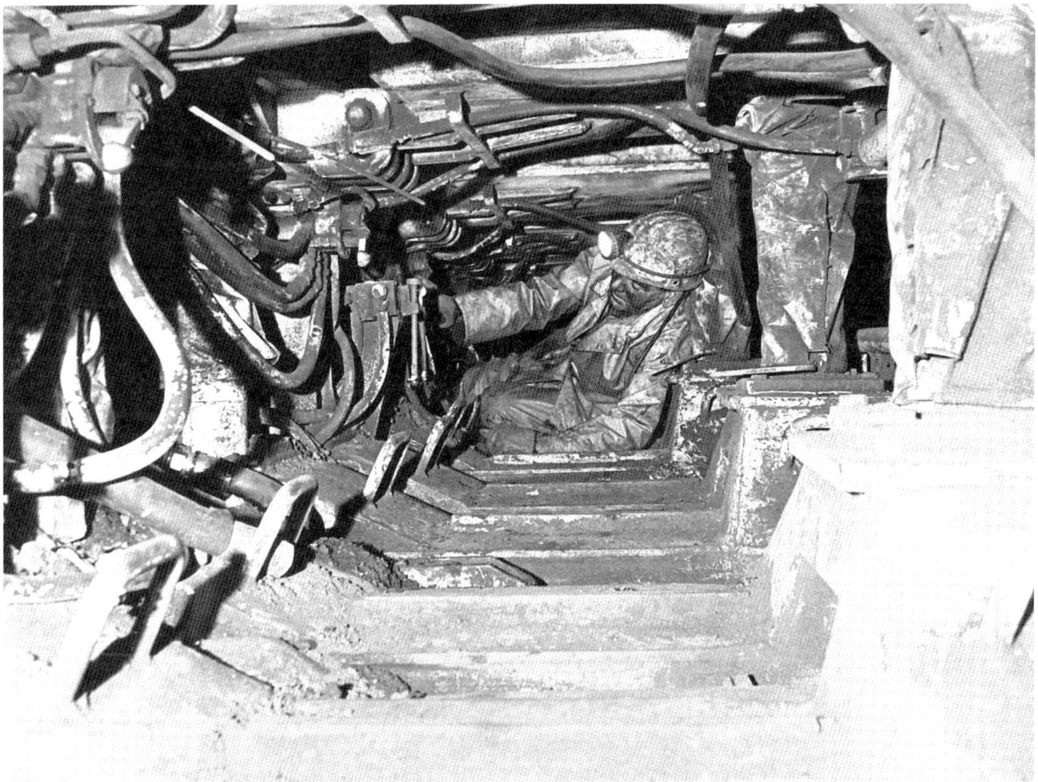

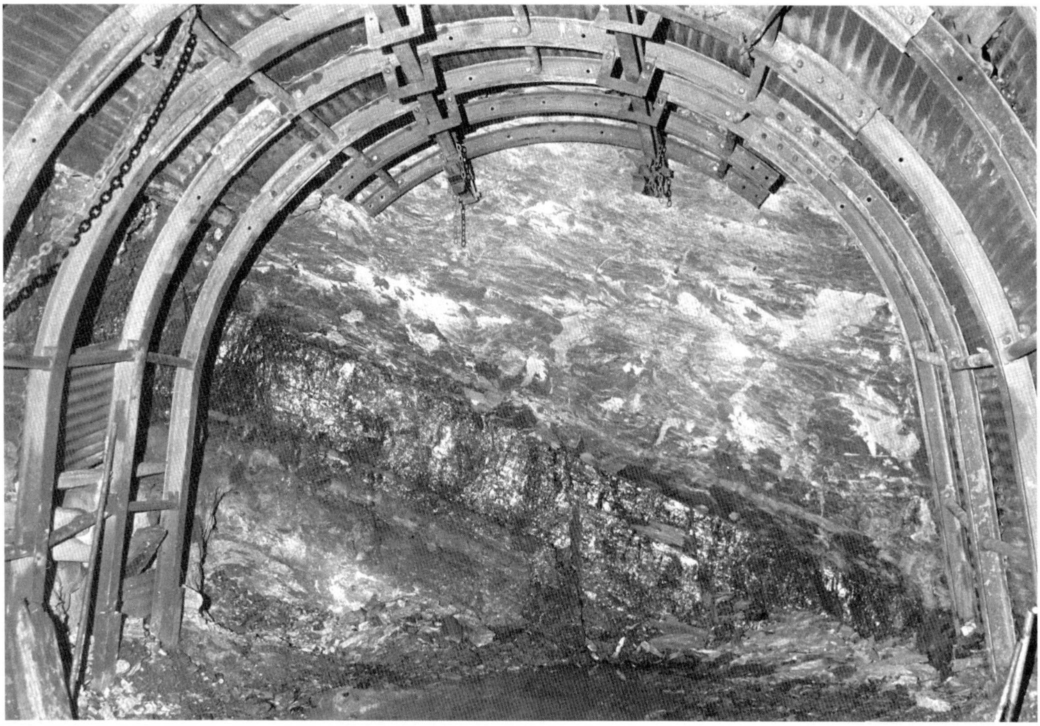

Above: View of a heading face with the Winpenny Seam dipping steeply to the right. MC
Below: Mine cars full of coal in the Victoria Shaft pit bottom. MC

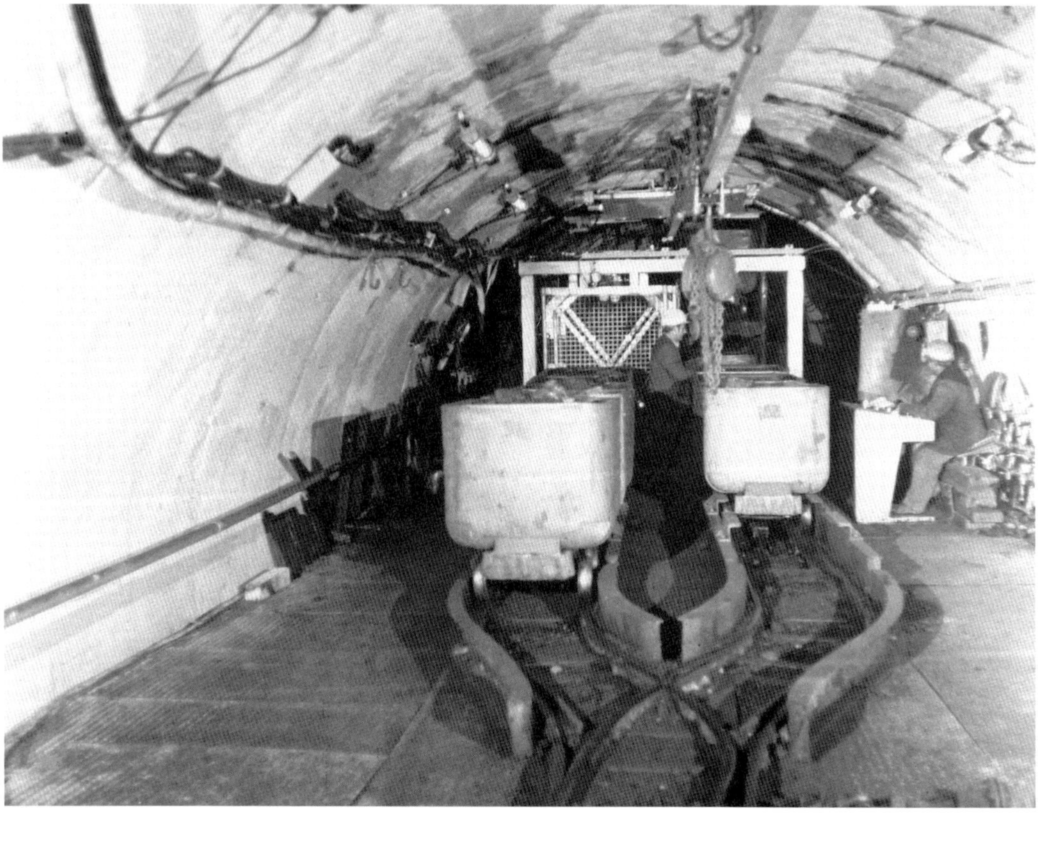

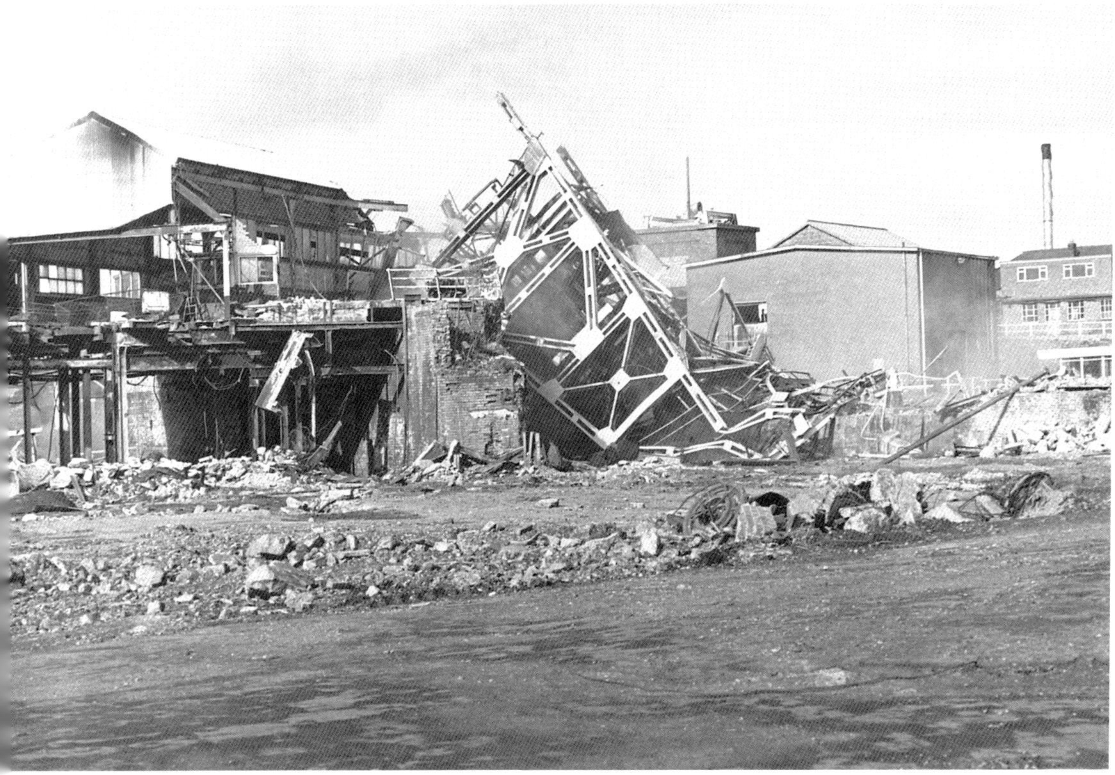

The headstocks on Victoria Shaft shortly after demolition, c.1983. MC

WOLSTANTON
Wolstanton Ltd.

Two shafts were sunk in the 1920s to work the ironstones of the Blackband series. It continued to work the ironstone until the 1940s, producing 250,000 tons of ironstone in 1940 alone. The mine also worked the Great Row coal seam. Shaft No. 2 was deepened in the 1950s-60s and a third shaft (No. 3) was also sunk at that time. The shaft depths were: No. 1, 568.4m; No. 2, 1,046m (being the deepest coal mining shaft in the British Isles); No. 3, 1,041.5m. The mine closed in October 1985.

After reconstruction the mine worked the rich coal resources below the Etruria Valley.

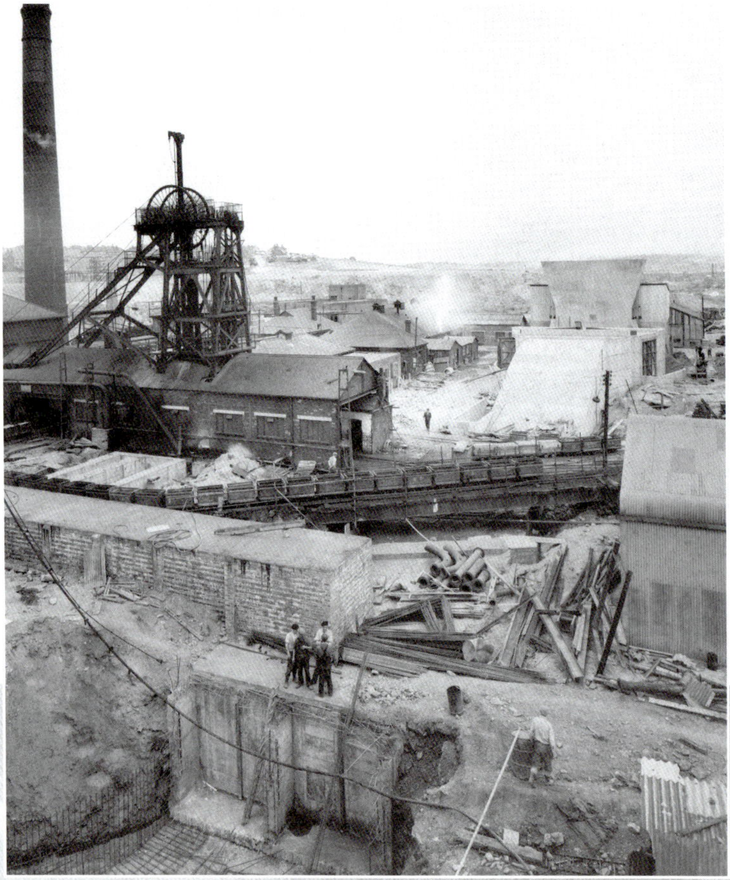

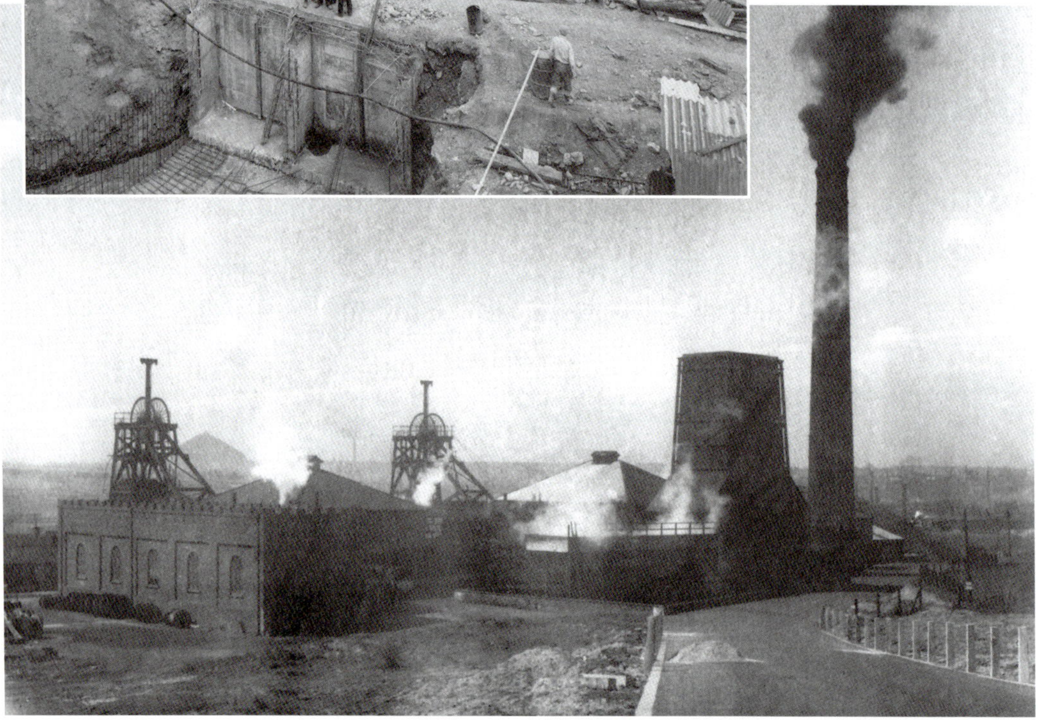

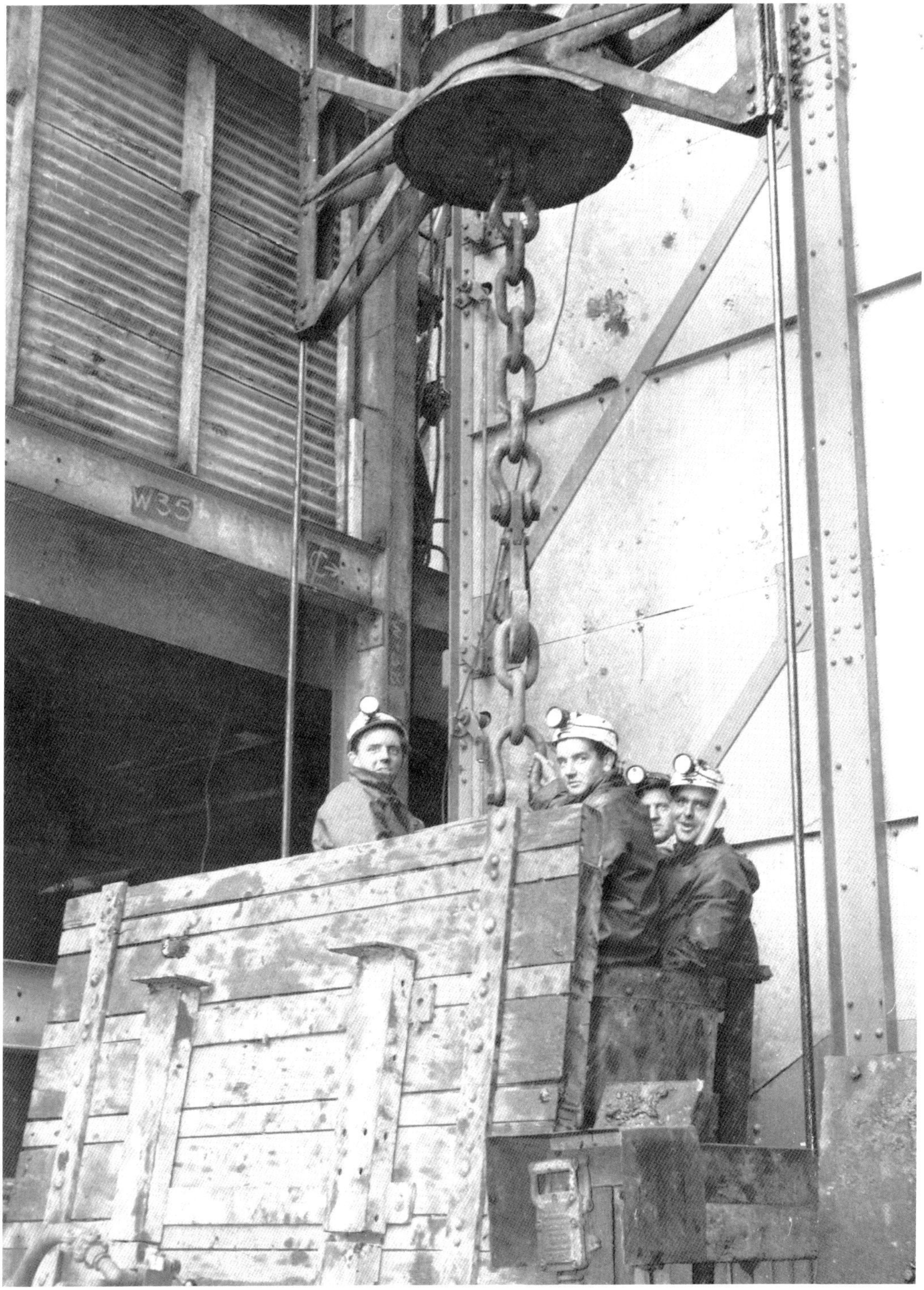

Above: A group of sinkers ready for descending the shaft in a kibble (bucket). MC

Opposite page; Top: A view of the Colliery during reconstruction. MC
Opposite page; Bottom: The Colliery in 1951 prior to reconstruction with the conical Sneyd Colliery tip in the background. MC

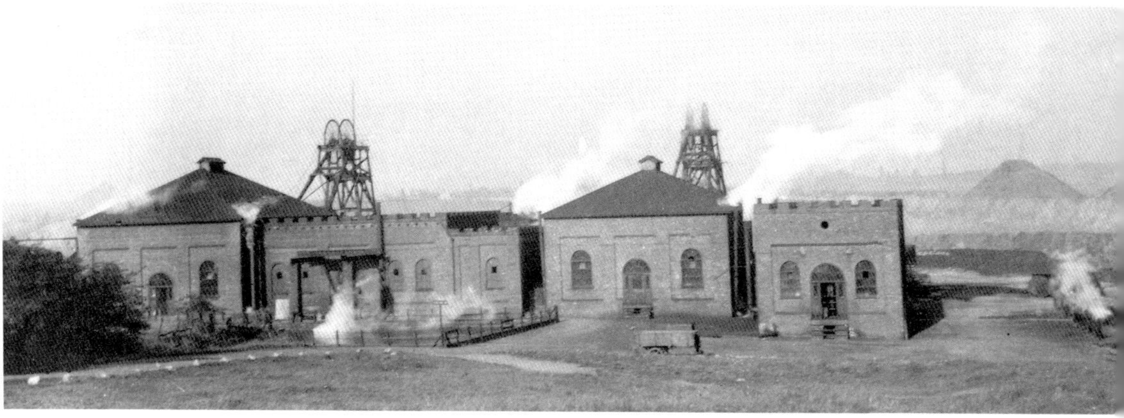

Above: The Colliery in 1951. The conical tip is at Racecourse Pit (now part of Festival Park). MC

Below: Looking over the coal preparation plant during construction. Note the conical tip at Whitfield in the distance with the Sneyd tip, on the right hand side, partially obscured by one of Shelton Bar's cooling towers. MC

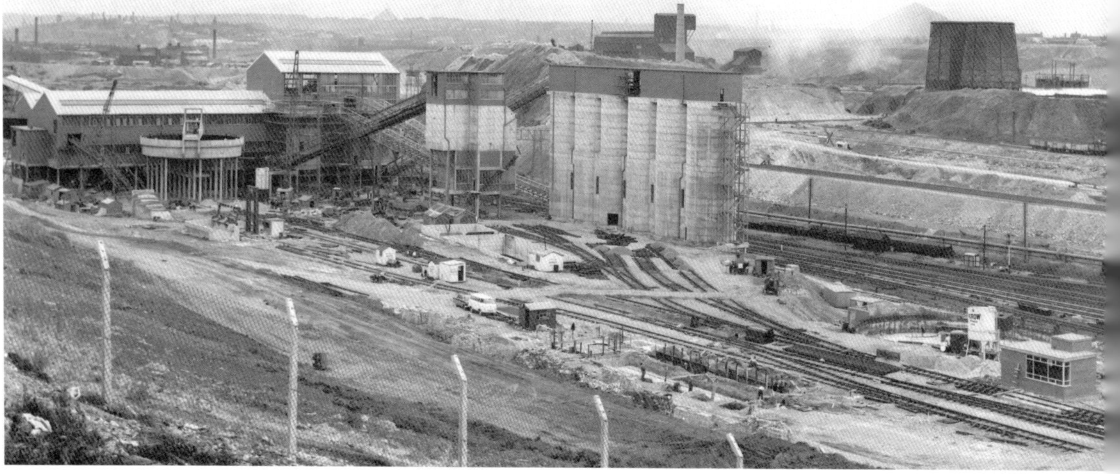

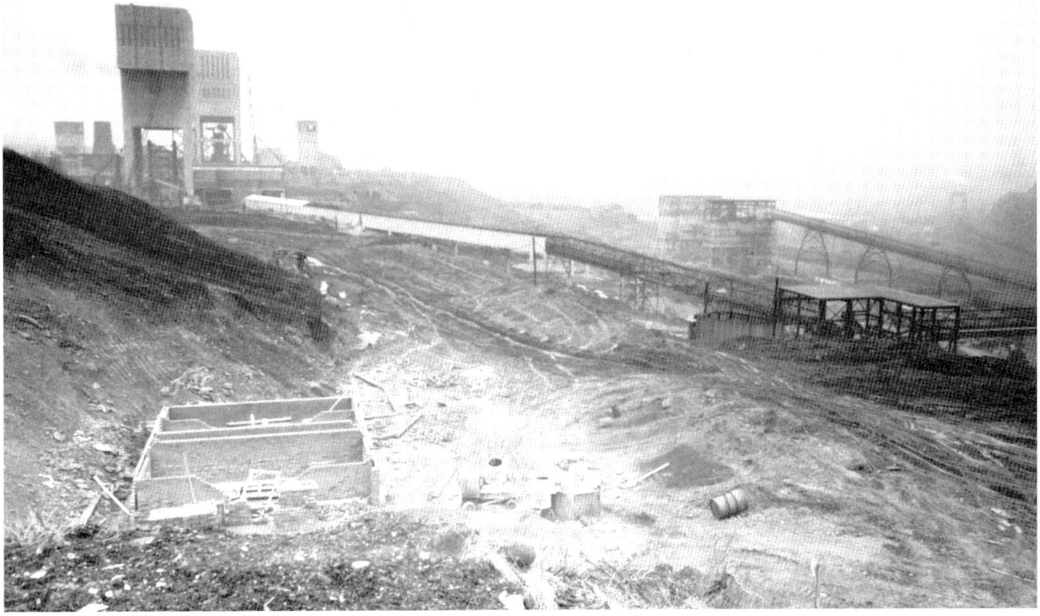

Above: A view of the Colliery looking north from what is now the D-road. No. 3 shaft is in the foreground with the winding tower of No. 2 shaft immediately behind it. MC

Below: A view looking south along the Etruria valley, taken from the No. 2 shaft tower. No. 3 shaft tower on the right hand side. Note Shelton Bar on the left and the gasometers in the south. The D-road now runs between the mine and the coal preparation plant, in the middle of the photograph. MC

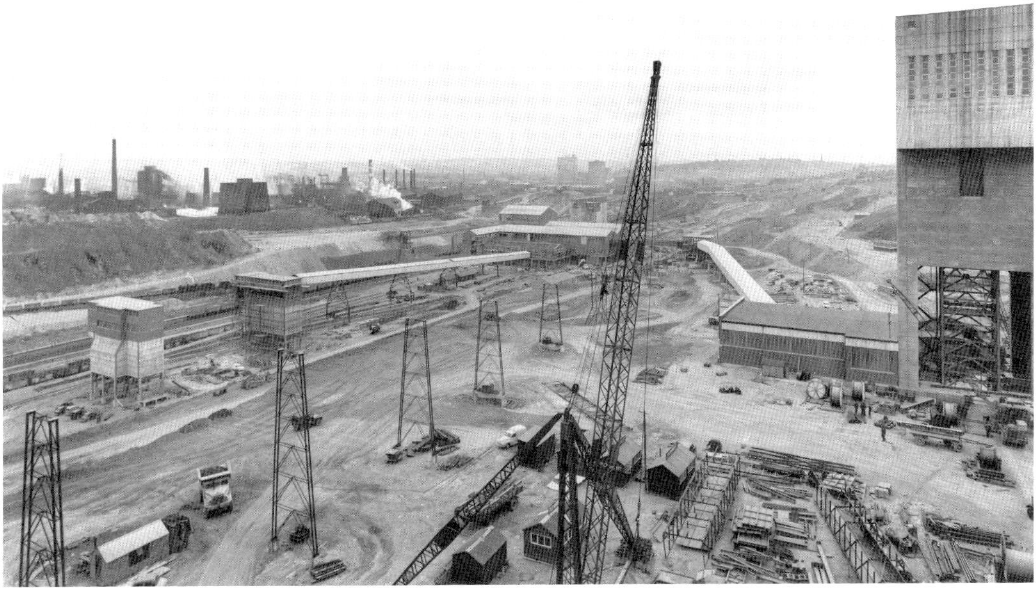

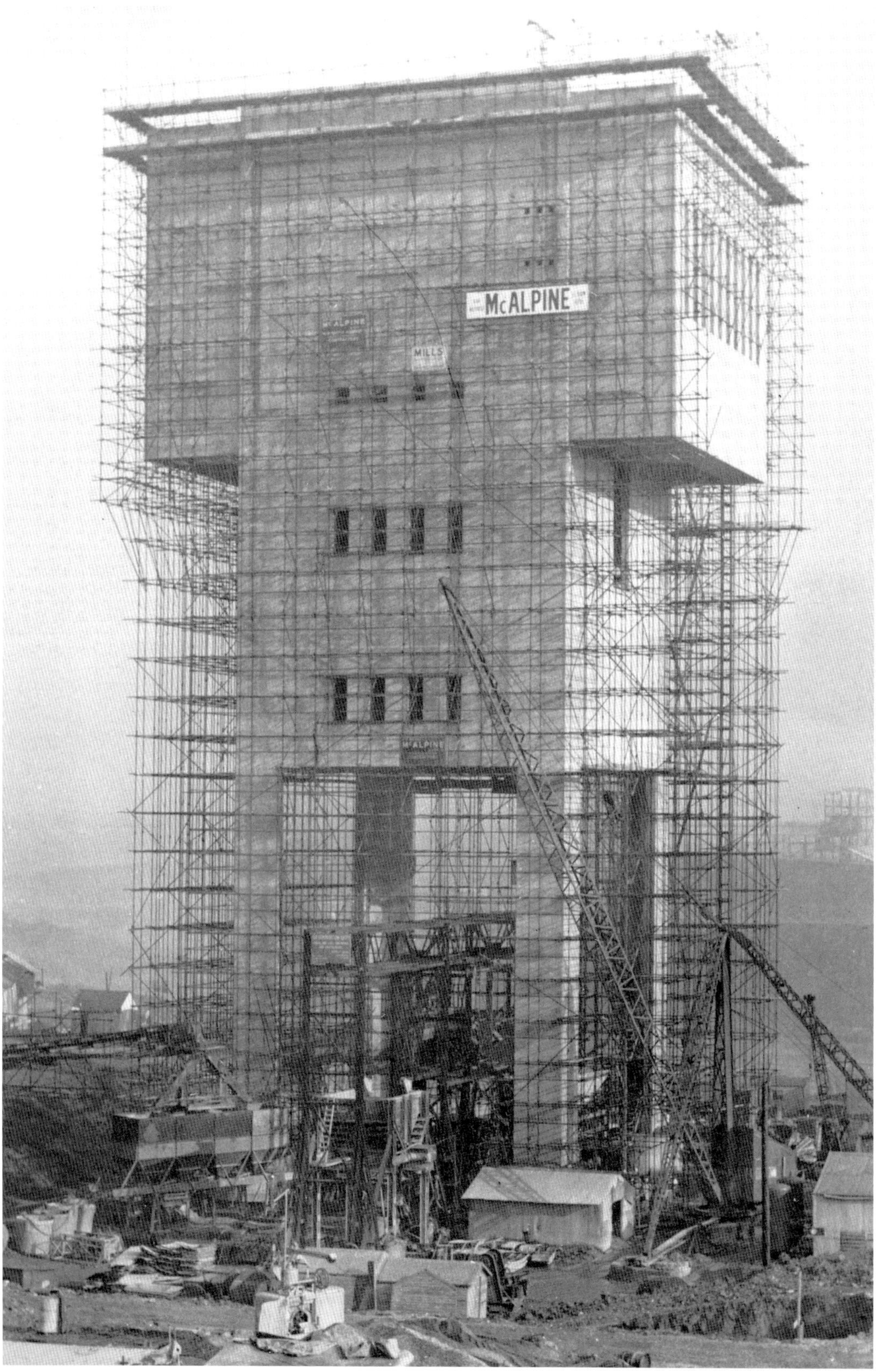
McALPINE
MILLS

Opposite page: The No. 3 shaft winding tower during construction with scaffolding still in place. MC

Right: A view of the winding tower on No. 3 shaft. Note the Shelton Bar Steel works in the background. MC

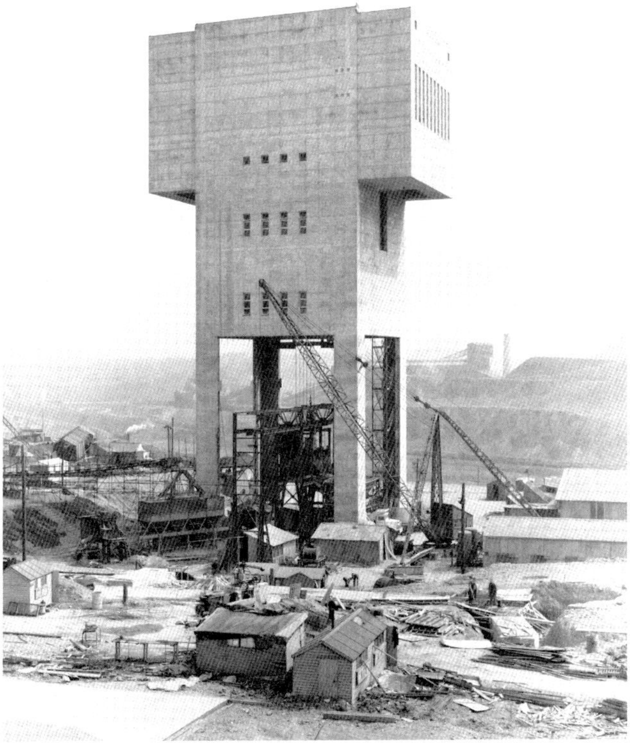

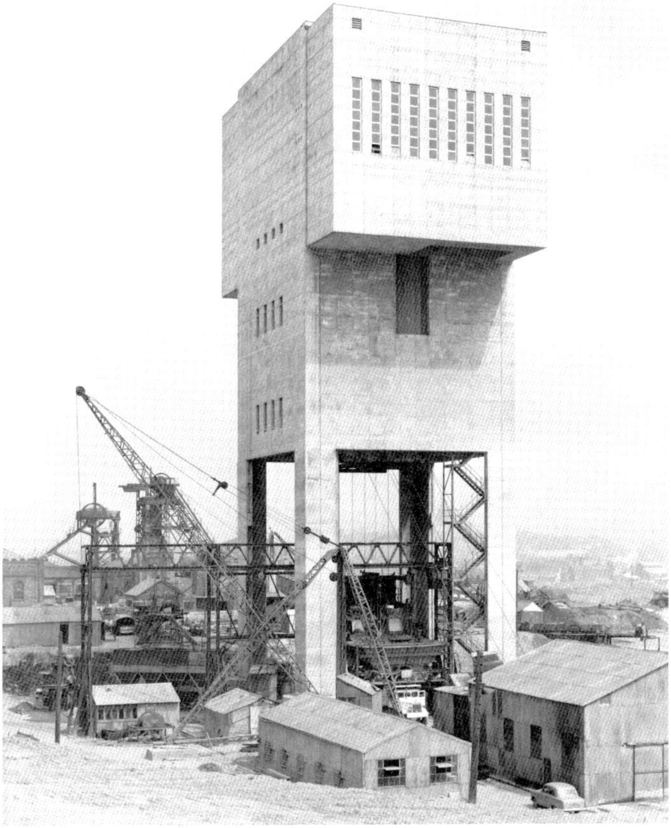

Another view of No. 3 winding tower and the headstocks on Shafts 1 & 2 in the background, 4th November 1959. MC

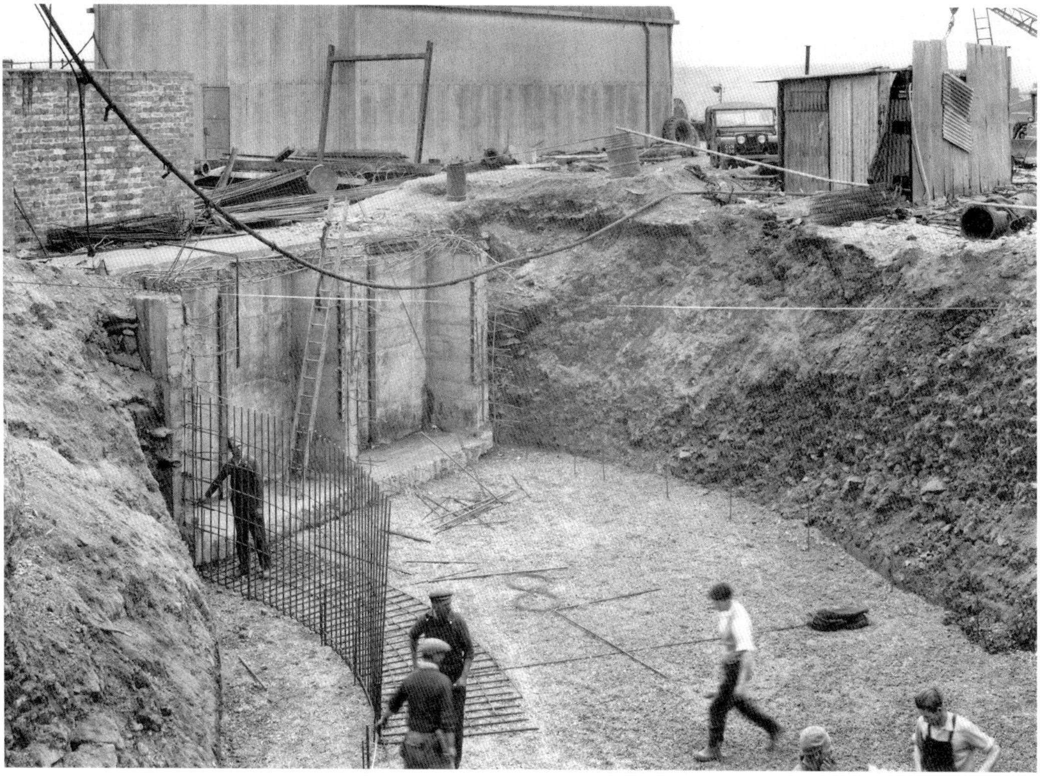

Above: Putting in steel reinforcing during the construction of the fan drift, 21st June 1961. MC Below: The top of No. 3 shaft showing the construction of the fan drift in the foreground. MC

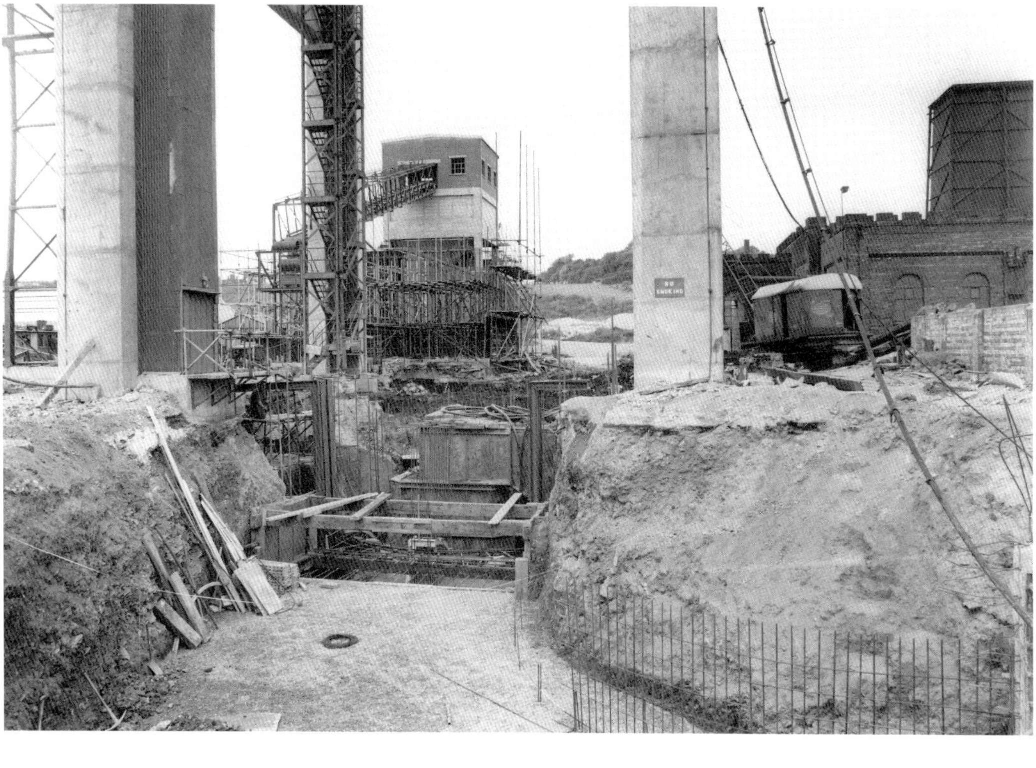

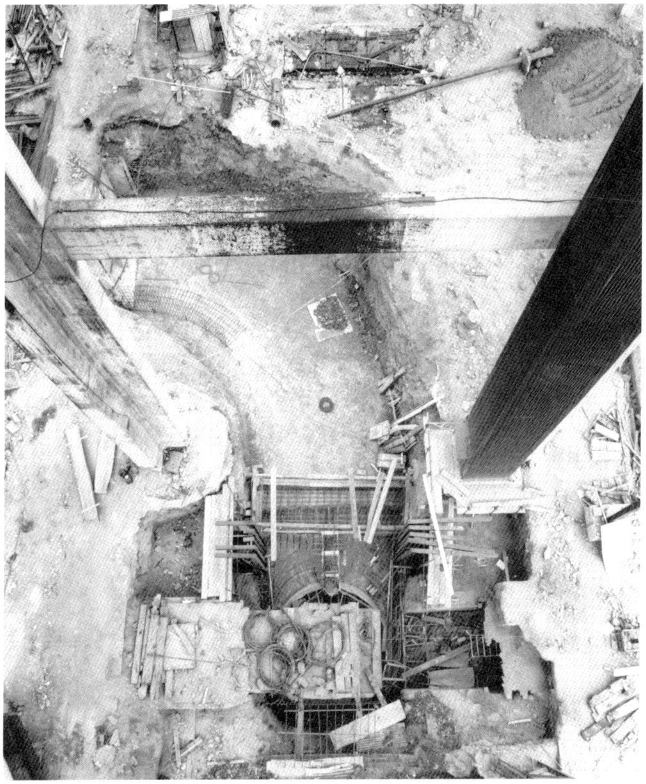

Looking vertically downwards from the winding tower to the shaft collar on No. 3 shaft, 21 June 1961. MC

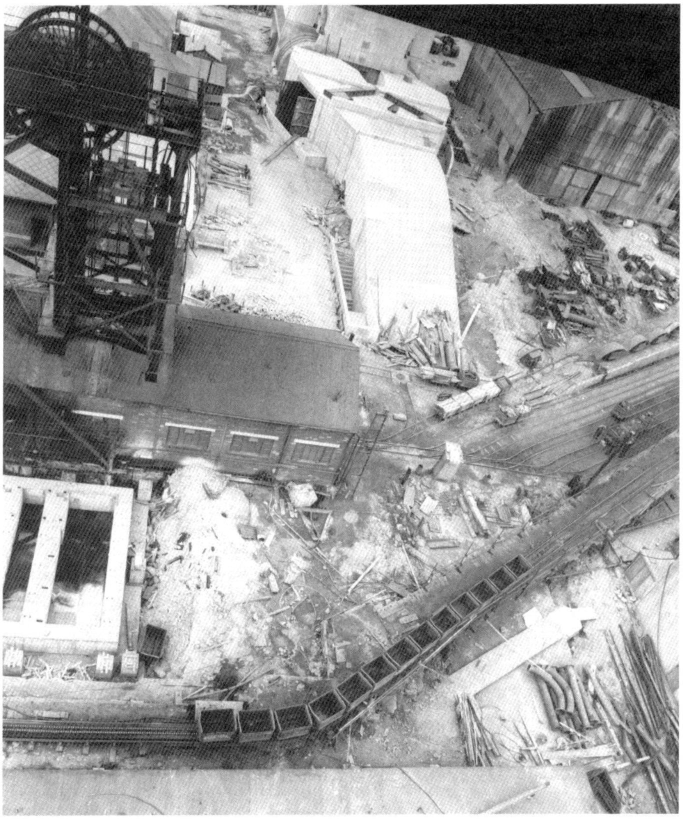

Looking down at the old headstocks and a Journey of 13 tubs, 21st June 1961. MC

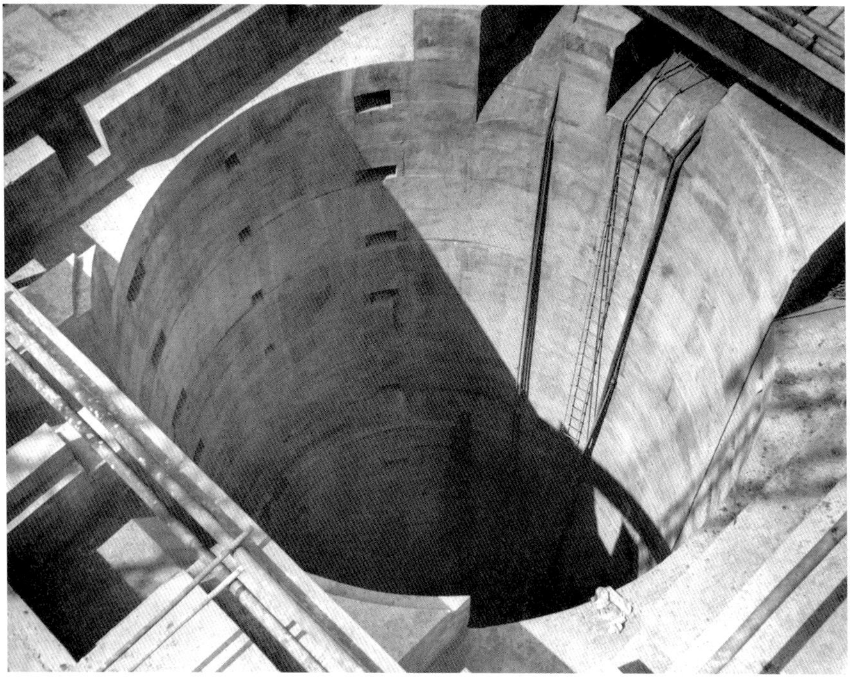

Above: A view of No. 3 shaft collar during construction. Note the rope ladder on the right hand side. MC

Below: Deflection pulleys on the friction winder. These are used to ensure that the ropes hang in the exact position in the shaft. MC

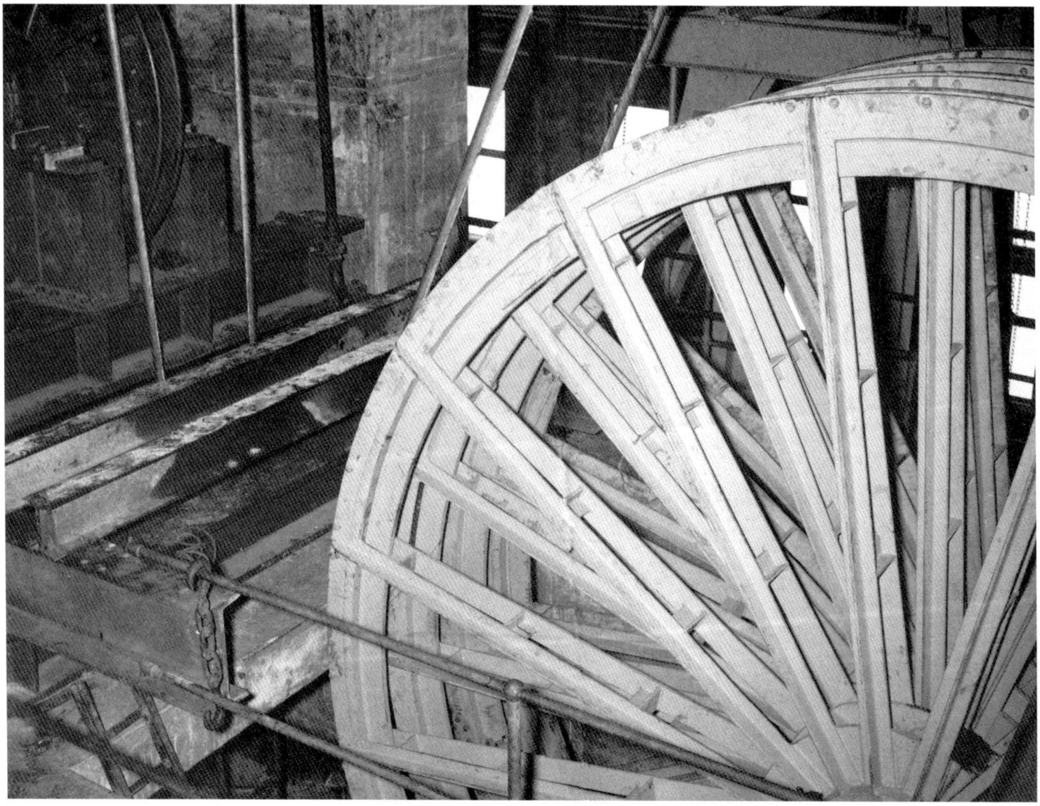

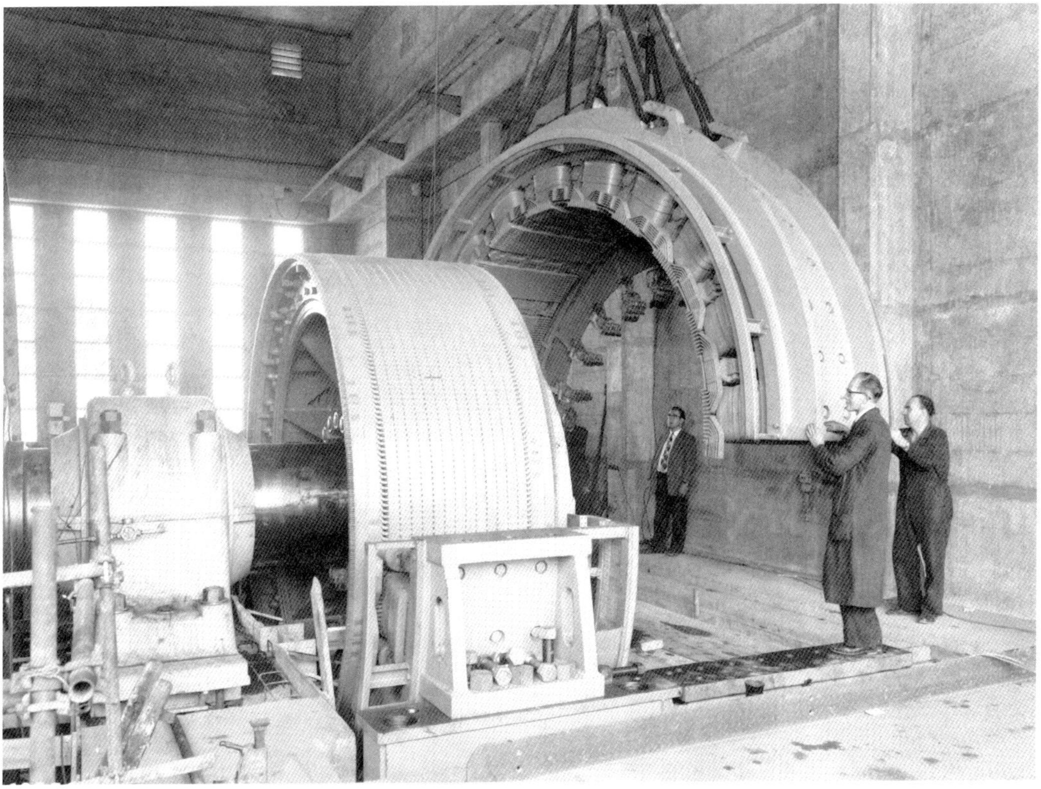

Completing the installation of the winding motor. Note the brake shoes in the winding drum in the background in the lower picture. MC

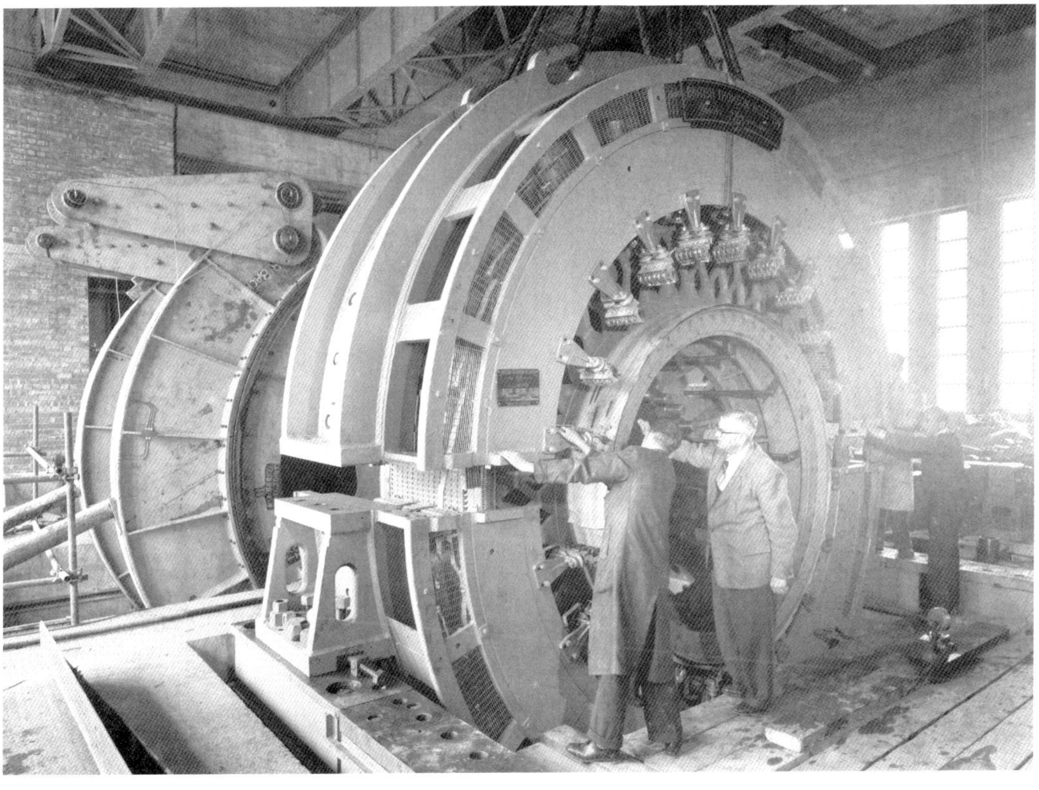

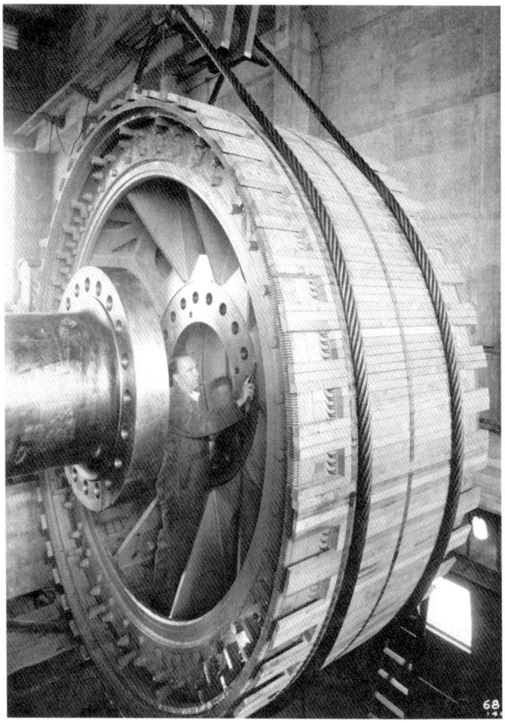

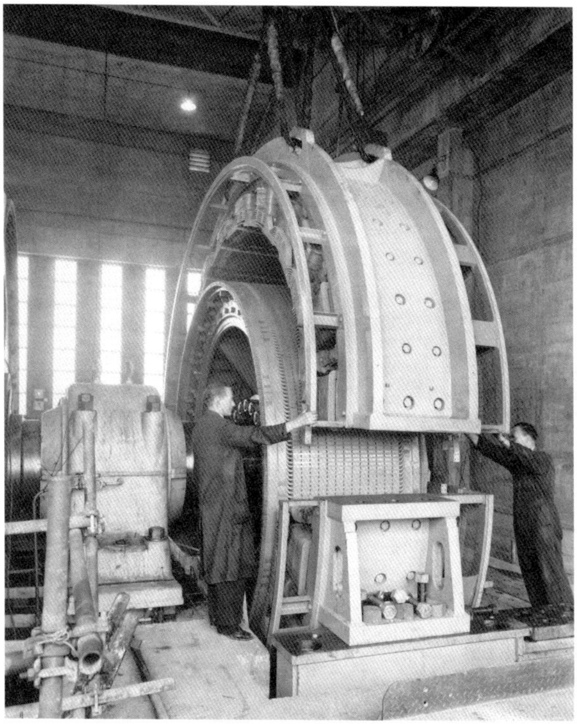

Above left: Fitting the winding engine armature to the drive shaft. MC

Above right: Completing the assembly of the winding engine motor. MC

Below: Rope men capping a rope. The men are opening up the wires for degreasing. The wires were then pulled into a steel cap, after which the cap would be filled with white metal. MC

Wolstanton Colliery is now a retail park, the shafts being situated in front of Asda.

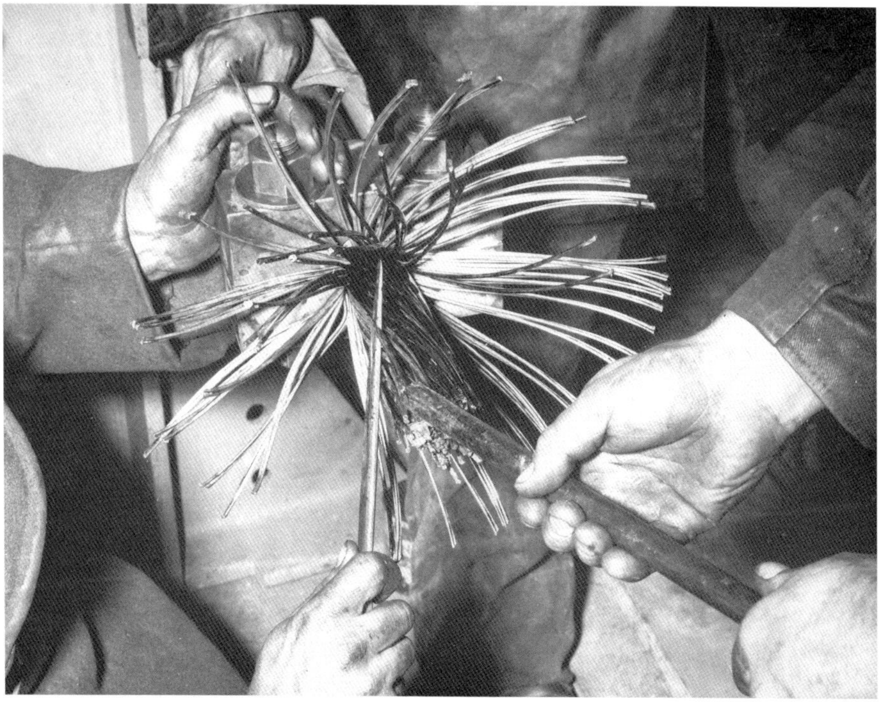

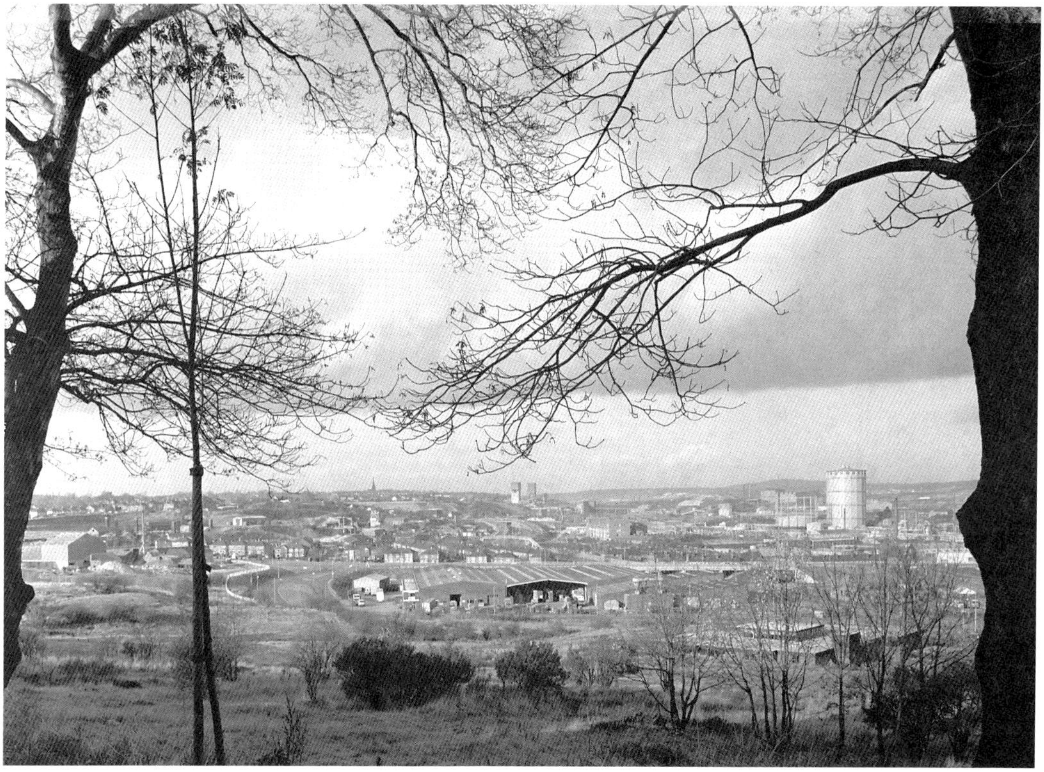

Above: Distant view of the Etruria Valley with the winding tower of Wolstanton Colliery on the horizon, 1976. PD

Below: View of the Etruria valley, 1965. PD

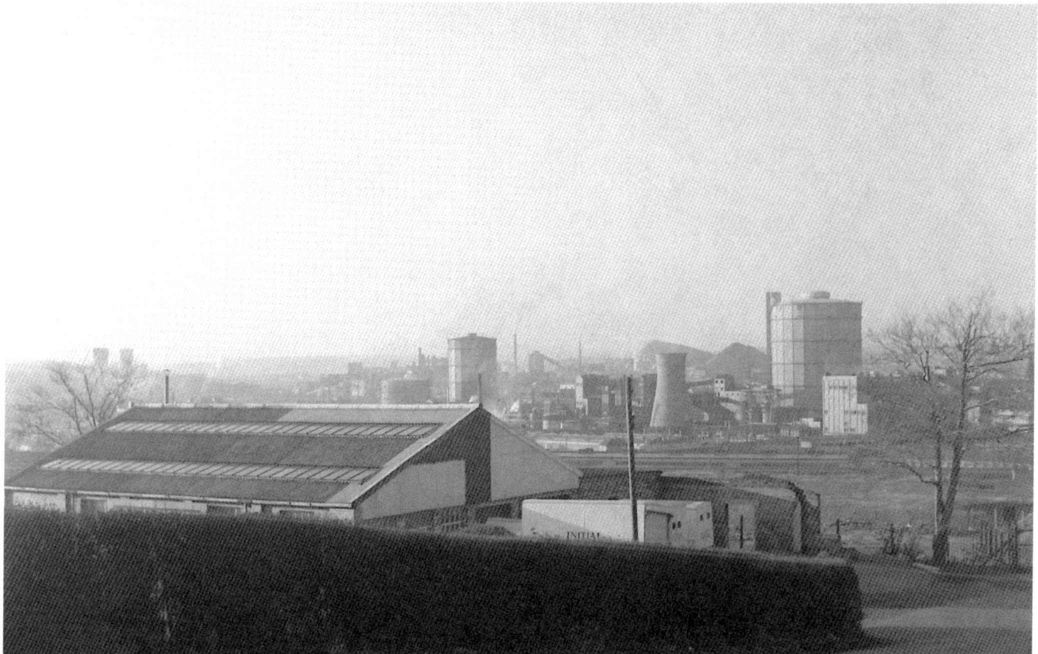

The Colliery Sites Today

All photographs taken 2004

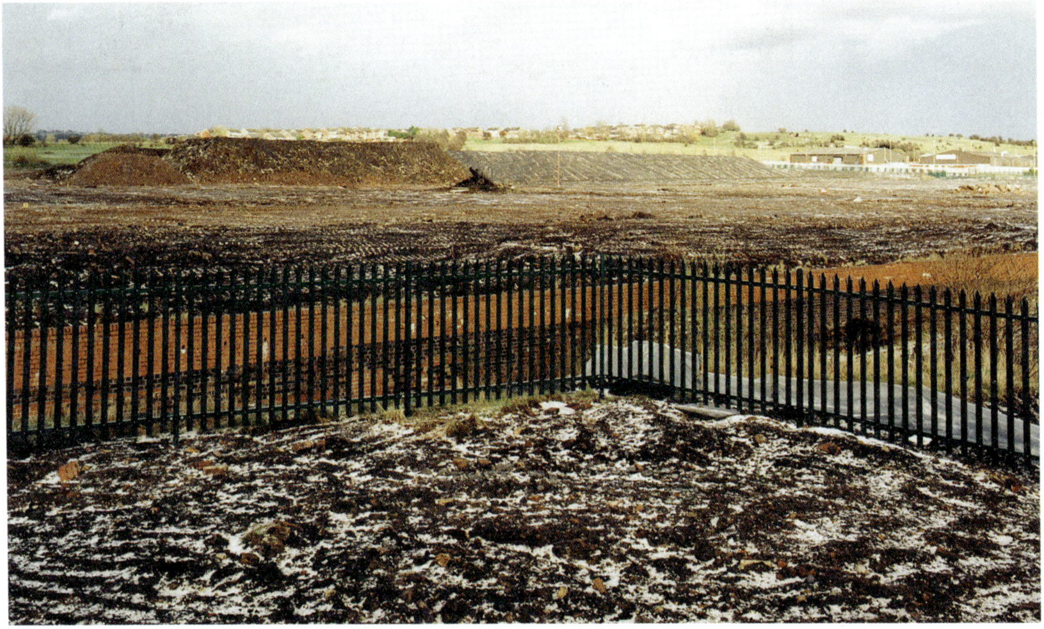

Above: Berry Hill.

Below: Chatterley Whitfield, awaiting a decision on its future.

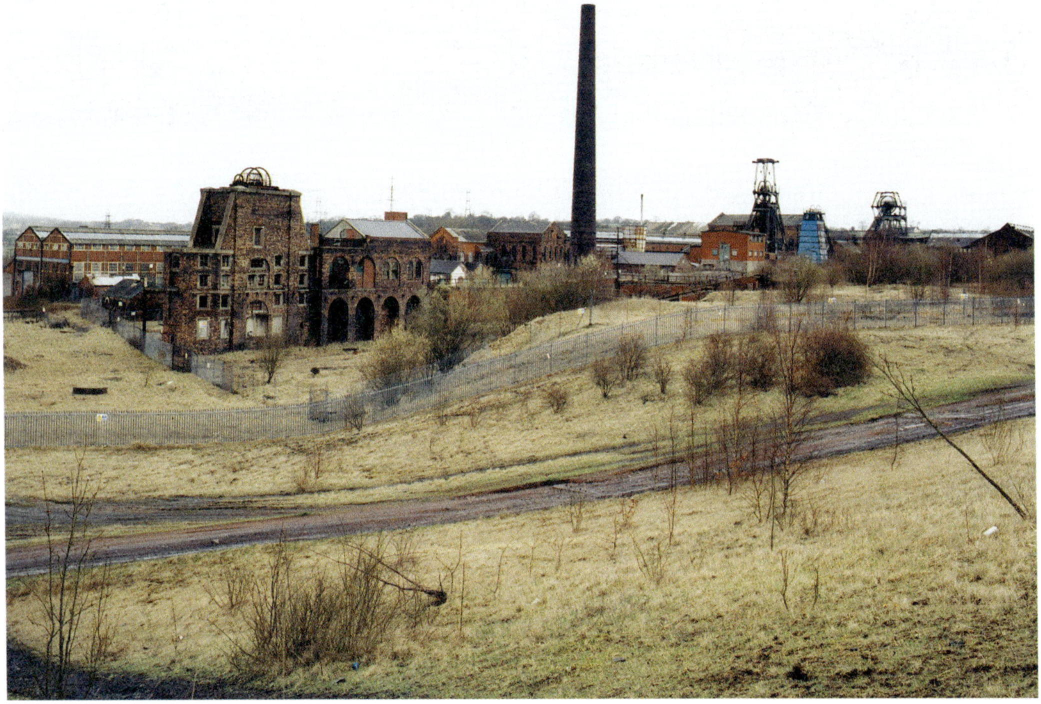

Right: Fenton (Glebe).

Middle: Florence.

Bottom: Florence; note the shaft ventilation pipe left of centre.

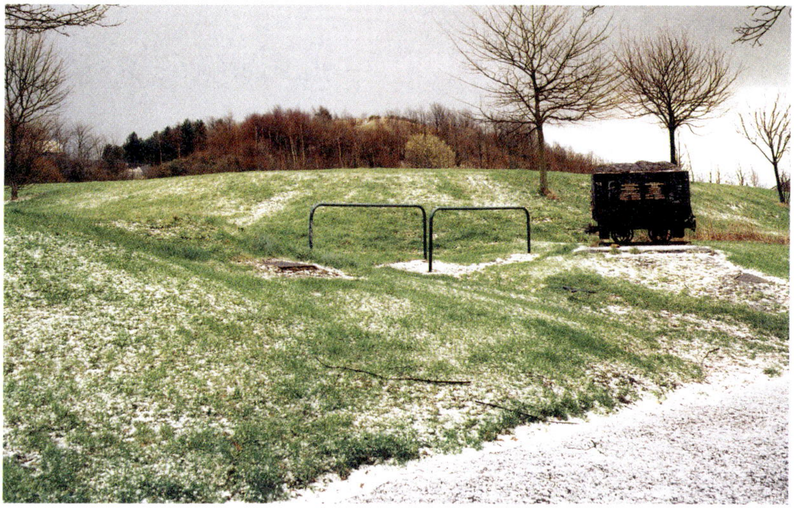

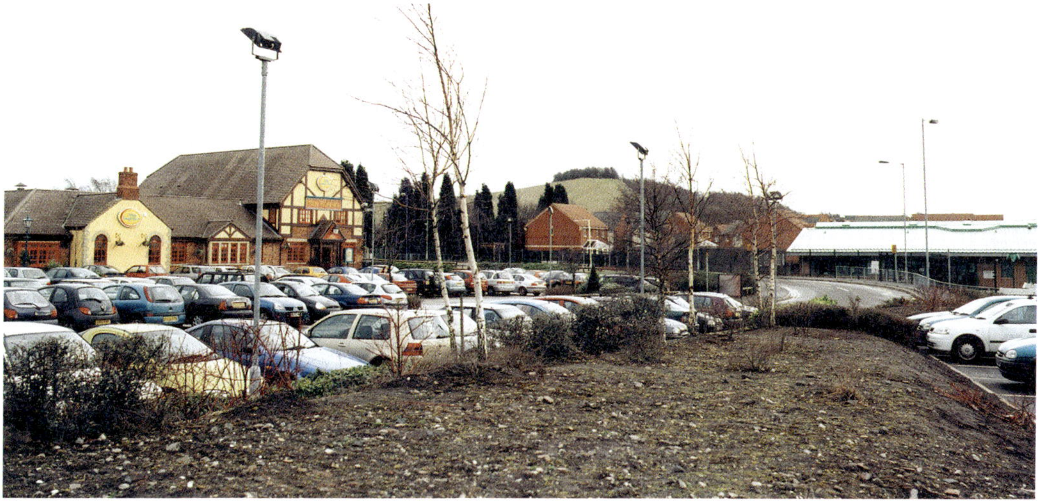

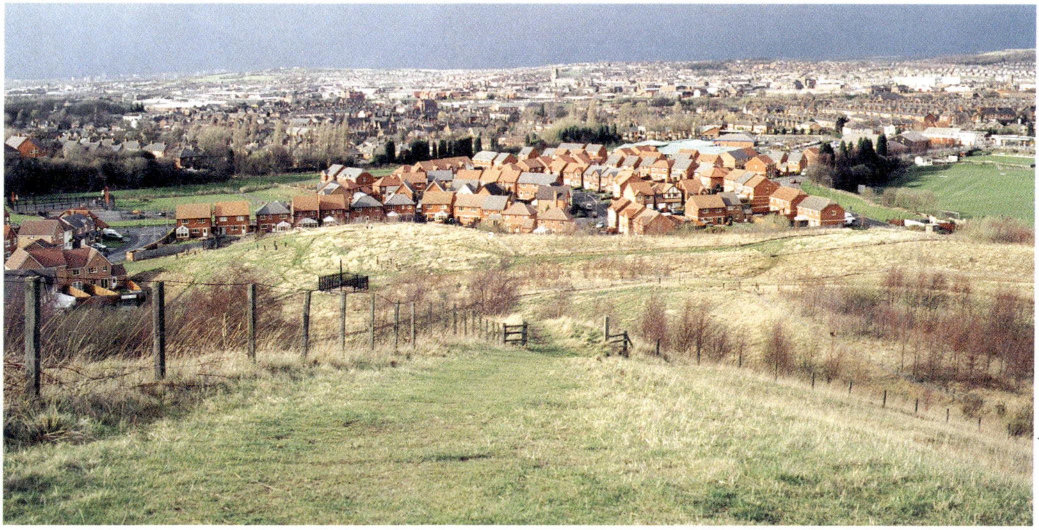

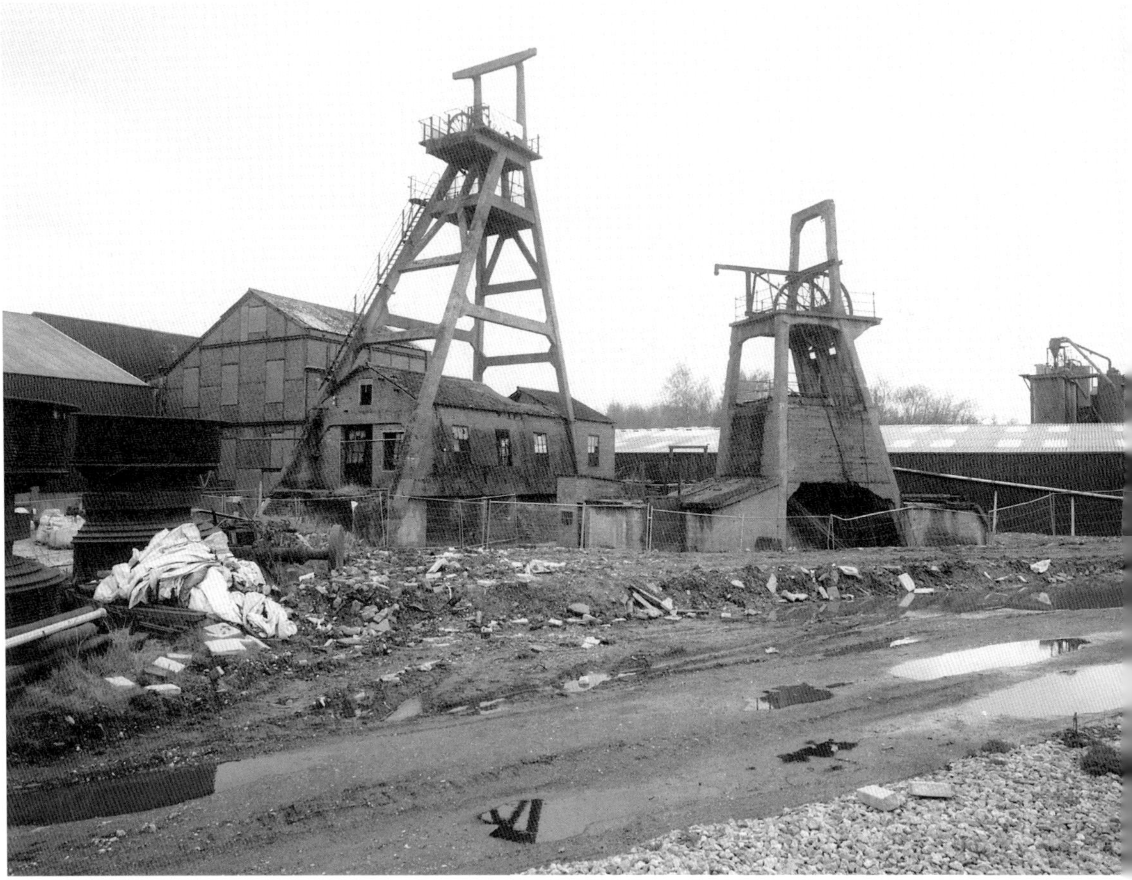

Above: Foxfield Colliery site from the Foxfield Railway

Below: Grange.

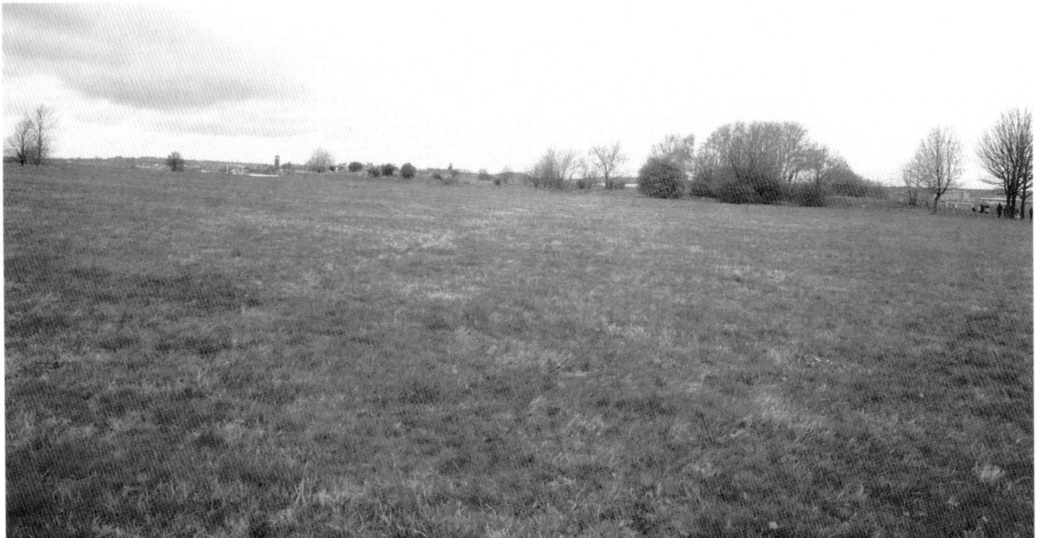

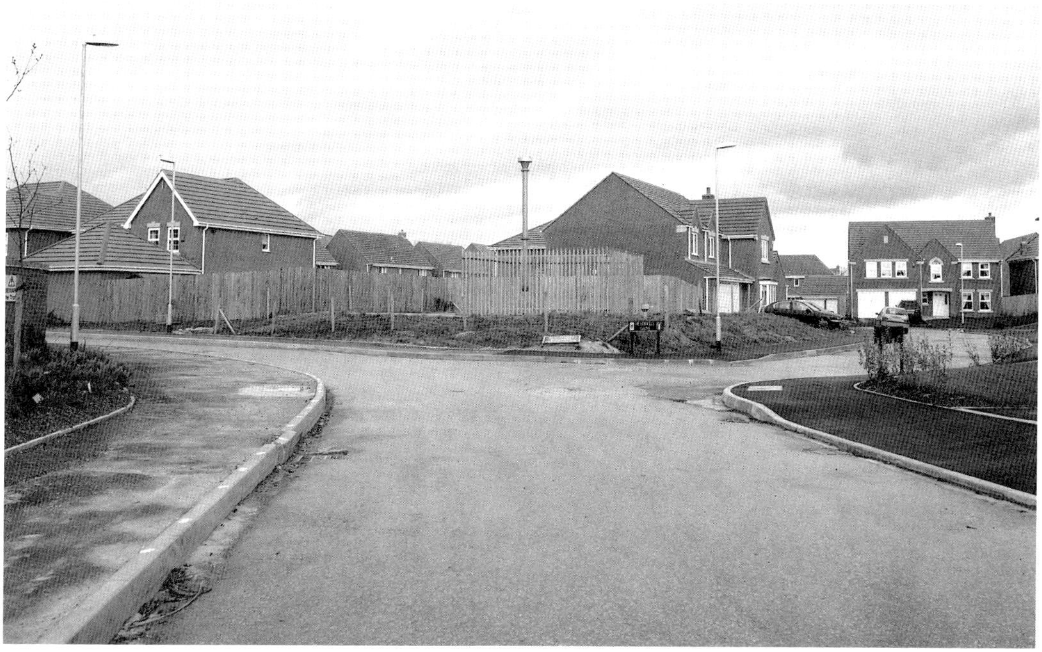

Above: Hem Heath Shaft No. 2.
Below: Hem Heath Adit.

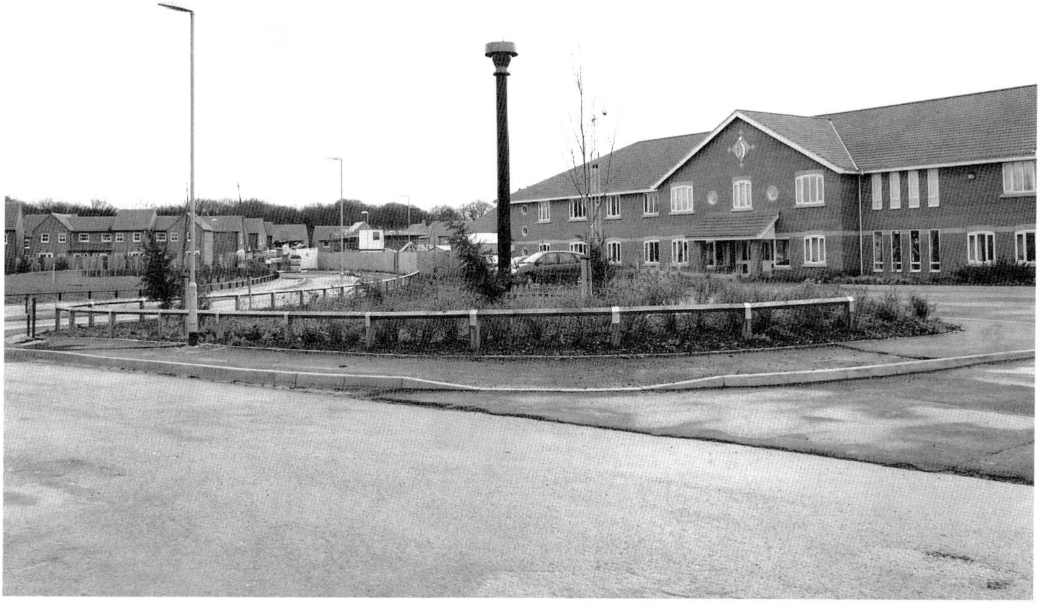

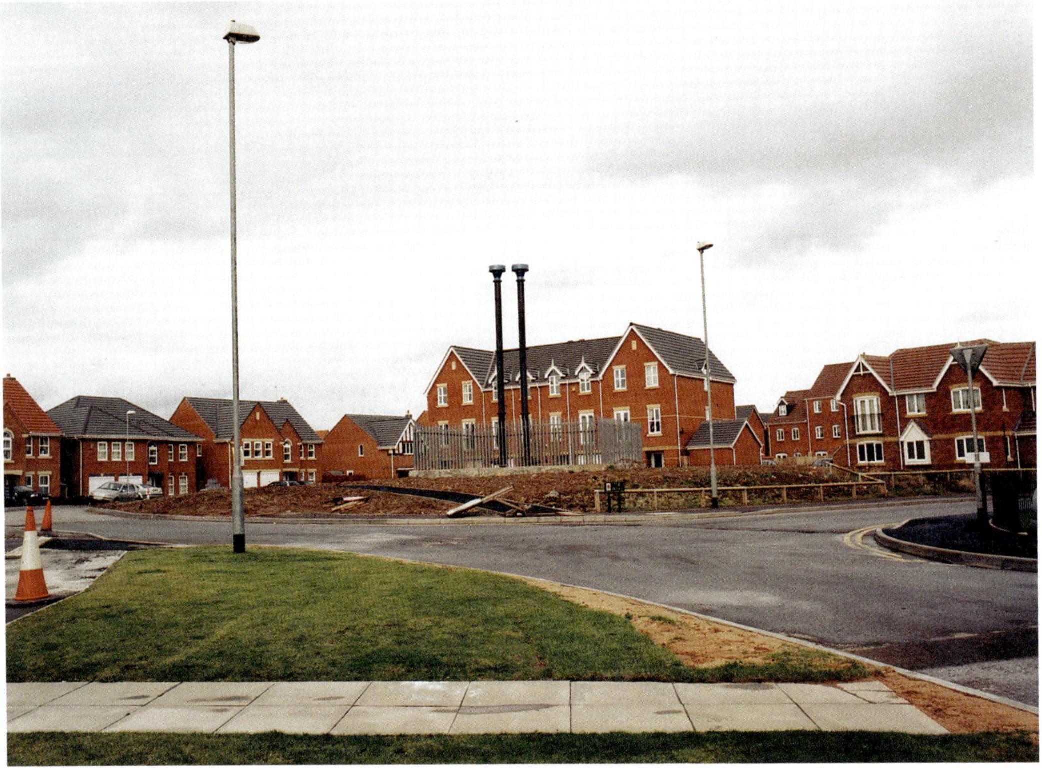

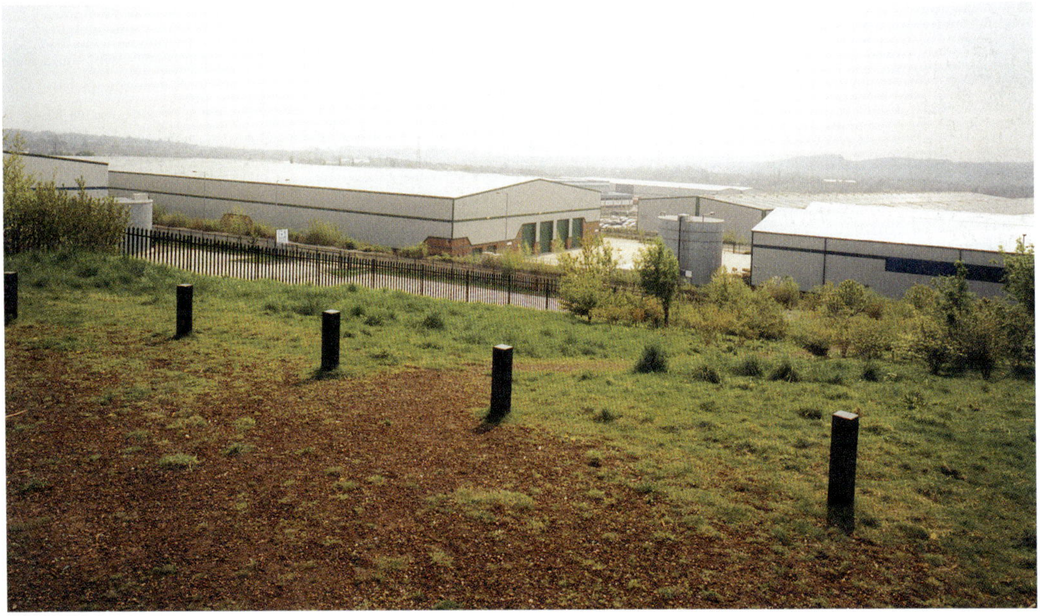

Above: Hem Heath No. 1 shaft.
Below: Holditch.

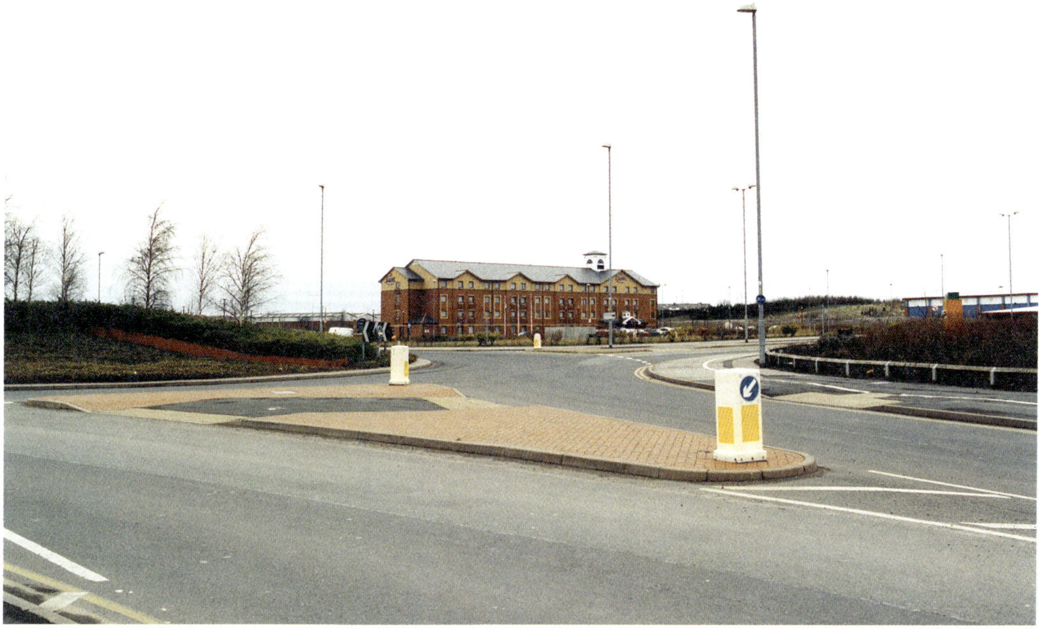

Above: Kemball Colliery site with the Holiday Inn.
Below: Madeley.

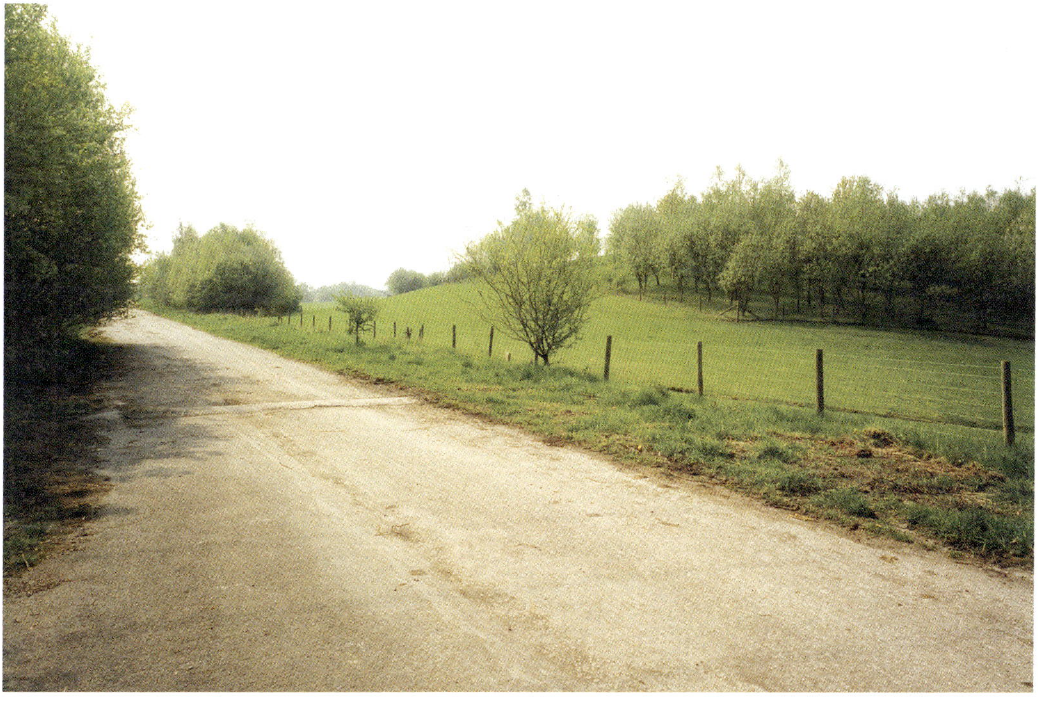

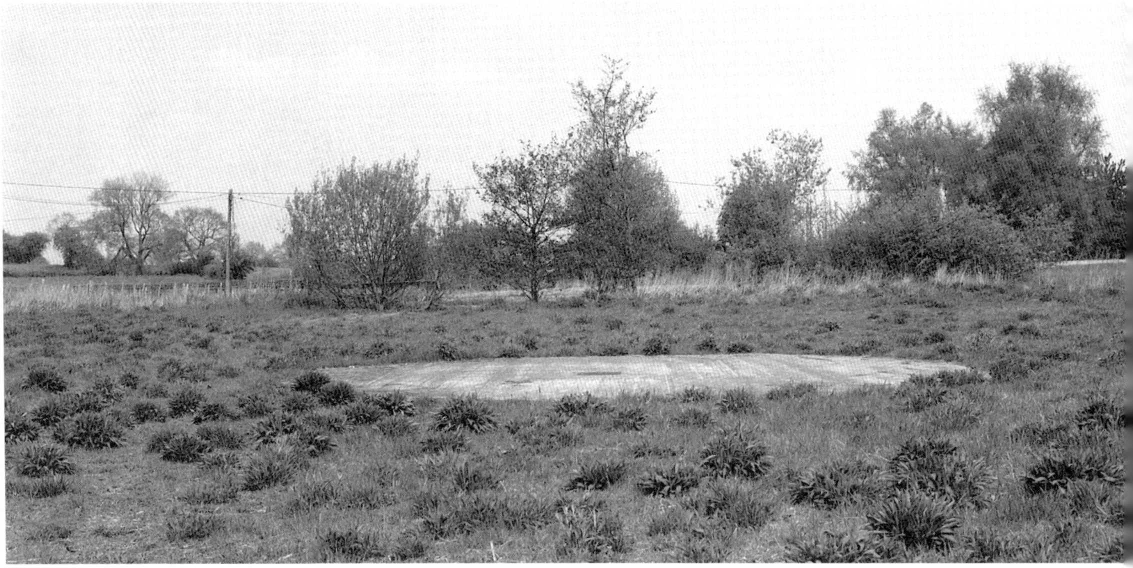

Above: Minnie Pit shaft cap.

Below: Minnie Pit; monument to the victims of the disaster of January 12th, 1918, that claimed 156 lives.

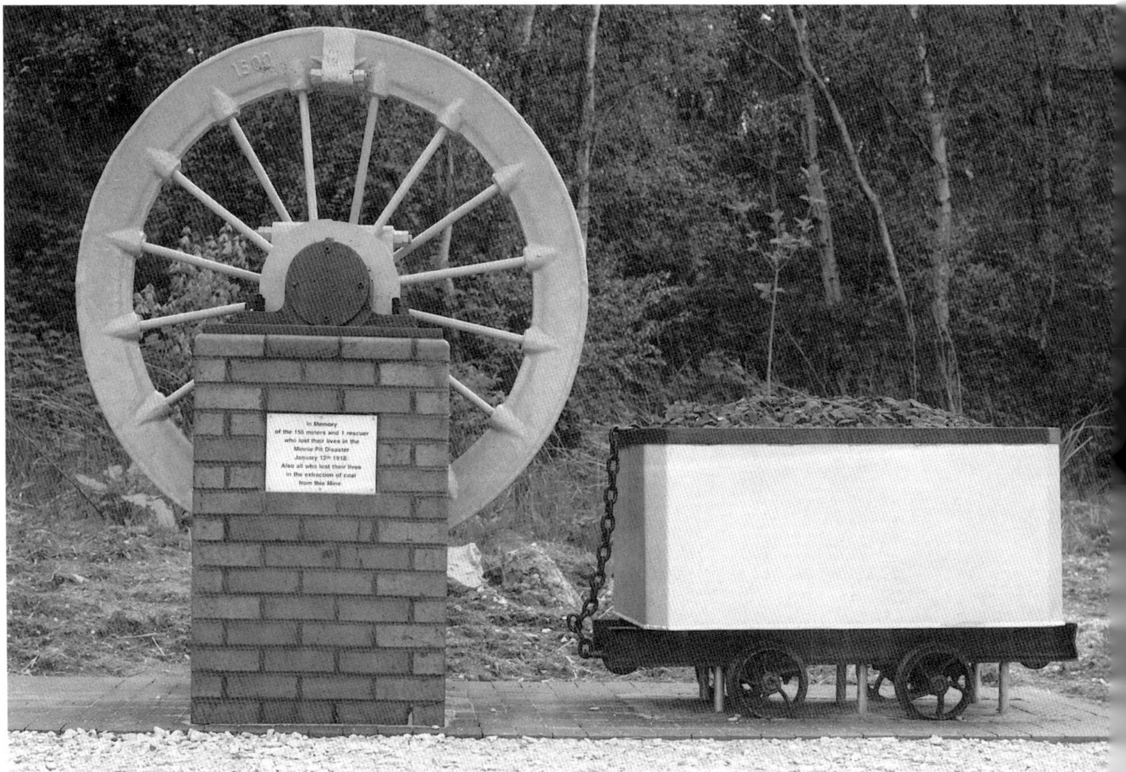

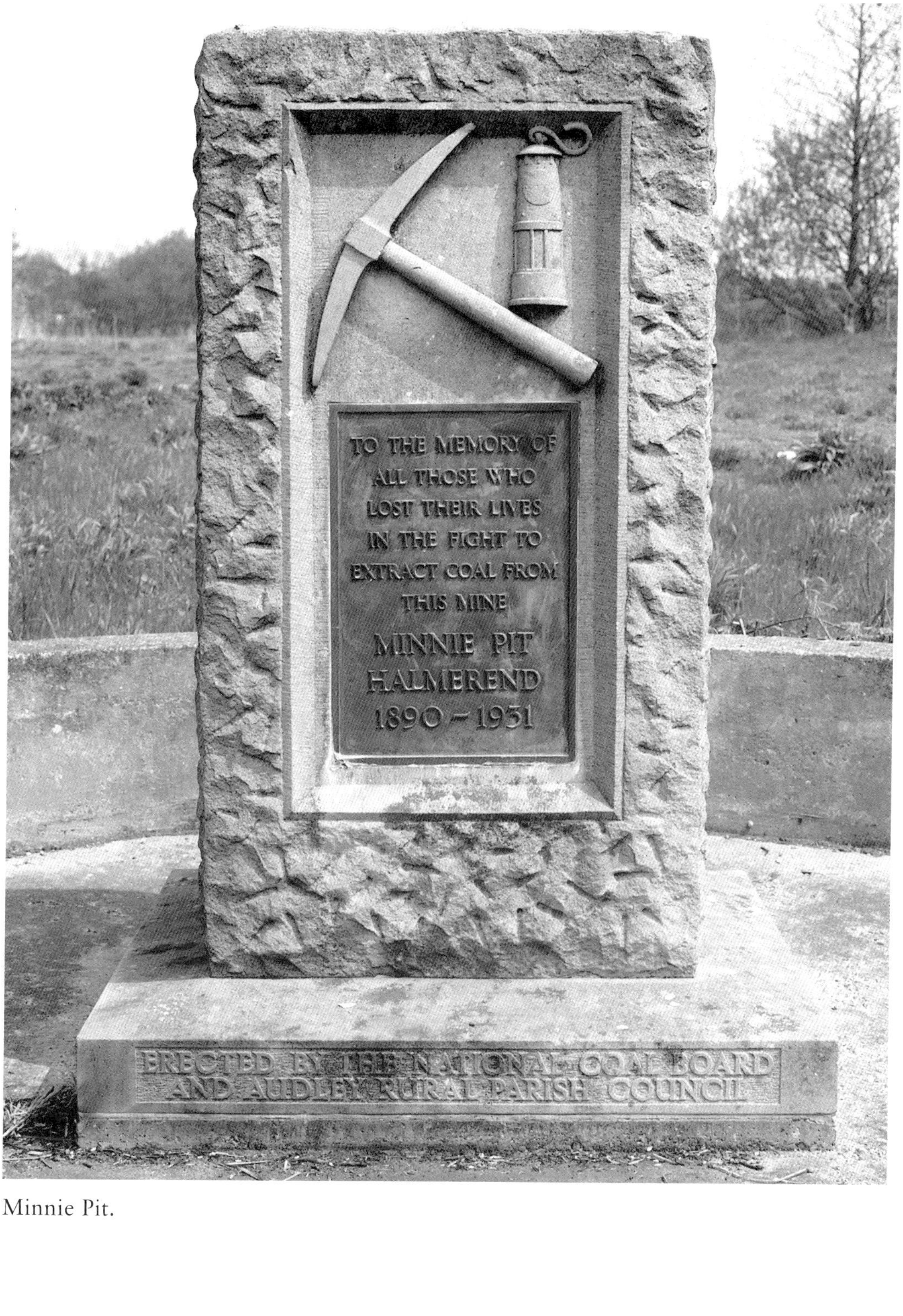
TO THE MEMORY OF
ALL THOSE WHO
LOST THEIR LIVES
IN THE FIGHT TO
EXTRACT COAL FROM
THIS MINE
MINNIE PIT
HALMEREND
1890 – 1931
ERECTED BY THE NATIONAL COAL BOARD
AND AUDLEY RURAL PARISH COUNCIL

Minnie Pit.

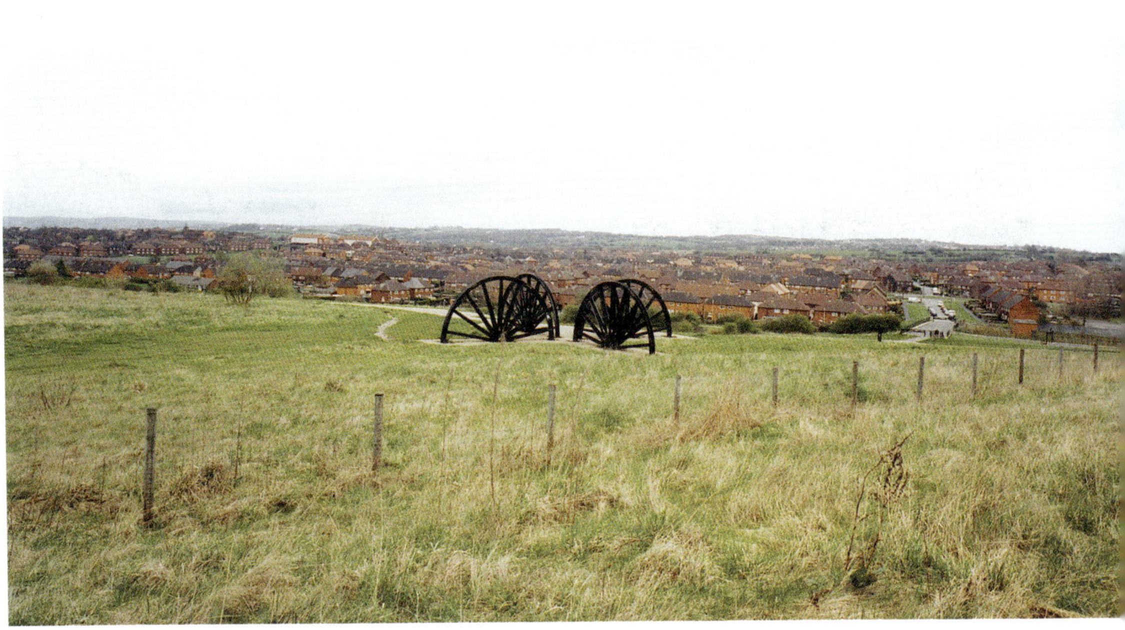

Above: Mossfield with the former winding wheels.

Below: Norton.

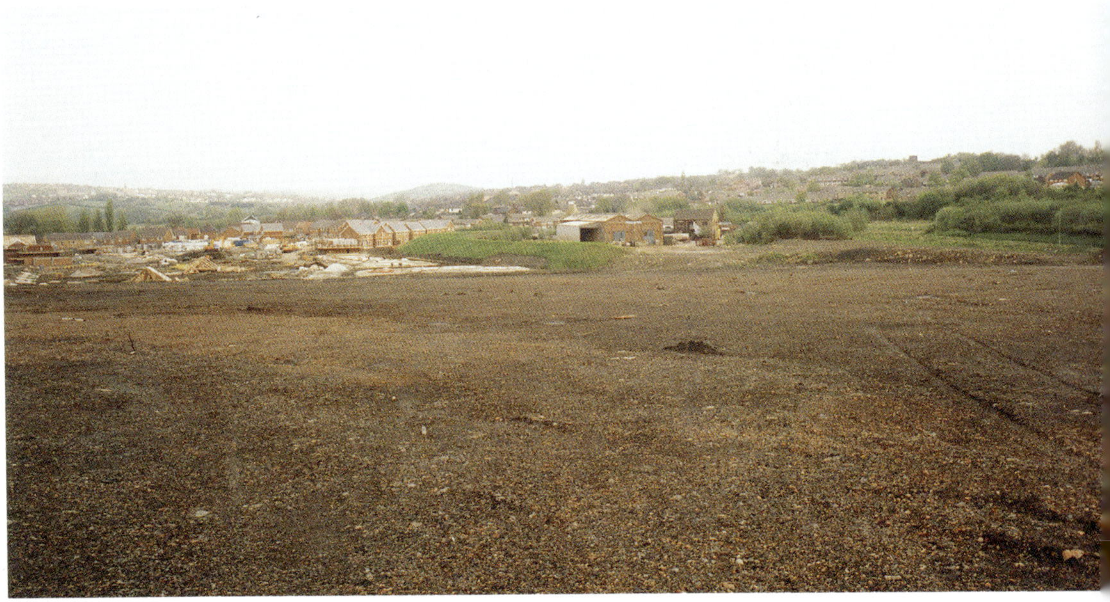

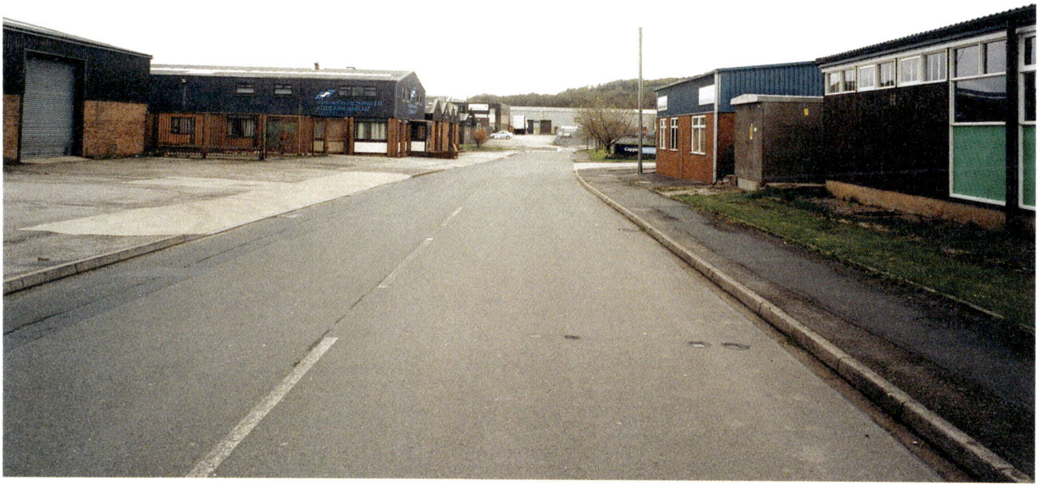

Above: Park Hall.
Below: Parkhouse.

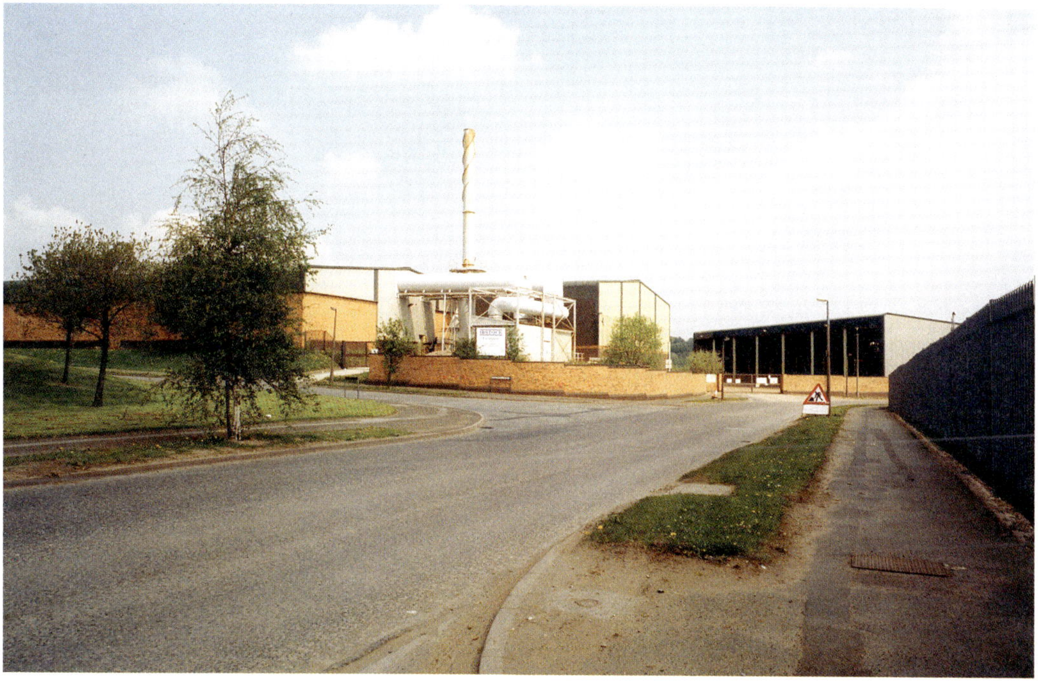

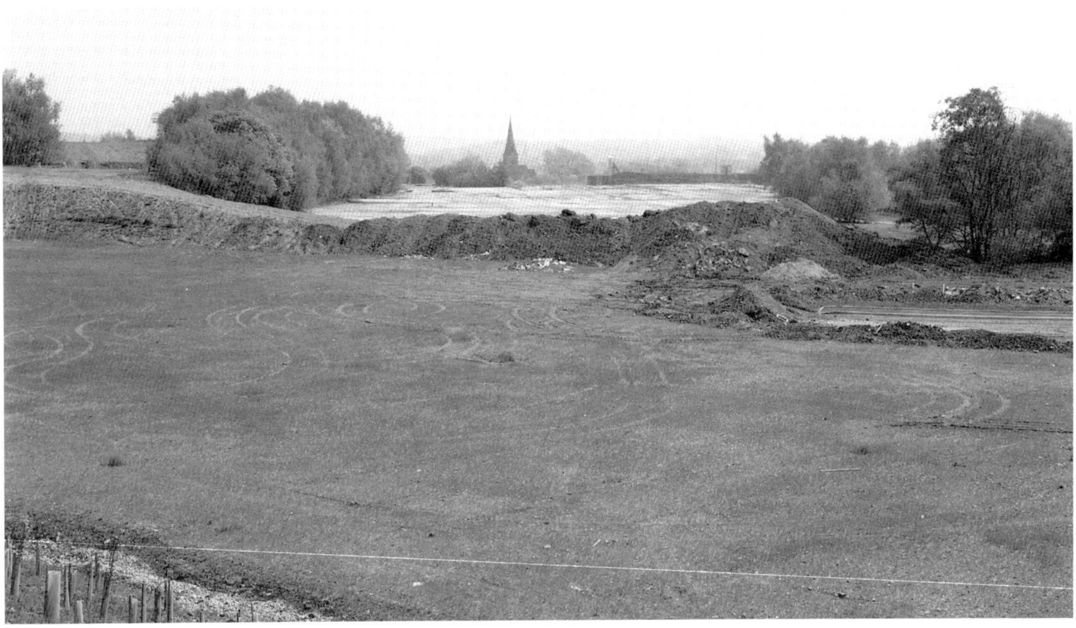

Above: Silverdale looking towards the church.

Below: Sneyd, taken off the colliery tip.

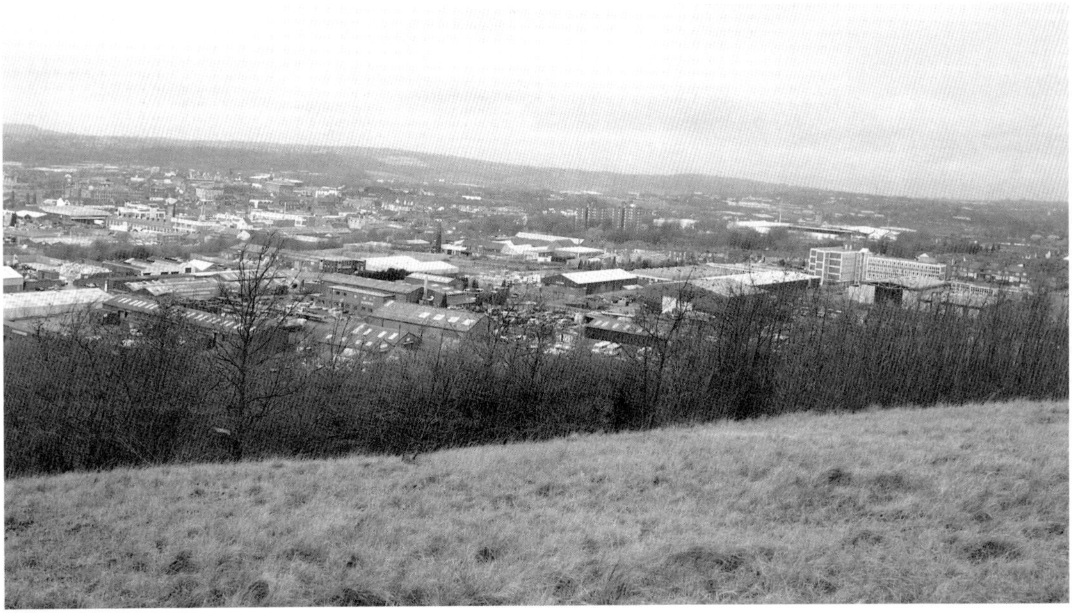

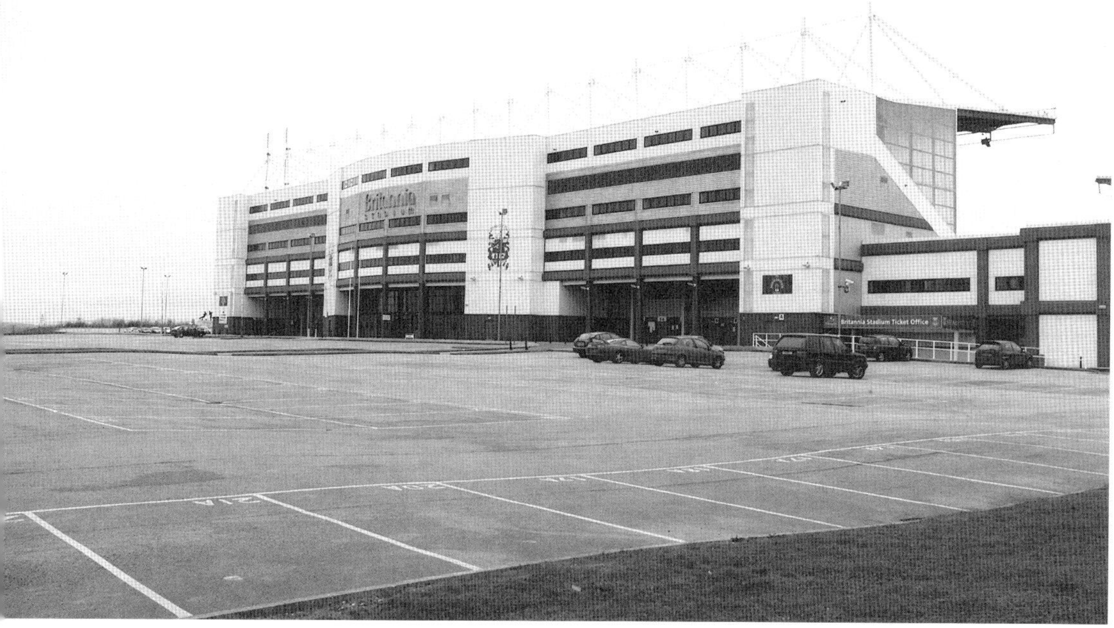

Above: Stafford Colliery site with the Britannia Stadium
Below: Victoria.

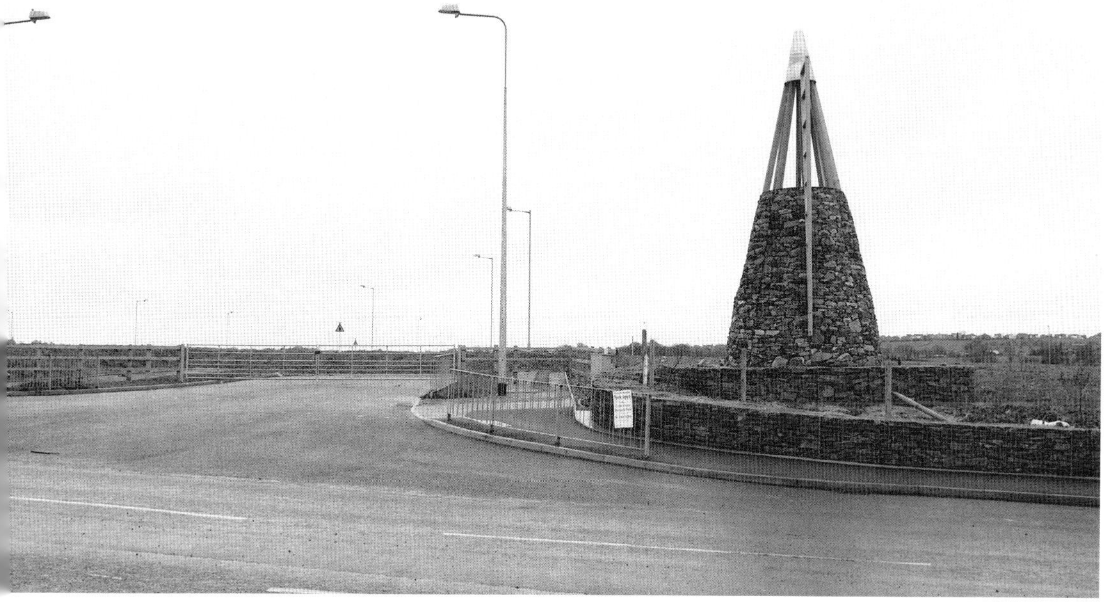

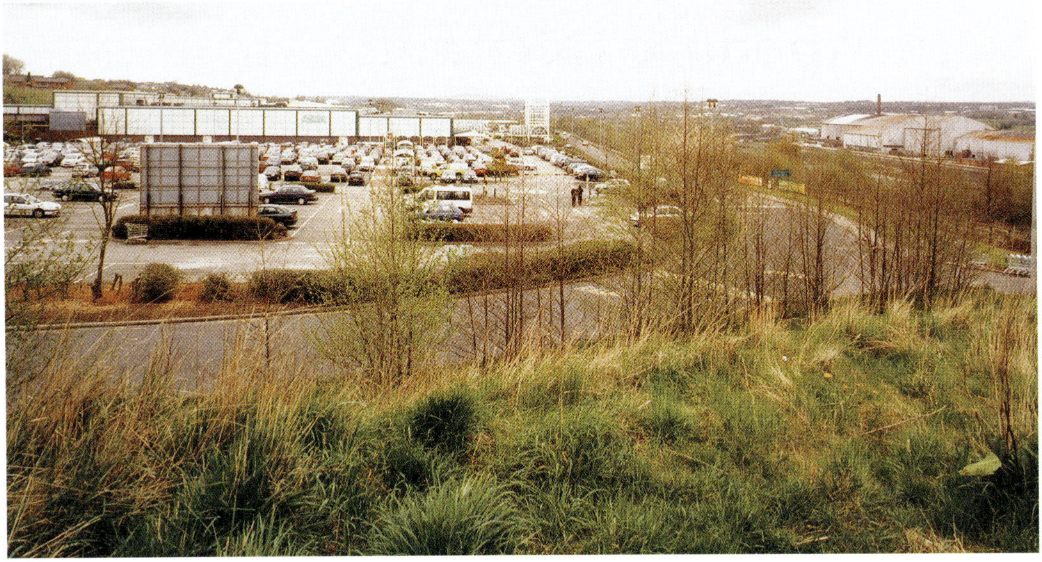

Wolstanton.

Men and Production in 1947

Colliery	Men U/G	Men surface	Approx Production
Norton	960	331	420 kt
Victoria	1059	352	475
Chatterley Whitfield	2000	630	900
Deep Pit	876	370	375
Glasshouse	98	23	40
Parkhouse	500	150	200
Sneyd	1000	460	450
Wolstanton	650	215	325
Rookery	233	111	50
Holditch	1150	335	400
Madeley	650	200	200
Silverdale	574	200	250
Glebe	550	200	160
Florence	1100	340	400
Stafford 1)			200
Stafford 2)	1050	517	325
Berryhill	850	230	320
Foxfield	390	150	200
Mossfield	720	330	360
Parkhall	460	200	240

Apedale Heritage Centre

This former working mine may now be visited underground as part of a tour to the Centre. Your tour also incorporates the Exhibition Hall and Museum where a reproduction miner's cottage evokes a fascinating insight into a past way of life. There is an impressive display of mining artefacts and memorabilia.

It is situated at the end of Loomer Road, Chesterton in the Apedale Community Country Park.

Underground visits are available on Saturdays and Sundays, 10.30am-3.00pm, booking not required. Weekday visits by appointment only.

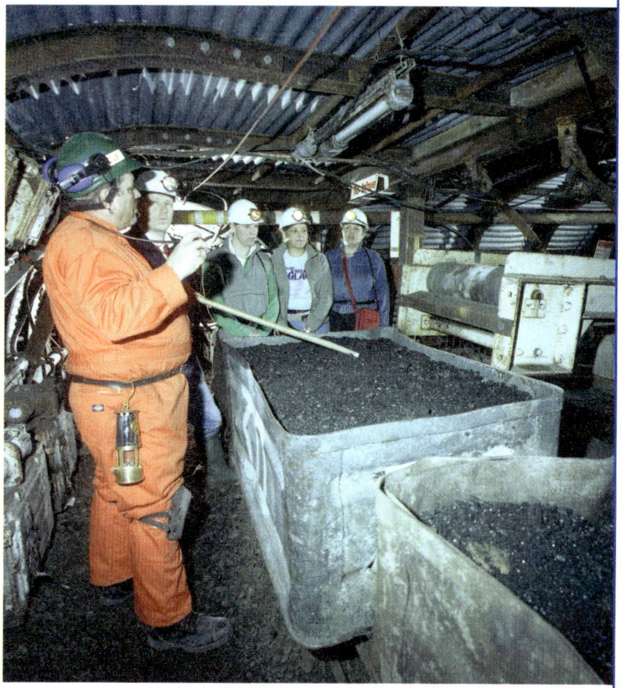

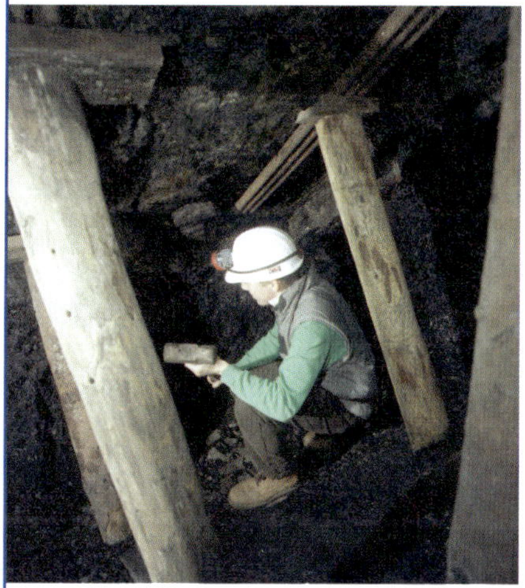

This new venture deserves to succeed, so give it your support!

For more details ring:
01785 813417 (daytime);
07940 192718 (evening).

Website:
www.apedale.heritage.cwc.net

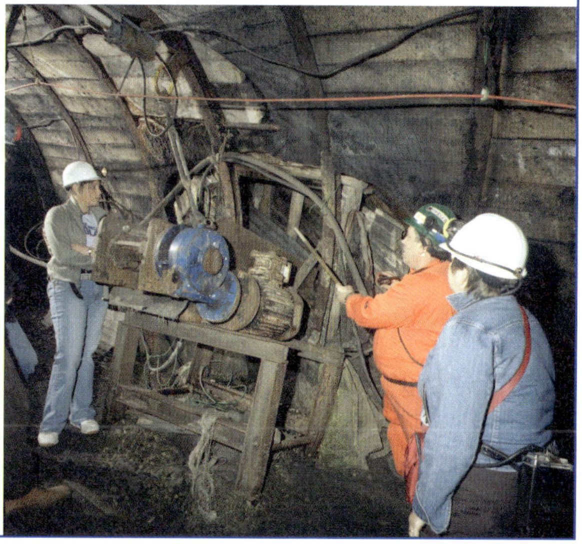